Novel Aspects of Insect–Plant Interactions

Novel Aspects of Insect–Plant Interactions

Edited by

PEDRO BARBOSA

Department of Entomology
University of Maryland
College Park, Maryland

and

DEBORAH K. LETOURNEAU

Board of Environmental Studies
University of California
Santa Cruz, California

WILEY

A WILEY-INTERSCIENCE PUBLICATION

JOHN WILEY & SONS

New York / Chichester / Brisbane / Toronto / Singapore

Library of Congress Cataloging in Publication Data:

Novel aspects of insect-plant interactions.

"A Wiley-Interscience publication."
Bibliography: p.
1. Insect-plant relationships.   2. Allelopathic
agents. I. Barbosa, Pedro, 1944-   . II. Letourneau,
Deborah Kay.
QL496.N68   1988   595.7'0524   88-5494
ISBN 0-471-83276-6

Printed in the United States of America

10 9 8 7 6 5 4 3 2 1

Preface

Throughout the history of ecological investigations of insect-plant interactions, controversies have arisen over the relative importance of one or another environmental factor in generating the patterns observed in nature. Clearly, a variety of factors, including temperature, light intensity, plant texture and morphology, relative availability of plants in time and space, and inter- and intrapopulation variation among herbivores, may play significant roles in ecological interactions. Nevertheless, the host plant an herbivore finds itself on, or in, can be critical in determining or altering dispersal, locomotory behavior, morph or biotype formation, pheromone production, morphogenesis of pheromone-producing organs, speciation, protection from insecticides, survivorship, as well as many of its other interactions and biological processes. Similarly, interactions across trophic levels may be influenced or mediated by aspects of the host plant upon which or within which an herbivore lives. This book, while recognizing the importance of many factors, focuses on the ways in which plant-derived allelochemicals mediate many interactions among organisms on higher trophic levels. Many of these interactions have been given little attention in Ecology. Others provide

v

insights that modify traditional explanations of the patterns we observe in nature. We hope the ideas and data presented will stimulate further cross-disciplinary research and open doors to even more novel concepts and perspectives.

PEDRO BARBOSA
DEBORAH K. LETOURNEAU

College Park, Maryland
Santa Cruz, California
June 1988

Acknowledgments

Discussions with and suggestions by P. Barbosa, E. A. Bernays, J. Eisenbach, and L. R. Fox were extremely valuable in the preparation of the section introductions. Some of the ideas for the book and the content of section introductions resulted from research conducted with the support of a UCSC Division of Social Sciences Faculty Research Grant and USDA competitive grant 85-CRCR-11590 to DKL (and L. R. Fox) and NSF grant BSR-8604303 to PB. Finally, we greatly appreciate the invaluable help and excellent technical assistance of S. Donkis, M. Hammer, P. Ponzini, and J. Lewis.

Contributors

M. A. Altieri
Division of Biological Control
University of California
Berkeley, California 94720

P. Barbosa
Department of Entomology
University of Maryland
College Park, Maryland 20742

M. R. Berenbaum
Department of Entomology
University of Illinois
Urbana, Illinois 61801

M. D. Bowers
Museum of Comparative Zoology
Harvard University
Cambridge, Massachusetts 02138

L. B. Brattsten
Department of Entomology
J. B. Smith Hall
Rutgers University/Cook College
New Brunswick, New Jersey 08903

M. Dicke
Department of Entomology
Agricultural University
Wageningen, The Netherlands

G. W. Elzen
USDA, Agricultural Research Service
Stoneville, Mississippi 38776

D. K. Letourneau
Board of Environmental Studies
University of California
Santa Cruz, California 95064

W. J. Lewis
USDA, Agricultural Research Service
Insect Biology and Population
Management Research Laboratory
Tifton, Georgia 31793

D. A. Nordlund
USDA, Agricultural Research Service
Cotton Insects Research Unit
College Station, Texas 77841

J. M. Pasteels
Laboratoire de Biologie Animale et Cellulaire
Université Libre de Bruxelles
B-1050 Brussels, Belgium

M. J. Raupp
Department of Entomology
University of Maryland
College Park, Maryland 20742

M. Rowell-Rahier
Zoologisches Institut der Universitat
Rheinsprung 9
4051 Basel, Switzerland

S. B. Vinson
Department of Entomology
Texas A&M University
College Station, Texas 77843

D. W. Whitman
Department of Biological Sciences
Illinois State University
Normal, Illinois 61761

H. J. Williams
Department of Entomology
Texas A&M University
College Station, Texas 77843

Contents

**PART I. CONCEPTUAL FRAMEWORK
OF THREE-TROPHIC-LEVEL
INTERACTIONS**   **1**

D. K. Letourneau

1. Allelochemical Interactions Among Plants,
    Herbivores, and Their Predators   11

    D. W. Whitman

    1. Introduction   12
    2. Synomones   14
    3. Antimones   18
    4. Kairomones   19
    5. Allomones   23
    6. The Future   37
    7. Caveat   40
       References   40

2. Influences of Plant-Produced Allelochemicals
    on the Host/Prey Selection Behavior of
    Entomophagous Insects   65

    D. A. Nordlund, W. J. Lewis and M. A. Altieri

xiii

1. Introduction   66
2. Plant-Produced Allelochemicals and Host/Prey
   Habitat Location   66
3. Effects of Plants on Herbivore Kairomones   73
4. Implications for the Use of Entomophagous
   Insects in Applied Biological Control   75
5. Conclusion   77
   References   79

**PART II.   MICROORGANISMS AS MEDIATORS OF
             INTERTROPHIC AND INTRATROPHIC
             INTERACTIONS   91**
D. K. Letourneau

3. Allelochemicals in Insect–Microbe–Plant
    Interactions; Agents Provocateurs in the
    Coevolutionary Arms Race   97
    M. R. Berenbaum

    1. Introduction   98
    2. Allelochemical Effects on Microbe–
       Insect Interactions   100
    3. Allelochemical Effects on Microbe–
       Plant Interactions   111
    4. Evolutionary Implications of Microbial
       Participation in Plant–Insect Interactions   115
       References   117

4. Microbial Allelochemicals Affecting the
    Behavior of Insects, Mites, Nematodes, and
    Protozoa in Different Trophic Levels   125
    M. Dicke

    1. Introduction   126
    2. Allelochemicals of Stored-Product
       Microorganisms   128
    3. Allelochemicals of Soil Microorganisms   130
    4. Microbial Allelochemicals Attracting
       Bacterivoral Protozoa   134
    5. Allelochemicals of Plant-Infesting Microorganisms
       and Their Influence on the Behavior of
       Microbivores   134

6. Allelochemicals of Aquatic Microorganisms   141
7. Allelochemicals of Insect-Associated
   Microorganisms that are Used as
   Pheromones by the Insect   142
8. Microbial Allelochemicals Affecting Natural
   Enemies of Microbe-Associated Insects   144
9. Consequences for Semiochemical Terminology   150
10. Prospects for Future Research   151
    References   154

PART III.   THEORY AND MECHANISMS: PLANT EFFECTS VIA ALLELOCHEMICALS ON THE THIRD TROPHIC LEVEL   165

D. K. Letourneau

5. Parasitoid–Host–Plant Interactions,
   Emphasizing Cotton (*Gossypium*)   171

   H. J. Williams, G. W. Elzen and S. B. Vinson

   1. Introduction   172
   2. Attraction of Parasitoids to Plants: Historical
      Background   174
   3. Chemical Mediation of *Campoletis
      sonorensis* Behavior   175
   4. Genetic Variability in Cotton Volatile Chemistry 184
   5. Effects of Allelochemical Toxins on Parasitoid
      and Host Development   185
   6. Summary   189
      References   191

6. Natural Enemies and Herbivore–Plant
   Interactions: Influence of Plant Allelochemicals
   and Host Specificity   201

   P. Barbosa

   1. Introduction   201
   2. Influence of Allelochemicals on Specialist and
      Generalist Herbivores   205
   3. Parasitoid Host Specificity and the Influence
      of Plant Allelochemicals   211

4. Interactions Among Herbivores, Allelochemicals,
   and Insect Pathogens   215
5. Counterintuitive Evolution or Unpredictable
   and Variable Selective Forces?   219
   References   223

PART IV. KEY ROLES OF PLANT
          ALLELOCHEMICALS IN SURVIVAL
          STRATEGIES OF HERBIVORES   231

D. K. Letourneau

7. Plant-Derived Defense in Chrysomelid Beetles   235

    J. M. Pasteels, M. Rowell-Rahier and M. J. Raupp

    1. Introduction: Diversity and Distribution of
       Chemical Defenses in Leaf Beetles   236
    2. Influence of the Host Plant on the Defense
       Secretion of Chrysomelinae   241
    3. Impact of Host Plant Variation on the Spatial
       and Temporal Distributions of Leaf Beetles   247
    4. Efficacy of the Different Groups of Chemical
       Compounds and Their Distribution Among
       Developmental Stages of Herbivores   252
    5. Protection Against Predators and Parasitoids   254
    6. Protection Against Competing Herbivores   262
    7. Summary and Conclusions   264
       References   266

8. Plant Allelochemistry and Mimicry   273

    M. D. Bowers

    1. Introduction   274
    2. Unpalatable Insects and the Chemical Bases of
       Unpalatability   275
    3. Allelochemical Variation and Implications
       for Mimicry   286
    4. Plant Allelochemistry and Predator Behavior   294
    5. Plant Allelochemistry and the Evolution
       of Mimicry   296
    6. Future Directions for Research on the Role
       of Plant Allelochemistry Mimicry   298
       References   301

9. Potential Role of Plant Allelochemicals in the
Development of Insecticide Resistance   313

   L. B. Brattsten

   1. Introduction   313
   2. Resistance and Induction: Similarities
      and Differences   314
   3. Resistance Mechanisms   318
   4. Induction of Insecticide-Metabolizing Enzymes   333
   5. Does Induction Help Insects Develop
      Resistance?   337
   6. Summary   340
      References   341

Species Index   349

Subject Index   357

Novel Aspects of Insect–Plant Interactions

PART I

CONCEPTUAL FRAMEWORK OF THREE-TROPHIC-LEVEL INTERACTIONS

In 1980, Price et al. put forth a compelling argument: The development of theory on insect–plant interactions, heretofore preoccupied with interactions between herbivores and their host plants, could not "progress realistically without consideration of the third trophic level." To introduce the first two chapters and to set the stage for the entire volume, I discuss here the importance of this argument and describe the impact it has had on ecological thought. Have we witnessed an actual change in ecological studies since the publication of this challenge by Price and coworkers? A comparison of papers published in the journal *Ecology* separated by a decade does show a clear trend toward greater consideration of complex multi-trophic-level interactions. Of 125 papers published in *Ecology* in 1976 and 181 in 1986, 23% and 31%, respectively, included studies of interactions between species from different trophic levels. In 1976, 13.8% of those represented studies of three-trophic-level interactions (Bentley 1976, Kushlan 1976, Morse 1976, Root and Chaplin 1976) and the rest were studies of resource exploitation between two trophic levels. In 1986, the proportion of trophic interaction studies that involved at least three levels was nearly double (26.8%). When only studies on terrestrial arthropods are considered, the percentage of studies on interactions involving three trophic levels was initially higher (i.e., 20%); however, the proportion in 1986 was still double, at 42.3%. If we can assume that trends in

1

topics in *Ecology* are real, not simply a change in editorial selection by the journal, there seems to have been a general shift toward more complex interactive studies. Among trophic-level studies, if two-level interactions (e.g., herbivore–plant, predator–prey, parasite–host) held central importance in the 1970s, a more holistic approach is the hallmark of ecology in the 1980s (Whitham 1983, Price et al. 1984, Levins and Lewontin 1985, Duffey and Bloem 1986).

Why has there been a propensity for two-trophic-level studies? In past years, investigations not only of plant–herbivore relationships, as pointed out by Price et al. (1980), but also studies of predator–prey interactions, involved primarily two species in laboratory or field studies or as mathematical constructs. The motivation for such simplification stems, at least in part, from systemic constraints of the scientific method. Controlled experiments are more feasible, analytical models more precise, and statistically sound experimental designs more tractable with one or two populations. Consider the differences, for example, between a field experiment to test the effects of diamondback moth larval density on parasitism rates and one to test the effects of host plant nitrogen content on larval density and parasitoid attack rates. Host plants encountered by ovipositing females are not likely to be homogeneous with respect to nutritional levels; and differences are likely within and among individuals as well as among populations. Not only do the total number of levels multiply (larval densities $\times$ N content), but the dissection of possibly confounding factors increases the complexity of the design. Is nitrogen distributed evenly among plant parts fed upon by larvae? Are parasitoids responding to plant factors, to larval quality, to larval density, or to some combination of factors? Clearly, though, the latter, multidimensional experiment is more realistic.

Restricted experimentation, then, is at once elegant, manipulable, analyzable, and incomplete. The implicit assumption within tests of hypotheses on predator–prey or herbivore–plant interactions is that the effects of the additional levels are constant across treatments. In patch-size experiments, for example, herbivore levels are routinely sampled and compared as response variables to the unit size of the host plant assemblage. The experiments are designed to answer questions about resource exploitation in heterogeneous environments, a vitally important set of questions for understanding a vast array of ecological patterns. To deduce that differences in herbivore densities within resource patches are due to patch size, the assumption must be that mortality factors are responding equally along the size gradient. If herbivores find, immigrate to, and emigrate from patches of different size at different rates, is it reasonable to assume that predators and parasitoids do not? Experiments that consider the role of at least one additional trophic level will improve the quality of theory derived from these and many other simplified systems such that the results will more accurately predict patterns in complex systems.

Many examples are provided throughout this volume for which the dismissal of plants as important factors in driving insect predator–prey or parasitoid-host interactions leads to misinterpretation of experimental results or to the failure of biological control programs. Vegetation is a primary habitat factor for hundreds of thousands of insects as well as a variable mixture of resource and obstacle for herbivores. New emphasis on the effects of plants on predation and parasitism of their associated herbivores has spurred a number of relevant studies.

For example, the searching efficiency of *Podisus maculiventris* predators calculated in an artificial arena with prey items is unlikely to reach values obtained from tests in which the feeding of its herbivore prey has caused the release of a plant allelochemical that attracts the predator (Greany and Hagen 1981). Without consideration of the third trophic level, a plant volatile that does not have a direct effect on the herbivore may be deemed unimportant in the plant–herbivore interaction. Similarly, the work of Whitman and coworkers on *Romalea* grasshoppers (Chapter 1) shows that predation rates on this polyphagous herbivore are not independent of the grasshopper's diet of food plants. Nordlund, Lewis, and Altieri (Chapter 2) summarize data from parasitism experiments using *Trichogramma pretiosum,* which show that attack rates of *Heliothis zea* eggs cannot be generalized across agroecosystems, but in fact, are dependent upon the crop composition. Similar effects of plant composition were found in lowland, tropical agroecosystems. *Diaphania hyalinata,* a squash specialist, suffered higher parasitism rates in general, and by *T. pretiosum* in particular, in the presence of maize than in pure stands of squash (Letourneau 1987). Recent work on plant resistance is also contributing a great deal of information on such discrepancies. For example, finely tuned recommendations of predatory mite releases on cotton will depend not only on predator–prey interaction studies, but also on the strength and timing of the induction of plant resistance and its effect on herbivorous mite populations (McMurtry and Scriven 1968, Karban and Carey

1984, Harrison and Karban 1986). Thus, the outcome of interactions between two species within a context of at least three trophic levels is qualitatively different from that gained as a by-product of data from two-level studies.

One of the reasons that two-level interactions taken separately cannot adequately describe interactions on three levels is that many processes of resource exploitation are coupled and plant–predator/parasitoid interactions involve responses to emergent factors associated with coupling of three trophic levels. To explore the variety of possible interconnections and their mechanisms, consider the steps in finding and exploiting a phytophagous host or prey item. The process of host/prey, finding can entail (1) coming into contact with appropriate habitats that contain hosts or prey, sometimes over long distances if ovipositing females are the searching stage; (2) contacting particular host plants if prey or hosts are likely to be associated with them; (3) moving along the substrate to encounter prey individuals; (4) assessing the suitability of prey items; and (5) capturing or handling the resource. Each of these steps is likely to be modified by plant characteristics. Synchrony, habitat location, and prey detection can be cued by plant volatiles during the initial phases. For example, gravid female green lacewings, *Chrysoperla carnea,* are arrested in flight by the detection of indole acetaldehyde from the honeydew of their aphid prey plus a volatile that is released during active plant growth of cotton (Hagen 1986). Habitat suitability for parasitoids or predators can also be determined by plant characteristics or microenvironments caused by the vegetation. *Artogeia* (= *Pieris*) *rapae* escapes contact with a major parasitoid, *Cotesia glomeratus,* when feeding on crucifer host plants that grow below the foliage of other plants. In undergrowth, it encounters a certain degree of "enemy-free" conditions because the parasitoid avoids plants growing under shady conditions, whether or not they contain hosts (Sato and Ohsaki 1987). Vegetation can also provide mechanical interferences to movement and thereby modify predator–prey or parasitoid–host interactions. Plant density and structural complexity themselves, independent of prey density, seem to increase the tenure time, and thus the foraging activity of the generalist predator, *Orius tristicolor,* in a patch (Letourneau and Altieri 1983; Letourneau, unpublished data). Finally, the activity of parasitoids and predators attacking similar herbivores can depend upon the properties of the host plant on which the herbivore is feeding. Different species of eulophid parasitoids, for example, attack *Liriomyza* leafminers at different rates on different host plants (Johnson and Hara 1987); thus herbivore mortality, as discussed previously, can depend upon the species composition of enemies, which, in turn, is influenced by plant species.

The expansion in studies of three trophic levels is not restricted to fields of insect behavior but is evident across a range of subdisciplines in ecology. Major advances are taking place in the development of analytical and simulation models (May and Hassell 1981, Pimm 1982, Gutierrez 1986), in biological control (Nordlund et al. 1981, Boethel and Eikenbary 1986, Kareiva 1986, Powell 1986), and in chemical ecology (Bell and Cardé 1984). Advances in theoretical ecology are addressing a variety of types of interactions, including mutualism and competition across trophic levels. Vandermeer (1980) put forth a conceptual model of indirect mutualism, suggesting that mutualism and competition may alternate as organizing forces in the community as one ascends trophic levels. Lawton (1986) expands the argument that competition between phytophagous insects can be regulated by shared parasitoids via many avenues, including patterns of host plant use. Pimm (1982) used "vertical communities," or species linked trophically, to assess the degree of compartmentalization in local food webs based on different resources in the first trophic level.

Of the wide range of mechanisms by which plants and insects interact, we have chosen to concentrate on only a subset in this volume of collected works—those mediated by allelochemicals. Behaviorally or physiologically active chemicals are ubiquitous and form the mechanism of a wide range of plant–insect interactions. "Allelochemics" (Whittaker 1970) may be defined as nonnutrient substances originating from an organism but affecting the behavior, physiological condition, or ecological welfare of organisms of a different species (Scriber 1984). Allelochemicals mediate interactions between plants, herbivores, and predators/parasitoids. As illustrated earlier in the discussion of host-finding and prey-finding processes, chemical effects work in concert with other cues: visual and auditory signals, physical conditions like temperature and humidity, and mechanical factors.

What place do chemically mediated interactions hold in the study of tritrophic level interactions between plants, herbivores, and their parasitoids and predators? As illustrated in the following chapters, interactions involving behavioral and physiological responses of receivers to chemicals emitted by other organisms are rich and abundant. Their recent emergence as a distinct and growing area of research may create an impression of dominance over the importance of other mechanisms such as visual, auditory, and tactile cues. A more direct source of such an impression is claims of researchers, themselves: "In insects the chemistry of the environment is a dominant modality mediating adaptive behavior, including the choice of food and feeding, avoidance of danger, location of a sexual partner and the choice of a habitat for the progeny" (Städler 1984). "The available information concerning host selection by insect parasitoids suggests tha host selection is regulated by a combination of factors, the most important of which appears to be chemicals" (Vinson 1976). There may, however, be no basis at this time for constructing a hierarchy that attributes a dominant role to chemicals. Bell and Cardé (1984), in fact, warn against ignoring or suppressing the importance of other modalities and thus losing track of vast quantities of information available to insects. The unique aspect of allelochemicals is that they affect organisms primarily through two sensory modalities (olfaction and gustation), but also through tactile senses (e.g., sticky exudates from trichomes) and via direct physiological pathways when allelochemicals are ingested or absorbed. The relative importance of chemical mediation is difficult to approach quantitatively (e.g., the proportion of an insect's sensory apparati, energy budget, or fitness that is devoted to or dependent upon receiving and processing chemicals versus visual stimuli) because receptors are not comparable and sufficient data may never be available. Certainly, the role of chemicals is very important, and the burgeoning field of study is contributing a great deal to basic knowledge and its applications.

Thus, this volume represents the forefront of two rapidly advancing areas of ecology: three-trophic-level interactions and the interdisciplinary field of chemical ecology. Although all possible categorical interactions, parasitism, mutualism, competition, and commensalism, are suitable targets of studies among multiple trophic levels, most of the chapters emphasize resource exploitation in the form of predation (predators and parasitoids) and herbivory. Chapters 1 and 2, comprising Part I, are both concerned with plants, insects, and allelochemicals. Whitman introduces the language of chemical ecology and emphasizes predators in interactive systems. Nordlund, Lewis, and Altieri review the role of plant chemicals as mediators of parasitoid–host interactions, and synthesize current knowledge on its application in the biological control of agricultural pests.

D. K. LETOURNEAU

REFERENCES

Bell, W. J., and R. T. Cardé (eds.). 1984. Chemical Ecology of Insects. Chapman & Hall, London.

Bentley, B. L. 1976. Plants bearing extrafloral nectaries and the associated ant community: interhabitat differences in the reduction of herbivore damage. Ecology 57:815–820.

Boethel, D. J., and R. D. Eikenbary (eds.). 1986. Interactions of Host Plant Resistance and Parasitoids and Predators of Insects. Wiley, New York.

Duffey, S. S., and K. A. Bloem. 1986. Plant defense-herbivore-parasite interactions and biological control. In M. Kogan (ed.), Ecological Theory and Integrated Pest Management Practice. pp. 135–184. Wiley, New York.

Greany, P. D., and K. S. Hagen. 1981. Prey selection. In D. A. Nordlund, R. L. Jones, and W. J. Lewis (eds.), Semiochemicals: Their Role in Pest Control. pp. 121–135. Wiley, New York.

Gutierrez, A. P. 1986. Analysis of the interactions of host plant resistance, phytophagous and entomophagous species. In D. J. Boethel and R. D. Eikenbary (eds.), Interactions of Plant Resistance and Parasitoids and Predators of Insects. Wiley, New York.

Hagen, K. S. 1986. Ecosystem analysis: plant cultivars (HPR), entomophagous species and food supplements. In D. J. Boethel and R. D. Eikenbary (eds.), Interactions of Plant Resistance and Parasitoids and Predators of Insects. Wiley, New York.

Harrison, S., and R. Karban. 1986. Behavioral response of spider mites (Tetranychus urticae) in induced resistance of cotton plants. Ecol. Ent. 11:181–188.

Johnson, M. W., and A. H. Hara. 1987. Influence of host crop on parasitoids (Hymenoptera) of Liriomyza spp. (Diptera: Agromyzidae). Environ. Ent. 16:339–344.

Karban, R., and J. Carey. 1984. Induced resistance of cotton seedlings to mites. Science 197:497–499.

Kareiva, P. 1986. Trivial movements and foraging by crop colonizers. In M. Kogan (ed.), Ecological Theory and Integrated Pest Management Practice. pp. 59–82. Wiley, New York.

Kushlan, J. A. 1976. Environmental stability and fish community diversity. Ecology 57:821–825.

Lawton, J. H. 1986. The effects of parasitoids on phytophagous insect communities. In J. Waage and D. Greathead (eds.), Insect Parasitoids. pp. 265–287. Academic Press, London

Letourneau, D. K. 1987. The enemies hypothesis: tritrophic interactions and vegetational diversity in tropical agroecosystems. Ecology 68:1616–1622.

Letourneau, D. K., and M. A. Altieri. 1983. Abundance patterns of a predator, *Orius tristicolor* (Hemiptera: Anthocoridae), and its prey, *Frankliniella occidentalis* (Thysanoptera: Thripidae): habitat attraction in polycultures versus monocultures. Environ. Ent. 12:1464–1469.

Levins, R., and R. Lewontin. 1985. The Dialectical Biologist. Harvard University Press, Cambridge, Mass.

May, R. M., and M. P. Hassell. 1981. The dynamics of multi-parasitoid host interactions. Amer. Nat. 117:234–261.

McMurtry, J. A., and G. T. Scriven. 1968. Studies on predator–prey interactions between *Amblyseius hibisci* and *Oligonychus punicae:* effects of host-plant conditioning and limited quantities of an alternate food. Ann. Ent. Soc. Amer. 61:393–397.

Morse, D. H. 1976. Variables affecting the density and territory size of breeding sprucewoods warblers. Ecology 57:290–301.

Nordlund, D. A., W. J. Lewis, and H. R. Gross. 1981. Elucidation and employment of semiochemicals in the manipulation of entomophagous insects. In E. R. Mitchell (ed.), Management of Insect Pests with Semiochemicals: Concepts and Practice, pp. 463–475. Plenum, New York.

Pimm, S. L. 1982. Food Webs. Chapman & Hall, New York.

Powell, W. 1986. Enhancing parasitoid activity in crops. In J. Waage and D. Greathead (eds.), Insect Parasitoids, pp. 319–340. Academic Press, London.

Price, P. W., C. E. Bouton, P. Gross, B. A. McPheron, J. N. Thompson, and A. E. Weis. 1980. Interactions among three trophic levels: influence of plants on interactions between insect herbivores and natural enemies. Annu. Rev. Ecol. Syst. 11:1141–1165.

Price, P. W., C. N. Slobodchikoff, and W. S. Gaud. 1984. A New Ecology. Wiley, New York.

Root, R. B., and S. J. Chaplin. 1976. The life-styles of tropical milkweed bugs, *Oncopeltus* (Hemiptera: Lygaeidae) utilizing the same hosts. Ecology 57:132–140.

Sato, Y., and N. Ohsaki. 1987. Host-habitat location by *Apanteles glomeratus* and effects of food-plant exposure on host-parasitism. Ecol. Ent. 12:291–297.

Scriber, J. M. 1984. Host-plant suitability. In W. J. Bell and R. T. Cardé (eds.), Chemical Ecology of Insects, pp. 159–202. Chapman & Hall, London.

Städler, E. 1984. Contact chemoreception. In W. J. Bell and R. T. Cardé (eds.), Chemical Ecology of Insects. pp. 3–36. Chapman & Hall, London.

Vandermeer, J. H. 1980. Indirect mutualism: variations on a theme by Stephen Levine. Amer. Nat. 116:441–442.

Vinson, S. B. 1976. Host selection by insect parasitoids. Annu. Rev. Ent. 21:109–133.

Whitham, T. G. 1983. Host manipulation of parasites: within-plant variation as a defense against rapidly evolving pests. In R. F. Denno and M. S. McClure (eds.), Variable Plants and Herbivores in Natural and Managed Systems. pp. 15–41. Academic Press, New York.

Whittaker, R. H. 1970. The biochemical ecology of higher plants. In E. Sondheimer and J. B. Simeone (eds.), Chemical Ecology. pp. 43–70. Academic Press, New York.

CHAPTER 1

Allelochemical Interactions Among Plants, Herbivores, and Their Predators

D. W. Whitman
Illinois State University
Normal, Illinois

CONTENTS

1 Introduction
2 Synomones
  2.1 Plant-produced synomones
    2.1.1 Floral scents, extrafloral nectaries, and food bodies
    2.1.2 Plant odors
    2.1.3 Guardian synomones
  2.2 Herbivore-released synomones
  2.3 Predator-released synomones
3 Antimones
4 Kairomones
  4.1 Plant-produced kairomones
  4.2 Herbivore-released kairomones
    4.2.1 Prey location and identification
    4.2.2 Plant detection of herbivores
  4.3 Predator-released kairomones
5 Allomones
  5.1 Plant-produced allomones
  5.2 Predator-released allomones
    5.2.1 Chemocryptic or mimetic allomones
    5.2.2 Prey disruption allomones
    5.2.3 Prey attraction allomones

     5.2.4   Prey subduction allomones
     5.2.5   Predator defense allomones
  5.3   Herbivore-released allomones
     5.3.1   Plant manipulation allomones
     5.3.2   Defense allomones
     5.3.3   Chemocryptic or mimetic allomones
     5.3.4   Plant-derived allomones
        5.3.4.1   The monarch
        5.3.4.2   The lubber grasshopper
        5.3.4.3   Monarchs and lubber grasshoppers: conclusions
6   The Future
7   Caveat
   References

1   INTRODUCTION

Paralleling the celebrated food web that links the various trophic levels of all communities is an allelochemical web. This vast network of interspecifically active substances is as significant as and in some ways more extensive than its trophic counterpart because it can unite organisms that are unbound nutritionally. In this sense, allelochemicals help to organize and structure a community. As with food webs, allelochemical associations can extend beyond two adjacent trophic categories to mediate interactions between three or more levels (Price et al. 1980). This chapter deals with those diverse allelochemical interactions important to insect herbivores, their food plants, and their predators. Tritrophic relationships are emphasized, especially those involving the direct and indirect effects of plant allelochemicals on predators.

In any tritrophic level food chain (i.e., where a carnivore feeds upon an herbivore, which feeds upon a plant), the allelochemical relationships can be organized according to benefit (Fig. 1.1; see also Chapter 4). A compound released by one organism which evokes a reaction in an individual of a different species that is favorable to the emitter but not to the receiver is termed an "allomone." A compound released by an organism which evokes a response beneficial to a member of another species but not to the emitter is termed a "kairomone." Substances released by organisms which benefit both the sender and receiver are termed "synomones." Finally, as proposed here, a substance produced or acquired by an organism that, when it contacts an individual of another species in the natural

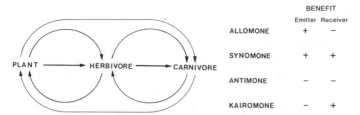

Figure 1.1   Allelochemical interactions between three trophic levels. Each arrow represents four theoretically possible ecological relationships: allomonal, synomonal, antimonal, or kairomonal.

context, evokes in the receiver a behavioral or physiological reaction that is maladaptive to both the emitter and the receiver is termed an "antimone."

The use of these terms is not without controversy (see Burghardt 1970; Blum 1974, 1980; Nordlund and Lewis 1976; Duffey 1977; Weldon 1980; Nordlund 1981; Pasteels 1982), since a substance produced by a species can serve a variety of functions simultaneously. For example, pine tree terpenoids may act as allomones, kairomones, or synomones. They deter herbivory and thus function as plant allomones (Smith 1963). However, bark beetles are attracted to these compounds, and thus they also perform as kairomones. Since certain bark beetle predators are also attracted to the tree terpenoids (Wood 1982), these chemicals can serve as synomones; that is, they are beneficial to both the host tree and the carnivore. Similarly, the roles of other compounds change sequentially as they move through the food chain. The cucurbitacins produced by the Cucurbitaceae are feeding stimulants for the cucumber beetle, *Diabrotica undecimpunctata,* and therefore serve as kairomones (Chambliss and Jones 1966). The same bitter substances are sequestered in the blood and tissues of the beetle, where they presumably serve as allomones against predators (Ferguson and Metcalf 1985, Ferguson et al. 1985). Some have argued against the concept of kairomones since detrimental substances should never evolve; maladaptive substances should be quickly selected against (Blum 1974, Pasteels 1982). However, organisms interact with many species; a substance having detrimental effects in one situation may provide overwhelming benefits in other situations and thus be evolutionarily favored. The multiple functionality of such allelochemicals makes it difficult to calculate their total net values. It is thus imperative that these substances be categorized only in reference to precisely defined situations.

Despite the controversy surrounding these terms, they are invaluable in teaching, and they help us to organize and understand these very real relationships. Furthermore, their use encourages consideration of their broader ecological and evolutionary significances. These terms therefore have legitimate value if their use is restricted to specific, defined interactions, and if there remains the realization that in different contexts, one substance may well serve divergent purposes.

2 SYNOMONES

Synomones are common in insects, mediating plant–herbivore and herbivore–predator interactions as well as tritrophic level interactions.

2.1 Plant-Produced Synomones

2.1.1 *Floral Scents, Extrafloral Nectaries, and Food Bodies*

The best known and most prevalent plant–herbivore synomonal interaction links flowering plants and their pollinators. Floral volatiles benefit plants by attracting pollinators and benefit insects by serving as nutrition and mate location sign posts (Barth 1985, Pellmyr and Thein 1986).

Flower scents also attract and/or arrest potential predators and parasitoids (natural enemies) of herbivores. Many such natural enemies are carnivorous only as larvae: as adults they require nutrition from flowers (Clausen 1940, House 1976, Hagen 1986). Plant pollen and nectar increase the life spans and fecundities of many parasitoids and predators, fostering greater herbivore mortality (Leius 1967, Sundby 1967, Streams et al. 1968, Altieri and Whitcomb 1979, Altieri and Letourneau 1982, Foster and Ruesink 1984).

The ability to allure pollinators and natural enemies holds important ramifications for plant fitness. Flowers differ intra- and interspecifically in their floral rewards and attractive capabilities (Streams et al. 1968, Shahjahan 1974, Baker and Baker 1983, Simpson and Neff 1983). These differences may determine visitation rates of appropriate entomophages and pollinators. In some plants, nectar and pollen production are synchronized with periods when herbivores are most damaging or susceptible to natural enemies. Indeed, an unexplored possibility is that availability of flower nutrients (nectar and pollen) remains high for periods much longer than necessary for fertilization, thereby attracting and feeding carnivores.

Sugar-rich extrafloral nectaries and protein- and lipid-laden food bodies are thought to serve as currency in a number of plant-carnivore interactions (Hagen 1962, 1986; Janzen 1975, Bentley 1977, O'Dowd 1980, Altieri 1981, Hespenheide 1985, Beattie 1985 but see Boecklen 1984, Rogers 1985). Many such plant–entomophage associations have evolved, producing highly dependent and specialized plants and predators (Janzen 1966) as well as herbivores (Heads and Lawton 1985). It has even been suggested that wild cherry and catalpa trees time extrafloral nectar production to coincide with susceptible periods of herbivores (Tilman 1978, Stephenson 1982a). If extrafloral nectaries have evolved to supplement natural enemies, then they may contain compounds that act as synomones.

2.1.2 Plant Odors

A growing body of information suggests that plant odors are extremely important in tritrophic level interactions. Although most examples involve parasitoids attracted to the host's food plant (Ullyett 1953; Monteith 1958; Arthur 1962; Read et al. 1970; Nettles 1980; Altieri et al. 1981, 1982; Lecomte and Thibout 1984; Vinson 1984), some cases of predator attraction are known. *Anatis ocellata,* the coccinellid predator of the pine aphid, orients to pine needle odors (Kesten 1969), and the carabid *Bembidion obtusidens,* is attracted by odors emanating from *Oscillatoria* algal mats (Evens 1982).

Particularly exciting is the possibility that some plants respond to feeding or tissue damage by emitting synomones attractive to herbivore natural enemies. The idea of plants sending a chemical "SOS" when under attack (the same way they "notify" other conspecifics) is an intriguing and potentially important one. For example, *Podisus maculiventris,* the spined soldier bug, orients to soybean plants damaged by *Trichoplusia ni* (Greany and Hagen 1981). The predators *Chrysoperla* (= *Chrysopa*) *carnea* and *Collops vittatus* are attracted by caryophyllene, a terpenoid released by damaged cotton leaves (Flint et al. 1979), and the eastern yellow jacket, *Vespula maculifrons,* is thought to use plant-released substances to locate leaf-feeding insects (Aldrich et al. 1985). The predaceous dolichopodid fly, *Medetera bistriata,* is attracted to its bark beetle prey only when both frontalin, the beetle's aggregation pheromone, and tree odors are present (Williamson 1971). Pine trees under attack liberate volatile terpenes attractive to numerous bark beetle predators (Wood 1982, Mizell et al. 1984). Herbivores may counter such plant responses by dispersing (Faeth 1985) or by severing damaged leaves so that they fall to the ground (Heinrich and Collins 1983).

Some cruciferous plants are normally relatively odorless. But in response to attacking insects they release enzymes which quickly convert inactive mustard oils into volatile parasitoid-attracting derivatives (Shmida and Auerbach 1983). Since these pungent compounds also serve as phagostimulants for numerous crucifer herbivores, it is advantageous to the plant to "hide" its odor until needed.

Substances released by some plants may profoundly influence ecological relationships in other plant species. Strong odors from one plant may lure or repel herbivores or natural enemies to or from the environment or mask odors of attractive plants (Tahvanainen and Root 1972, Raros 1973, Shahjahan and Streams 1973, Thiery and Visser 1986). Onions, garlic, and marigolds are commonly interspersed with vegetables in home gardens, where they are thought to repel herbivores. Wild mustard and nasturtium serve as trap crops for crucifer pests, apparently because of their high sinigrin levels (Hunter 1964, Altieri and Gliessman 1983, Altieri and Schmidt 1986). Monteith (1960) found that tachinid parasitism of a larch-feeding sawfly was reduced when pines were nearby, concluding that pine odors masked the attractive scent of the sawfly and its host. Read et al. (1970) concluded that the proximity of collards to sugar beets may enhance biological control of beet pests, because collards attracted braconid parasitoids to the general area. In some instances, diversified agroecosystems have suffered less herbivory than monocultures (Altieri and Whitcomb 1979, Altieri et al. 1981, Risch et al. 1983, Altieri and Letourneau 1984, Letourneau 1986), but this is not always the case (van Emden and Williams 1974, Sheehan 1986).

2.1.3 Guardian Synomones

The current overwhelming body of evidence indicates that the primary evolutionary purpose of plant secondary substances is allomonal: protection from herbivores, disease, and competition. But in many cases, herbivores turn the use of these substances to their own profit, by sequestering them for defense, a situation seemingly detrimental to the plant (see Sections 4.1 and 5.3). An intriguing hypothesis regarding the function of such compounds has recently been offered by Rothschild (1985). She suggests that plants may afford protection to their pollinators by providing them with food and secondary substances. Such a system might operate with the tree, *Catalpa speciosa*, and its major herbivore, the catalpa sphinx caterpillar, *Ceratomia catalpae*. This insect obtains deterrent iridoid glycosides from the tree, making it unpalatable to birds (Bowers and Puttick 1986). The adult moth, a principal pollinator, is unaffected by the toxic iridoid glycosides in the nectar, which incapacitate nonadapted nectar thieves (Stephenson 1982b).

There is no doubt that reproduction is extremely important for plants, as attested by the high percentage of metabolic resources directed to flower and seed production. The loss of a small portion of tissue to a "dedicated" pollinator may be cost-effective. Such mutually beneficial relationships are seen with fig wasps (Galil and Eisikowitch 1969, Ramirez 1969) and yucca moths (Powell and Mackie 1966). Elucidation of the pollination roles of allomone-sequestering herbivores is a promising area for future investigation.

2.2 Herbivore-Released Synomones

Synomone mediation of herbivore–predator interactions is illustrated by the nonnutritive ant attractants, arrestants, and feeding stimulants produced by aphids, treehoppers, and lycaenid butterfly larvae (Way 1963, Atsatt 1981, Pierce and Mead 1981, Boucher 1982, Fritz 1982, Pierce 1984). Strong, mutualistic associations have evolved between certain ant and aphid species whereby the ants protect the aphids from predators and parasitoids, and even move them from senescent to fresh plants or, in the winter, to subterranean locales (Way 1963). In return, aphids provide ants with carbohydrate and amino acid–containing honeydew. Some ants even respond to the alarm pheromone of their aphid symbionts, by orienting toward the pheromonal source and presumably attacking aphid predators (Nault et al. 1976).

2.3 Predator-Released Synomones

Predators may also emit synomones. In the mutualistic relationship between the ant, *Pheidole bicornis,* and the plant *Piper cenocladum,* high numbers of nutritious plant-produced food bodies appear only when *P. bicornis* is present. When other *Pheidole* species inhabit the plant, no food bodies are produced (Risch and Rickson 1981). A chemical factor from the ant may stimulate food body production; the actual mechanism is not yet known.

3 ANTIMONES

Substances released by organisms occasionally produce reactions in receivers which are detrimental to both the emitter and the receiver. For example, substances released by pathogenic microorganisms sometimes cause abnormal behavior or premature death in infected hosts, reactions often detrimental to both species. Selection against such antimonal effects is thought to be the cause of the attenuation process observed in many diseases. Numerous examples of this type of interaction can be found.

In some situations an allomonal defense may prove deleterious to a plant because it deters natural enemies of the plant's herbivores. For example, many plant families are characterized by the presence of glandular trichomes. These tiny fragile hairs cover stems and leaves, rupturing at the slightest pressure to release a variety of herbivore-deterring allomones (Levin 1973, Rodriquez et al. 1984). Trichomes are especially prevalent among the Solanaceae. Wild tomatoes (*Lycopersicon*) release the methyl ketone, 2-tridecanone, a contact poison for many tomato herbivores (Williams et al. 1980, Dimock and Kennedy 1983). Trichomes of some wild tobacco (*Nicotiana*) species contain nicotine, nornicotine, or anabasine, alkaloids known to be toxic to a variety of insects, including parasitoids of tobacco herbivores (Thurston and Fox 1972, Barbosa et al. 1986). Potato (*Solanum*) trichomes exude the alkaloids veratrine, tomatine, and acotine and deter feeding by herbivores (see Levin 1973). Studies demonstrate that herbivore-deterring trichomes can have a negative mechanical, and perhaps chemical, impact on natural enemies (Ekbom 1977, Obrycki 1986). For example, predatory coccinellid and chrysopid larvae and adults are sometimes repelled or inhibited by glandular trichomes, presumably to the benefit of phytophages and to the detriment of the plant (Gurney and Hussey 1970, Elsey 1974, Belcher and Thurston 1982, Shah 1982, Obrycki and Tauber 1984, Obrycki et al. 1985).

Other plant products may, in certain cases, prove detrimental to both the plant and beneficial insects. Many plants restrict floral access to specialized pollinators through morphological, phenological, or chemical adaptations. Plants in the latter category produce pollen or nectar that contains alkaloids, saponins, unsaturated lactones, and other substances toxic to generalist flower visitors which usually accomplish little or no pollination (sometimes referred to as "nectar thieves"), but not to specialized coevolved species (Baker and Baker 1975, Janzen 1977, Barker 1978, Guerrant and Fiedler, 1981). For example, nectar thieves that imbibe *Catalpa speciosa* nectar exhibit abnormal behaviors, including regurgitation and loss of locomotion, due to toxic iridoid glycosides (Stephenson 1982b). However, in cases where specialized pollinators are scarce, the poisoning of generalist pollinators could be detrimental to the plant. Honeybees feeding on California buckeye (*Aesculus californica*), death camas (*Zigadenus venenosus*), and titi (*Cyrilla racemiflora*) flowers that contain the sugars mannose or galactose, or on certain poisonous extrafloral nectars, can be killed directly or can carry toxins back to the hive which destroy the entire colony (Robinson and Oertel 1975, Barker 1978). Homoptera feeding on poisonous plants sometimes produce honeydew toxic to honeybees (Barker 1978). Toxic floral or extrafloral nectars might also harm natural enemies that require or supplement their diet with pollen or nectar. Beneficials can also be negatively infuenced by mechanical properties of plant products. The leaves and stipules of some pea varieties are so waxy that coccinellid larvae slip off and are unable to feed on attendant pea aphids (Edwards and Wratten 1980). Bees are occasionally trapped in herbivore-deterring gums (Barker 1978). The sticky trichomes, resins, and slippery waxes on flower pedicels, which deter nectar thieves, also inhibit floral access for beneficial natural enemies.

Plant odors are instrumental in both attracting or repelling entomophages (Monteith 1955, 1958; Arthur 1962; Greenblatt and Lewis 1983). In the latter case these substances would serve an antimonal role, being detrimental to both the plant and the natural enemy, but beneficial for attendant herbivores. The allelopathic substances released by plants may also have antimonal side effects when they eliminate herbivore-repellent or natural enemy-conserving allospecifics (Altieri and Letourneau 1982).

Antimone production seems maladaptive (as does the release of kairomones); however, this does not negate the fact that for organisms participating in these specific interactions, antimonal effects are real. The existence of antimones is often explained by the overwhelming benefits imparted by these compounds in other contexts.

4 KAIROMONES

4.1 Plant-Produced Kairomones

The plant-produced substances that serve as attractants, arrestants, and oviposition and feeding stimulants for herbivores (Kogan 1976, Städler 1976, Schoonhoven 1981, Ahmad 1983, Miller and Strickler 1984) are undoubtedly the best known kairomones. Many of these diverse compounds function as allomones toward nonadapted herbivores (Rosenthal and Janzen 1979, Städler and Buser 1984, Kogan 1986). These plant-derived substances also influence organisms of the third trophic level (sometimes to the detriment of the plant) when they are sequestered by herbivores and used as allomones against natural enemies. For willows (*Salix* spp.), the production of a toxic phenol glycoside, salicin, can be detrimental (see Chapter 7). The herbivorous beetle, *Chrysomela aenicollis,* is not only undeterred by salicin, but utilizes this compound to produce a defensive secretion. Beetles generate more secretion and suffer less predation on willows with high salicin content. Furthermore, leaf damage is linearly and positively related to leaf salicin level (Smiley et al. 1985).

4.2   Herbivore-Released Kairomones

4.2.1   Prey Location and Identification

Carnivores utilize a variety of stimuli to locate and identify their prey, including chemical cues emanating from their prey. Numerous kairomones have been documented for vertebrate predators (Curio 1976) and insect parasitoids (Vinson 1984). Parasitoids respond to the host's body odor (Vinson 1976, Strand and Vinson 1982, Loke and Ashley 1984, Noldus and van Lenteren 1985a), sex pheromones (Sternlicht 1973, Nordlund et al. 1983, Kennedy 1984), epideictic pheromones (Corbet 1971, Prokopy and Webster 1978), aggregation pheromones (Wood 1982, Aldrich et al. 1984), salivary constituents (Vinson 1968, Corbet 1971, Loke and Ashley 1984, Mudd et al. 1984), excretory products (Lewis and Jones 1971, Loke and Ashley 1984, Stafford et al. 1984, Nordlund and Lewis 1985), webbing (Wesloh 1976, Loke and Ashley 1984), honeydew (Carter and Dixon 1984), body scales (Gueldner et al. 1984, Loke and Ashley 1984), and eggs (Jones et al. 1973, Strand and Vinson 1983, Noldus and van Lenteren 1985b).

Kairomone use by insect predators is less well known, although many examples exist. The pioneering work of Tinbergen (1958, 1972) demonstrated that the bee wolf, *Philanthus triangulum,* a predatory sphecid wasp, utilized olfaction to recognize its host, *Apis mellifera.* Bee-mimicking syrphid flies were not attacked, unless they were rubbed against a bee and thus made to smell like one.

Other studies have shown prey body odors to elicit searching, orientation, or attack in predaceous dyticids (Tinbergen 1951), carabids (Bauer

1982, Ernsting et al. 1985), mites (Sabelis and van de Baan 1983), chrysopid larvae (Fleschner 1950), coccinellids (Colburn and Asquith 1970), and
myrmecophilous insects (Holldobler 1969). Aphid odors are both an arrestant and an oviposition stimulant for the syrphid fly, *Syrphus corollae*
(Bombosch and Volk 1966), and a short-range attractant for the cecido-
myiid predator, *Aphidoletes aphidimyza* (Wilbert 1974). Predaceous
Geocoris bugs and coccinellids are attracted to solutions of ground aphids
(Ben Saad and Bishop 1976b). Larvae of *Chrysoperla carnea* respond to
kairomones emanating from *Heliothis zea* eggs as well as to scales left
behind by the ovipositing moth adult (Lewis et al. 1977, Nordlund et
al. 1977). Predaceous mites respond to house fly kairomones (Farish and
Axtell 1966, Jalil and Rodriguez 1970). An interesting example comes
from McLain (1979), who observed three species of predatory penta-
tomids following the terrestrial trails of caterpillars. The predators also
responded to artificial trails made with *Trichoplusia ni* frass or hemolymph drawn on paper.

Other prey products can influence insect predators. The odor of frass
or webbing from herbivorous mites is highly stimulatory to predatory
mites (Hislop and Prokopy 1981). Sugar-rich honeydew is an attractant or
arrestant for numerous aphid natural enemies (Zoebelein 1956). Syrphid,
chrysopid, and coccinellid adults feed on natural honeydews and deposit
eggs from which aphid-consuming larvae hatch (Hagen 1962, 1986; Hagen
and van den Bosch 1968). Carter and Dixon (1984) observed increased
foraging and prey capture by coccinellid larvae in the presence of honeydew. Artificial honeydews have been used to manipulate predator populations (Hagen et al. 1971, Ben Saad and Bishop 1976a, Nichols and Neel
1977, Greany and Hagen 1981). Proteinaceous artificial honeydews
sprayed onto foliage attracted chrysopid and syrphid adults and arrested
the movement of a number of coccinellid larvae that proceeded to feed on
the honeydew and search for prey. Among the volatiles in artificial honeydew attractive to *Chrysoperla carnea* adults was a tryptophan degradation product, indole acetaldehyde (Hagen et al. 1976, van Emden and
Hagen 1976).

Predators are also known to respond to prey pheromones (Wood 1982).
Thanasimus dubius, a clerid predator of bark beetles, is attracted to frontalin, a pheromone of its host, *Dendroctonus frontalis* (Mizell et al. 1984,
Payne et al. 1984). Some ants, including many slave-making species, exploit their host's communicatory substances. The formicid *Megaponera
foetens,* an obligate predator of termites, locates its prey via kairomones,

which possibly also serve as pheromones within the termite colony (Longhurst and Howse 1978, Howse 1984).

Ants themselves fall prey to the decoding and exploitation of their pheromones by inquilines and predators. The blind snake, *Leptotyphlops dulcis,* follows the foraging trails of the army ant, *Neivamyrmex nigrescens,* then feeds on colony members (Watkins et al. 1969). The ephydrid fly, *Rhynchopsilopa nitidissima,* which mounts and feeds on *Crematogaster* spp. ants, presumably utilizes ant chemical communication cues to locate its host (Freidberg and Mathis 1985). Numerous inquilines, including roaches, beetles, and lycaenid caterpillars, follow their host's trail pheromone (Akre and Rettenmeyer 1968, Wilson 1971, Kistner 1982, Schroth and Maschwitz 1984). Greany and Hagen (1981) relate an intriguing case of a coccinellid beetle that locates its aphid prey by following the trails of aphid-tending ants.

Finally, kairomonal interactions between herbivores and their enemies may be influenced indirectly by host plants. Many parasitoids exhibit differential responses to hosts reared on different plants (Monteith 1958, Mueller 1983, Elzen et al. 1984, Loke and Ashley 1984). Apparently, plant substances influence the body or frass odor of herbivores, inciting or deterring natural enemy attack (Monteith 1958, Hendry et al. 1976, Sauls et al. 1979, Elzen et al. 1984, Nordlund and Lewis 1985). Indeed, it may benefit plants to "tag" herbivores with appropriate substances attractive to natural enemies.

4.2.2 Plant Detection of Herbivores

Plants may also respond to herbivore-produced substances. Some plants counteract herbivory through a variety of facultative defenses (increased growth, differential growth, or allomone production) (Ryan and Green 1974; Baldwin and Schultz 1983; McNaughton 1983a,b; Rhoades 1983, 1985; Ryan 1983). Although this can occur as a consequence of direct plant tissue removal (Haukioja and Niemela 1977, Haukioja and Neuvonen 1985), increasing evidence suggests the involvement of herbivore kairomones. Although the mechanism is unknown, cotton seedlings become more resistant following attack by mites (Karban and Carey 1984). Buds, flowers, fruits, and leaves infested by insects are often aborted (Coakley et al. 1969, Faeth et al. 1981, Stephenson 1981, Addicott 1982, Chabot and Hicks 1982, Kahn and Cornell 1983, Pritchard and James 1984, Bultman and Faeth 1986). Plants respond to the saliva of some insects by fruit abortion (Carter 1939) and increased growth (Detling and Dyer 1981). Even the production of predator- and parasitoid-attracting extrafloral nectaries near insect galls (Felt 1940, Mani 1964) may represent kairomonally induced countermeasures by plants against the gall maker.

4.3 Predator-Released Kairomones

Most examples of invertebrate predator-produced kairomones come from marine ecosystems (von Uexkull 1921, Hoffman 1930, Tinbergen 1951, Atema and Burd 1975, Phillips 1977), although some insect cases have been documented. For example, many ant, wasp, and bee species that are themselves prey or competitors of ants respond to formicid kairomones (Spangler and Taber 1970, Chadab 1979). *Camponotus* and other ants perform startle reactions in response to army ant trails 1 to 2 cm distant (Schneirla 1971). Fire ants (*Solenopsis invicta*) release a kairomone that elicits alarm recruitment in the ant, *Pheidole dentata* (Wilson 1975). The evacuating behavior of some ants in response to army ant attack (Alloway 1979, LaMon and Topoff 1981) may be an adaptive response to army ant kairomones, allowing at least some of the host colony to escape.

Although our knowledge of the use of predator-produced kairomones by insect prey is scanty, more cases will be discovered. For example, the lengthy antennae of many nocturnal Orthoptera and Thysanura, long thought to serve primarily a mechanoreceptive function, may also play an important role in chemodetection of predators. Cricket antennae certainly possess chemosensory capabilities, since they respond to food and conspecific odors (Otte and Cade 1976, Rence and Loher 1977).

5 ALLOMONES

5.1 Plant-Produced Allomones

The primary plant defense against herbivory is the possession of toxic or repugnant allomones. These substances repel, deter, or harm many potential or actual phytophages (Rosenthal and Janzen 1979). The direct effects of plant allomones on herbivores are well established, but a growing body of information suggests that plant allomones can indirectly, yet substantially, affect the herbivore–carnivore trophic level as well: an impact that often holds important consequences for plant survival. Plant allomones can indirectly influence herbivore–carnivore relations by affecting prey vigor, longevity, and development rate. The slowing of development or reduction in vigor of a herbivore species may favor natural enemy populations, whose numerical and functional response could increase faster than the herbivore population (Dahms 1972). For example, predators kill twice as many *Heliothis zea* larvae on resistant cotton varieties than on susceptible varieties (Lincoln et al. 1971). Price et al. (1980) noted that predation by the pentatomid, *Podisus* sp., on the Mexican bean beetle, *Epilachna varivestis,* was much higher when beetles were reared on high-tannin soybean varieties. Similarly, *Lycosa* spiders consumed more *Nilaparvata lugens* planthoppers on resistant rice varieties than on nonresistant ones (Kartohardjono and Heinrichs 1984). Of course, plant resistance is derived from a variety of features, but defensive allomones are prominent (Maxwell 1972).

Plant allomones may also indirectly affect herbivores living in temporally restricted habitats. For a herbivore whose generation time barely fits a narrow phenological window, a slight increase in developmental time concomitant with increasing plant resistance could result in elimination of that herbivore population. Likewise, multivoltine species would produce fewer generations per year.

Some plants may have even broken through the specific and complex semiochemical codes utilized by associated herbivores. Wild potato, *Solanum berthaultii,* for example, repels the aphid, *Myzus persicae,* by releasing the aphid's alarm pheromone, (E)-β-farnesene (Gibson and Pickett 1983). Many plants, however, are associated with specialized herbivores that not only are unharmed by the plants' secondary substances, but utilize them as attractants and feeding stimulants to the detriment of the plant (Harborne 1982). Other herbivores circumvent plant chemical defenses through mass attack, gregarious feeding, or by isolating foliage from the vascular system of the plant (Carroll and Hoffman 1980, Wood 1982, Heinrich and Collins 1983, Tallamy 1985, Compton 1987).

5.2   Predator-Released Allomones

Predators (also competitors and inquilines) release a variety of substances detrimental to other organisms. Many of these are manipulation compounds that serve to deceive, confuse, attract, or subdue prey.

5.2.1   Chemocryptic or Mimetic Allomones

Although some predator-released substances warn prey, other substances released by predators may act to hide or camouflage their true identity.

Many of the inquilines sharing ant and termite nests, to the detriment of their hosts, would undoubtedly be attacked were it not for the release of mimetic or appeasement substances (Holldobler 1971, Wilson 1971, Kistner 1982, Garnett et al. 1985). The staphylinid termitophile, *Trichopsenius frosti*, synthesizes a complex of surface hydrocarbons nearly identical to that of its host, *Reticulitermes flavipes* (Howard et al. 1980). Instead of synthesizing its own surface compounds, the myrmecophilous scarab beetle, *Myrmecaphodius excavaticollis*, acquires the hydrocarbon blend of its host, *Solenopsis*, allowing it to survive within the ant nest. When transferred to the colony of another species, it loses its former odor and procures that of its new host (Vander Meer and Wojcik 1982). Other inquilines may survive simply because they possess no odor, or emit substances that deaden the chemosensory capabilities of their hosts. Such a mechanism might explain why Argentine ants (*Iridomyrmex humilis*) easily enter honeybee hives, whereas the odors of most other ant species elicit strong anti-formicid defense (Spangler and Taber 1970).

5.2.2 Prey Disruption Allomones

Some predators produce substances that disrupt or confuse other insects. When the slave-making ant, *Formica subintegra*, raids other colonies, it releases decyl, dodecyl, and tetradecyl acetates from the Dufour's gland. This "propaganda substance" causes confusion and dispersion within the prey colony (Regnier and Wilson 1971). Workers of another slave-making ant, *Harpagoxenus sublaevis*, secrete a substance onto the host's brood, making the brood unattractive to their adult sisters (Buschinger et al. 1980). Prey ants that grapple with or contact the slave makers are themselves attacked by conspecifics (Alloway 1979). Similarly, the workerless inquiline ant, *Leptothorax kutteri*, produces a substance that causes host workers, *L. acervorum*, to attack each other. Thus, host nestmate recognition is overridden (Allies et al. 1986). Finally, the stingless bee, *Lestrimelitta limao*, successfully raids other colonies to steal food stores by releasing citral from the mandibular glands. This substance acts as an allomone, eliciting nonoriented dispersion in the host colony (Blum 1966, 1970).

5.2.3 Prey Attraction Allomones

Some arthropod predators have evolved the ability to attract their prey chemically. Instead of building a web, the bolas spider, *Mastophora*, dangles a sticky ball of silk which it flings at passing insects. Eberhard (1977) found that prey captured by bolas spiders were invariably male noctuid moths of two related species. Apparently, the spider produces an allomone similar to the pheromone of the female moth, thus luring males to a sticky death (Stowe et al. 1987). An assassin bug, *Apiomerus pictipes*, releases a substance attractive to the stingless bee, *Trigona fulviventris*. Approaching bees are seized and fed upon (Weaver et al. 1975). Instead of producing its own prey attractant, the assassin bug, *Salyavata variegata*, draws termites into capture range by presenting the carcass of its last termite meal (McMahan 1983).

5.2.4   Prey Subduction Allomones

Prey subduction substances are widespread in insects and arachnids and are represented by an extensive array of primarily organic compounds delivered into the prey via fangs, probosci, stings, and modified ovipositors (Maschwitz and Kloft 1971, Blum 1981, Schmidt 1982). Even more interesting are those prey subduction allomones that are not injected into the prey, but released into the environment. *Lomamyia latipennis*, a berothid, is an obligate predator of termites. Larvae of these rare neuropterans roam freely in galleries of *Reticulitermes hesperus*. When ready to feed, a larva backs toward a termite's head, lifts and points its abdominal tip, and releases a volatile substance which incapacitates the termite within 2 to 5 minutes, allowing the berothid to dine without objection from its prey (Johnson and Hagen 1981). Subduction does not involve contact and the host makes no effort to flee.

Nymphs of the chigger, *Womersia strandtmanni*, secrete a substance that attracts collembolans. Feeding on this secretion produces temporary paralysis in the collombolan, allowing the mite to attack. Nonattacked hosts recover after a short period (Huber 1979). A similar example is found in the predatory hemipteran, *Ptilocerus ochraceus*. This assassin bug releases a secretion from ventral trichomes which is attractive to ants. Upon feeding, the ants become narcotized and are easy prey for the bug (Jacobson 1911).

5.2.5   Predator Defense Allomones

Even predators must defend themselves from other carnivores. Accordingly, many predators (e.g., ladybeetles, ants, stink bugs, and ground beetles) synthesize defensive compounds (Blum 1981). Some, however, obtain their defensive substances from their herbivore prey. When feeding on the cochineal insect, *Dactylopius confusus*, larvae of the pyralid moth, *Laetilia coccidivora*, acquire the anthraquinone, carminic acid, and use it to drive away predaceous ants (Eisner et al. 1980). Female *Photuris* fireflies do not produce the noxious lucibufagins characteristic of most other lampyrids. Instead, they acquire these deterrent substances by feeding on male fireflies of the related *Photinus,* which they attract by imitating the flash signal of *Photinus* females (Eisner 1980).

5.3 Herbivore-Released Allomones

Herbivores obtain or synthesize a variety of substances which when released into the environment prove advantageous for the herbivore but detrimental for other organisms. These substances can act against both lower (plant) and higher (natural enemy) trophic levels.

5.3.1 Plant Manipulation Allomones

Many insects manipulate plants through the use of allomones. Gall insects secrete substances that cause abnormal or cancerous growth in plants (Mani 1964). These growths provide the gall inhabitants with a nutritious, moist, and stable habitat as well as some protection from generalist predators and parasitoids, all at the plant's expense. Some gall makers may benefit further by eliciting higher levels of nutritive or predator/parasitoid-deterring defensive substances within the gall (Weis and Kapelinski 1984). Spines and spicules on galls may deter vertebrate and invertebrate herbivory and thus protect the gall maker. The sugary and sticky exudates associated with certain galls elicit some intriguing questions (Mani 1964). Do these exudates benefit the plant by attracting and feeding parasitoids and predators of gall makers, or do they benefit the gall maker by trapping natural enemies or attracting parasitoid-deterring ants (Bequaert 1924, Felt 1940)?

Other insects, such as leaf miners, may manipulate plants by inducing early leaf abscission (Kahn and Cornell 1983). The speculation that plant growth regulators occur in insect saliva implies that even generalist herbivores may manipulate plant growth patterns (Hori 1976, Detling and Dyer 1981). Went (1970) speculates that spider mites inject growth inhibitors into plants with the result that "less sugar is used up by the plant and more is available to the spider mite." Such herbivore-induced reactions in plants could, however, serve the plant (see Part 4).

5.3.2 Defense Allomones

The staggering variety of defensive substances utilized by insects, and the many ways in which they are employed, stand as a testament to the driving force of natural selection. Summarizing this broad field would require an entire volume and has been accomplished in part by others (see Eisner 1970, 1972; Weatherston and Percy 1970; Duffey 1977; Blum 1981; Pasteels et al. 1983; Rothschild 1985; see also Chapter 7). Therefore, I will concentrate only on the more recent or theoretical advances in the area.

5.3.3   Chemocryptic or Mimetic Allomones

Many predators hunt visually, and in response insects have evolved a myriad of cryptic appearances. Since many predators, especially nocturnal ones, utilize olfaction during hunting, it follows that some prey may have developed a chemical-based crypsis. Such chemoconcealment could operate via crypsis, mimicry, or predator sensory overload. Hypothetically, some insects may escape predation simply by possessing no body odor at all. Thus, predator chemosensory organs would not be stimulated, and in the absence of prey clues such as motion, the prey would not be detected. Other insects may mimic environmental odors, analogous to visual crypsis. For example, insects might escape predation by incorporating and releasing host plant odors. Instead of acting as irritating repellents, these would make the herbivore indistinguishable from the host plant. Olfactorally hunting predators contacting such prey would be unable to distinguish them from other nonprey items that produce similar odors. The algal gardens grown on the elytra of some beetles and the leaf fragments that some insects attach to their backs may function in this manner (Gressitt 1966a,b).

Some insects have successfully cracked the chemical communication code of their natural enemies, advantageously turning the predator's own pheromonal system against them. In certain situations, the predators become the prey, as with the inquilines discussed earlier which mimic ant body odors. Given the great number of insects species known to produce defensive allomones, it is not surprising that numerous insect exudates contain substances identical to predator semiochemicals. Whether these cases represent mere coincidences, or adaptive responses to specific predators, requires examination. Blum (1974, 1980) lists a variety of insects that incorporate well-known social insect pheromones in their defensive secretions.

Predator sensory overload represents another possible defense strategy. As Blum and Brand (1972) and Blum (1981) have suggested, insects may occasionally escape predation by hiding under a chemical "smoke screen." Indeed, the defensive secretions of some insects contain an immense variety of substances which may function in this manner. For example, the metathoracic exudate of the grasshopper, *Romalea guttata,* contains over 50 compounds (see below), while the Dufour's gland secretion of the ant, *Camponotus ligniperda,* contains almost that many (Bergstrom and Lofqvist 1971). Such a defensive potpourri may act as the chemical equivalent of "white noise," generating an uncoded array of spikes in sensory neurons, thereby "jamming" a predator's sensory abilities (Blum 1974, Duffey et al. 1977).

5.3.4 Plant-Derived Allomones

Of all the possible tritrophic level allelochemical interactions, only one has been demonstrated repeatedly and studied extensively. This is the relationship between plants, allomone-sequestering herbivores, and predators. Although the storage of plant secondary products by insects had long been suspected (Slater 1877, Haase 1893; reviewed in Brower and Brower 1964), it was not until 1967 that Reichstein first identified plant cardenolides in the monarch butterfly (Reichstein 1967). Knowledge in this area expanded rapidly, and only six years later Rothschild (1973) listed 42 additional species known to sequester plant allelochemicals. That number has grown substantially (Duffey 1977, 1980; Blum 1981, 1983; Rothschild 1985), and today it is generally accepted that sequestering insects gain protection against carnivores.

Two examples illustrative of plant-derived allomone sequestration for defense will be discussed: (1) the stenophagous, holometabolous monarch butterfly, and (2) the polyphagous, hemimetabolous lubber grasshopper.

5.3.4.1 The Monarch

Every incipient biologist is familiar with the monarch story, in which *Danaus plexippus* larvae feed on cardenolide-containing milkweeds, storing these compounds for defensive use against predators. Birds feeding on the adult butterflies are poisoned by the sequestered cardenolides and develop a conditioned aversion toward subsequent monarch feeding (Brower 1969).

However, this story is not as simple as it first appears, for the literature is replete with cases of monarch predation by mammals and birds (Clark 1927; Poulton 1932; Lane 1957; Urquhart 1960, 1976; Reichstein et al. 1968; Kammer 1970; Rothschild and Kellet 1972; Smithers 1973; Shapiro 1977; Tuskes and Brower 1978; Rothschild et al. 1984; Brower and Fink 1985; Brower et al. 1985). Peterson (1964) reported that 110 of 112 wingless monarchs were eaten by brown thrashers, grackles, robins, cardinals, and sparrows coming to his feeder during a 12-day period. In a subsequent experiment, scrub and piñon jays ate almost 80 winged monarchs in a 1-week period. Brower and Calvert (1985) estimated that black-backed orioles and black-headed grosbeaks killed an average of 15,000 overwintering Mexican monarchs per day. Finally, Fink et al. (1983) recorded nine species of birds attacking them in nature, and four additional species eating them in cages.

Clearly, numerous predators, including jays, can and do eat monarchs: How is this discrepancy accounted for? The answer lies in a series of studies conducted by Lincoln Brower, Miriam Rothschild, and numerous other researchers over the last 15 years.

The monarch is indigenous to the temperate and tropical new world. Its stenophagous larvae feed mainly on milkweed species in the large genus *Asclepias* (Urquhart 1960). Although most milkweeds possess cardiac glycosides (powerful vertebrate heart poisons thought to deter herbivory), the qualitative and quantitative composition of these toxins varies greatly between and within species, and between individual plant parts, as well as seasonally (Duffey and Scudder 1972, Roeske et al. 1976, Nelson et al. 1981, Nishio et al. 1983, Brower et al. 1984a). The cardenolide content of adult monarchs, to a limited extent, reflects that of their food plant. Thus, while *A. eriocarpa* leaves average 421 μg of cardenolide per 0.1 g of dry leaf, *A. speciosa* shows only 90 μg per 0.1 g of dry leaf. Monarchs reared on the former contain twice as much cardenolide and are typically 10 times more emetic to blue jays as monarchs reared on the latter (Brower et al. 1982, 1984a).

Upon this host plant variability is imposed insect variability. Not only do different monarch "subspecies" differ in cardenolide-sequestering abilities (Rothschild and Marsh 1978), but siblings raised side by side on the same plant can acquire slightly different concentrations and have different emetic properties.

The monarch story is made even more complex by the fact that larvae selectively sequester only certain cardenolides (Seiber et al. 1980, Brower et al. 1984a). Furthermore, cardenolide acquisition is regulated. Larvae feeding on cardenolide-poor milkweeds concentrate the toxin, while monarchs feeding on cardenolide-rich plants egest a portion (Seiber et al. 1980, Brower et al. 1982). Although cardenolide uptake so mirrors the plant cardenolide profile that individual adults can be chemically "fingerprinted" as to their host plant (Brower et al. 1982, 1984a,b), some cardenolides are metabolically altered within the insect's body (Roeske et al. 1976; Seiber et al. 1980; Brower et al. 1982, 1984b). In some cases the insects become more toxic with time (Dixon et al. 1978, Rothschild and Marsh 1978). Once in the monarch's body, cardenolides are not distributed uniformly, but are concentrated in the wings and abdomen; moreover, females frequently possess higher cardenolide levels than males (Brower and Glazier 1975).

Adding to this variability is the possibility that monarchs synthesize additional toxins *de novo,* since artificial diet-reared monarchs contained cardioactive substances (Rothschild and Marsh 1978). Furthermore, there is evidence that adult *Danaus* acquire an additional, entirely different class of deterrent substances, pyrrolizidine alkaloids, by imbibing plant juices, nectar, and aphid honeydew from certain species in the Boraginaceae and Asteraceae (Rothschild and Edgar 1978, Rothschild and Marsh 1978). Rothschild and coworkers believe that these alkaloids provide the insect with a powerful warning scent capable of deterring predators (Rothschild et al. 1984; but see Brower 1984). Finally, the long-distance migration exhibited by this butterfly results in the mixing of adults from widely divergent locales and host plants.

The ultimate consequence of this variability in sequestration potential is what Brower (1968) and Brower et al. (1972) call a palatability spectrum. In any monarch population, those individuals containing particularly potent cardenolides, or high concentrations, are toxic to avian predators. Those containing less toxic cardenolides or low concentrations are edible. The latter represent palatable, Batesian mimics, or "automimics" of their more unpalatable conspecifics. Brower (1969) and Pough et al. (1973) calculated that a monarch population could contain up to 75% automimics and remain protected from avian predation. However, certain predators have overcome monarch defenses. Some birds measure cardenolide content by taste, rejecting cardenolide-rich individuals and consuming cardenolide-poor ones. Others differentiate between various body areas, consuming only the most palatable parts (Calvert et al. 1979, Fink and Brower 1981). Clearly, the monarch story is complex and still not completely understood. Apparently though, a multitude of intrinsic and extrinsic factors work to create a highly variable chemical defense.

5.3.4.2   The Lubber Grasshopper
Another insect with many similar features but a very different life history is the polyphagous hemimetabolous lubber grasshopper, *Romalea guttata.* This large aposematic grasshopper, native to the southeastern

United States, overwinters in the egg stage, emerges in spring, reaches adulthood in midsummer, and mates and oviposits in late summer and fall.

Romalea is the quintessence of a chemically defended insect. It is large, flightless, gregarious, diurnal, sluggish, and possesses a multisensory aposematic display. Included in this display are not only visual warnings, but acoustic, olfactory, and mechanical aposematic elements. When disturbed, *Romalea* first flashes bright red hind wings, made more conspicuous by a sideways rocking motion of the body. The abdominal tip twists about as if to sting. When touched, a defensive odorous froth erupts from paired metathoracic spiracles with a hissing sound. Intruders that press the attack face an armory of spiny hind legs raised over the head as well as the unpleasant products of regurgitation and defecation.

Potent emetic allomones back up these aposematic threats. In feeding trials, naive blue jays, mocking birds, robins, and thrashers all vomited after consuming *Romalea* nymphs, and developed a strong food aversion toward subsequent offerings. Some birds, however, were not affected by lubber toxins (Whitman, unpublished data).

Romalea defensive secretion is particularly interesting. An exceedingly complex mixture, with over 50 components, it consists largely of phenolics and quinones in an aqueous solution, dominated by phenol, hydroquinone, catechol, guaiacol, *p*-benzoquinone, 4-methoxybenzaldehyde, *p*-cresol, isophorone, verbenone, and romallenone (Meinwald et al. 1968, Eisner et al. 1971, Jones et al. 1986). The secretion composition varies intraspecifically; secretions from females of the same age and population differed in component concentrations as much as 70-fold, with some compounds absent in certain individuals (Jones et al. 1986).

The extreme irregularity in secretion composition was puzzling since insect defensive secretions are generally considered species stereotypic, varying occasionally as a result of sex, age, caste, race, season, population, or secretory history (Eisner et al. 1967, Wallbank and Waterhouse 1970, Tschinkel 1975, Owen 1978, Blum 1981, Prestwich 1983, Goh et al. 1984). But these factors could not explain the variation in *Romalea,* since tested individuals were of the same age, sex, race, population, and so on. An earlier discovery by Eisner et al. (1971) of the incorporation of 2,5-dichlorophenol into the secretion of Florida *Romalea* provided a clue. Since chlorinated compounds are rare in nature, and since 2,5-dichlorophenol is quite similar to the widely used herbicide 2,4-dichlorophenol, Eisner suggested a dietary origin for this substance, the first case of the incorporation of a synthetic compound into a defensive secretion.

Could diet, then, be the source of the variability observed in *Romalea* secretions? Analyses of the secretions of grasshoppers reared on a variety of single-plant and artificial diets showed that diet strongly affected secretion composition. Secretions of insects reared on restricted diets contained lower concentrations of fewer components and markedly altered relative compositions of compounds compared to insects reared on natural, multiple-plant diets (Jones et al. 1988). Furthermore, secretions of grasshoppers reared on certain single-plant diets contained specific host plant compounds. Most striking was the secretion from wild onion-reared grasshoppers. Not only did it contain a variety of characteristic onion compounds (1-propanethiol, methylthirane, methyldisulfide, isopropylsulfide, and propyldisulfide), but it possessed a pervasive onion odor. These compounds and their characteristic smell were absent from the secretions of grasshoppers reared on other diets (Jones et al. 1988). Plant substances were found in grasshoppers reared on other single-plant diets as well; the secretion from catnip (*Nepeta cataria*)-reared lubbers contained nepeta lactone breakdown products (Blum, unpublished data). These experiments demonstrated that *Romalea* could indeed sequester plant allomones.

Diet alone, however, did not explain all the observed variation in the secretion. In fact, sex, age, instar, and defensive history did have some influence (Jones et al., unpublished data). In addition, all major secretion components were present in at least some individuals reared on single host plants or artificial diets, indicating *de novo* origin (Jones et al. 1987).

It is clear that for *Romalea,* secretion composition has both an individual and a dietary origin. Probably any nutritionally balanced diet would result in the presence of the major components. However, certain plants, if consumed in large quantities, may supply additional phenolics or even idiosyncratic allomones.

What makes the *Romalea* story intriguing is the polyphagous nature of the insect. Allomone sequestration by herbivores, as a rule, occurs in monophagous or stenophagous insects. *Romalea* not only is polyphagous, but exhibits an obligatory host plant switching whereby favored plants become less favored following feeding (Whitman, unpublished data). The result is that *Romalea* seeks a wide variety of host plants and ingests an equally wide variety of plant allelochemicals. In all, 104 plants from 38 families are known to be consumed, including many containing potent allomones. The fate of most of these substances vis-à-vis *Romalea* is yet to be determined, but it is possible that some of the 50 or so minor components present in *Romalea* defensive secretion owe their occurrence to the insect's polyphagous habits.

How does such polyphagy relate to the defense of *Romalea?* Eisner et al. (1971) demonstrated the repellent efficacy of *Romalea* secretion against ants, and Whitman et al. (1985) showed that the defensive secretion of the closely related grasshopper, *Taeniopoda eques,* was aversive to grasshopper mice, *Onychomys torridus.* Secretions from onion-reared *Romalea* were twice as effective in repelling fire ants, *Solenopsis invicta,* as secretions from multiple-host-plant-reared lubbers and four times as effective as artificial-diet–reared *Romalea* (Jones et al. 1988). This demonstrated a predator's ability to sense diet-related differences in *Romalea* secretion and suggested that diet influenced defense capacity. Secretion variation may or may not benefit lubbers. Many predators are inhibited by novel prey stimuli (Coppinger 1970, Eisenberg and Leyhausen 1972) and show a preference for familiar prey (Holling 1961, Pasteels and Gregoire 1984). Idiosyncratic secretions may disrupt olfactory search image formation and food-attraction conditioning in natural enemies. Conversely, variation could impede food-aversion learning in predators.

5.3.4.3   Monarchs and Lubber Grasshoppers: Conclusions
The monarch and lubber demonstrate the extreme complexity inherent in allelochemical relationships. This complexity, arising from a number of sources, is not fully appreciated by researchers. These areas of misunderstanding include the origin, action, and identification of insect allelochemicals.

Insects may (1) sequester plant allomones directly from host plants, unchanged; (2) selectively sequester only certain compounds; (3) metabolically alter, then sequester, plant compounds; or (4) synthesize allelochemicals *de novo.* The exact relationships between these four processes are not clearly understood. Several may function simultaneously, and the occurrence of one may influence the operation of another. Some insects may even synthesize compounds identical to those present in their host plant. Clearly, for many insects, dietary history influences the synthesis and storage of allelochemicals. In the lubber, forced monophagy increases plant allomone sequestration and, with some diets, reduces *de*

novo synthesis. For other insects, dietary history might influence sequestration through induction of allomone-denaturing gut microflora or mixed-
function oxidases (Brattsten 1979, but see Gould 1984).

Our knowledge of the noxious principles in many chemically defended
insects is inadequate, as evidenced by recent findings of new, purportedly
deterrent substances in well-studied insects such as the monarch (Rothschild and Marsh 1978, Rothschild et al. 1984), Colorado potato beetle
(Daloze et al. 1986), and Mexican bean beetle (Eisner et al. 1986). For
lubber grasshoppers, only a portion of the secretion components are
known, and the internal emesis-inducing principle remains unidentified.

The mode of action for many allomones is also unknown. As Brower
(1984) relates, a repellent substance may act as either an unconditioned
negative stimulus (class I) or as a conditioned negative stimulus (class II).
Class I substances are noxious by themselves and cause immediate rejection or, following feeding, delayed effects such as poisoning. Class II
compounds are innocuous by themselves but become repellent because of
their conditioned association with some unpleasant effect such as poisoning by a class I substance. Confusion arises because predators sometimes
acquire conditioned aversions to class I substances. These compounds
can also act as class II substances when they are present in low concentrations. The phenolic secretion of lubber grasshoppers and the pyrrolizidine scent of monarchs may not in themselves be harmful, but act as
olfactory aposematic stimuli, reminding experienced predators of the unpleasant consequences of ingesting these prey.

Another source of confusion is the variability in insect defenses; probably no two individuals are exactly alike in their chemical defense composition. The tiny first instars of many chemically defended insects, especially those that derive their defense from plants, may be relatively
innocuous compared to later instars, which have had time to build up
allomone stores (Rowell-Rahier and Pasteels 1986). Conversely, many
chemically defended larvae, such as the lackey moth (*Malacosoma neus-
tria*) and goat moth (*Cossus cossus*), are apparently palatable as adults
(Rothschild 1985). The monarch and lubber grasshopper exemplify how
sex, diet, age, past defensive history, and inherent individual differences
influence allomone quantity and quality. Other studies demonstrate the
importance of season, race, population, instar, caste, and physiological
condition (see Jones et al. 1986).

As we have seen, there is also a great variability in allomone concentrations within individuals. As a result, predators may be quite selective in their feeding. When attacking monarchs, black-backed orioles, *Icterus abeillei,* preferentially consumed the cardenolide-poor flight muscles (Fink and Brower 1981). Birds commonly remove the head (and attached gut) of grasshoppers and caterpillars before swallowing the body (Vince 1964, Eisner 1970). Grasshopper mice prefer the head of *Taeniopoda eques* over the thorax, which contains the defensive gland (Whitman et al. 1985). The significance of preferential feeding is that it allows predators to circumvent arthropod defenses.

Another concept which is often unrecognized is that no defense is absolute. For any chemically defended organism, predators capable of breaching the defense can be found. Grosbeaks and orioles are avid feeders of monarchs, and while most birds are strongly affected by lubber toxins, force-fed chickens exhibit no ill effects. Some predators may even be adapted to feed on chemically defended prey. Toads are notoriously indiscriminate in their prey choice (but see Eisner 1960; Brower and Brower 1962, 1965), and grasshopper mice, *Onychomys,* appear only partially deterred by noxious or dangerous prey (Whitman et al. 1986). Other predators, such as bee eaters (Meropidae) and horned lizards (*Phrynosoma*), specialize on defended prey (Morse 1978). Chemical defense in many insects may have evolved mainly to thwart avian predation (Pasteels et al. 1983, Brower 1984, Rothschild 1985). In this respect, arthropod predators may be less deterred. There is a real dearth of knowledge on the broad-spectrum efficacy of insect chemical defenses against various taxonomic groups of predators.

Just as allomonal defenses vary in prey populations, conspecific predators vary in their capacity to overcome defended prey. For example, when a series of hand-reared robins were offered *Romalea,* some refused to attack, some attacked and rejected the grasshopper, and others fed and vomited. Some developed food aversion conditioning following one trial; others required multiple prey contacts (Whitman, unpublished data). Similar predator variability was observed in blue jays (Brower and Fink 1985).

Undoubtedly, given any defended prey, some predator species are always and some are never deterred, but the vast majority are probably facultatively deterred. For these, a variety of factors such as age, size, hunger, and prior experience influence whether or not an individual will attack a prey (Brower 1984). Such predator variability is seldom pointed out in studies on the efficacy of chemical defense, yet it holds important ramifications for the evolution of defense. It suggests that even a slight increase in a prey's defensive capabilities could have profound effects on that broad group of facultative predators.

6 THE FUTURE

Plants, herbivores, and natural enemies are clearly interconnected through an intricate array of chemical links. Substances released by plants can have direct and indirect, beneficial or detrimental effects on herbivores or carnivores. Likewise, substances from herbivores and carnivores can influence other trophic levels. Elucidating these complex allelochemical relationships not only furthers our basic understanding, but provides greater potential for the manipulation of these communication systems.

We have already seen the widespread commercial and scientific use of insect sex and aggregation pheromones for trapping, monitoring, and disrupting various insect pests (Mitchell 1981). The knowledge gained from studying allelochemical interactions might likewise be applied. For example, the potential for attracting and stimulating natural enemies by the use of kairomones has been suggested for predatory mites (Hislop and Prokopy 1981) and has been tested for parasitic wasps (Beevers et al. 1981, Lewis et al. 1975). The same techniques might be applicable for manipulating certain specialist predators, such as those attacking aphids or bark beetles. Artificial honeydews have been used experimentally to attract, arrest, and feed aphidophagous insects, with subsequent reduction in pest populations (Hagen et al. 1971, Ben Saad and Bishop 1976a, Nichols and Neel 1977, Hagen and Bishop 1979).

A knowledge of herbivore-produced kairomones could also aid biological control programs through the incorporation of feeding or oviposition stimulants in artificial diets, or by the application of conditioning, stimulating, or arresting kairomones to natural enemies prior to release (Greany et al. 1984, Lewis and Nordlund 1985). Food packets containing attractive kairomones might be dispersed to build up natural predator populations. Arthropod allomones such as the disruptants some insects use against ant, termite, or bee species (Blum 1980) might also lend themselves to profitable exploitation.

The manipulation of plant-produced substances such as odors, nectars, and allomones is another potentially successful approach to the control of insect pests. Plant-derived kairomones utilized by herbivores to locate their hosts could be synthesized or extracted from crop residues and released at appropriate times to attract pests to nonagricultural areas, or to poisoned bait, as has been suggested by Boppre et al. (1984) for the control of certain pestiferous grasshoppers. Plant breeders have successfully applied these concepts, creating hybrid trap crops that are superattractive to certain pests (Rhodes et al. 1980).

Pest monitoring is another application already in use for some fruit flies. Methyl eugenol and other floral-derived scents are widely used for the detection and evaluating the density of these pests (Metcalf et al. 1975, Sivinski and Calkins 1986). Synthetic apple volatiles have successfully attracted apple maggot flies in commercial apple orchards (Reissig et al. 1985). Likewise, deterrent plant allomones might be applied to crops to act as long- or short-distance herbivore repellents or contact-feeding deterrents (Altieri et al. 1977, Schmutterer et al. 1981, Schoonhoven 1982). Extracted plant synomones have already been used to attract, arrest, or stimulate carnivores (Altieri et al. 1981, Nordlund et al. 1985).

Intercropping of repellent or attractive plants with other economically important ones has long been suggested as a means for controlling certain herbivores (Raros 1973, Altieri and Letourneau 1984). Diversification of agroecosystems has demonstrated both pest population suppression and enhancement through direct effects on herbivores (Tahvanainen and Root 1972, Risch 1981, Altieri and Gliessman 1983, Risch et al. 1983, Letourneau 1986) or indirect effects on the third trophic level (Letourneau and Altieri 1983, Andow 1986). In the latter case, diversification provides increased microhabitats, refuges, alternative prey, supplementary nutrition sources (floral and extrafloral nectar, pollen, and honeydew), and disruptive or stimulating odor sources for the third trophic level (Monteith 1960, Leius 1967, Read et al. 1970, Shahjahan and Streams 1973, van Emden and Williams 1974, Bach 1980, Altieri et al. 1981, Horn 1981, Risch et al. 1983, Altieri and Letourneau 1984, Nordlund et al. 1985).

Plant breeders must also consider the tritrophic level consequences of their work, since some forms of plant resistance may be incompatible with natural herbivore control (Bergman and Tingey 1979, Campbell and Duffey 1979, Boethel and Eikenbary 1986). New plant varieties may influence natural enemies directly by altering levels of attractive, arrestant, or repellent substances, such as trichome toxins (Jones et al. 1985), or indirectly through physiological effects on the herbivore (Price 1980), or predator (Strickler and Croft 1985, Rogers and Sullivan 1986). Plant antibiosis may even extend through four trophic levels (Orr and Boethel 1986).

Natural enemy success is often keyed to the host's plant variety (Wyatt 1965, 1970; Starks et al. 1972; Mueller 1983). The anticarnivore implications of herbivore sequestration of plant allelochemicals are well documented (Smith 1957, Rothschild 1966, Thurston and Fox 1972, Barbosa and Saunders 1985). Ideally, plant breeders should strive to reduce substances attractive to herbivores while increasing substances attractive to natural enemies. This presents a quandary when the same substance serves both roles, as does allyl isothiocyanate, which is stimulating to both the cabbage aphid and its parasitoid (Read et al. 1970). Crop plants could be made to influence more favorably the survival of carnivores by breeding for increased flowers, nectar, blooming period, or extrafloral nectaries.

Exciting possibilities have surfaced with the dawning of biotechnology. After plant-produced synomones or herbivore-released kairomones attractive to natural enemies are identified, the techniques of genetic engineering might be used to incorporate the production of such substances into plants. Crops could be designed to release powerful predator-attracting synomones at sites of herbivory. Likewise, biotechnology might someday build plants which "tag" herbivores with substances attractive to parasitoids and predators. Some plant compounds, which appear in the frass after passing through herbivores' bodies, act as kairomones (Hendry et al. 1976, Elzen et al. 1984, Nordlund and Lewis 1985). Crops might be fabricated to produce higher levels of such substances.

Likewise, the incorporation of allomones via genetic engineering into predators and parasitoids could reduce mortality in this group, making them more effective natural controllers of herbivores. Eventually, predators and parasitoids may even be constructed to produce new or higher concentrations of allomone-detoxifying enzymes, allowing them to better utilize chemically defended hosts.

Utilizing biotechnology, powerful attractants or phagostimulants could be incorporated into living poisonous plants. Such plants, designed against specific crop pests, could be planted as trap crops, attracting the pest, inducing it to feed, and eventually poisoning it. Cases are known of herbivores feeding or ovipositing on plants toxic to them (Chew 1977, Janzen 1978, Wiklund 1981). In the same vein, flowers used by natural enemies could be made to produce more copious nectar, with higher carbohydrate, lipid, or amino acid concentrations. Conversely, flowers of crops attacked by dedicated pollinators could be designed to produce lethal allomones.

7 CAVEAT

Many of the interactions suggested in this chapter are admittedly speculative; others are represented by few examples. Just as we have yet to see widespread application of certain pest control techniques (e.g., food sprays, mass trapping, antifeedants, insect hormone disruption, autocidal elimination), the control of pests through either the manipulation of intraspecific chemical communication or via biotechnology is no panacea. As with the use of pheromones, the sterile male techniques, and biological and cultural controls, there will be successes and failures (Wall 1984, Mitchell 1986). Still, the acquisition of new weapons in the arsenal of integrated control is welcomed.

Although many of the interactions suggested in this review are hypothetical or poorly represented, we can expect to see progress as this field continues to advance. Two things are certain: (1) the future will hold new surprises as well as new opportunities to use these interactions for the benefit of humankind, and (2) insects will continue to be formidable adversaries, meeting our challenges with surprising flexibility.

ACKNOWLEDGMENTS

I am indebted to the following colleagues for their assistance in preparing this chapter: Murray Blum, Lincoln Brower, Stephen Buchmann, James Cane, William Hargrove, Clive Jones, Deborah Letourneau, Jule Macie, Larry Orsak, Justin Schmidt, Susan Watkins, and Kathy Smith-Whitman. Financial support was provided by the Mary Flagler Cary Trust and NSF grants DEB-8117943 and DEB-8117999. This is a contribution to the program of the Institute of Ecosystem Studies, the New York Botanical Garden.

REFERENCES

Addicott, F. T. 1982. Abscission. University of California Press, Berkeley, Calif.

Ahmad, S. (ed). 1983. Herbivorous Insects: Host-Seeking Behavior and Mechanisms. Academic Press. New York.

Akre, R. D., and C. W. Rettenmeyer. 1968. Trail-following by guests of army ants (Hymenoptera: Formicidae: Ecitonini). J. Kans. Ent. Soc. 41:165–174.

Aldrich, J. R., J. P. Kochansky, and C. B. Abrams. 1984. Attractant for a beneficial insect and its parasitoids: pheromone of the predatory spined soldier bug,

Podisus maculiventris (Hemiptera: Pentatomidae). Environ. Ent. 13:1031–1036.

Aldrich, J. R., J. P. Kochansky, and J. D. Sexton. 1985. Chemical attraction of the eastern yellow jacket, *Vespula maculifrons* (Hymenoptera: Vespidae). Experientia 41:420–422.

Allies, A., A. Bourke, and N. Franks. 1986. Propaganda substances in the cuckoo ant *Leptothorax kutteri* and the slave-maker *Harpagoxenus sublaevis*. J. Chem. Ecol. 12:1285–1293.

Alloway, T. M. 1979. Raiding behavior of two species of slave-making ants, *Harpagoxenus americanus* (Emory) and *Leptothorax duloticus (Wesson)*. Anim. Behav. 27:202–210.

Altieri, M. A. 1981. Weeds may augment biological control of insects. Calif. Agric. 35:22–24.

Altieri, M. A., and S. R. Gliessman. 1983. Effects of plant diversity on the density and herbivory of the flea beetle, *Phyllotreta cruciferae* Goeze, in California collard (*Brassica olaracea*) cropping systems. Crop Prot. 2:497–501.

Altieri, M.A., and D. K. Letourneau. 1982. Vegetation management and biological control in agroecosystems. Crop Prot. 1:405–430.

Altieri, M. A., and D. K. Letourneau. 1984. Vegetation diversity and insect pest outbreaks. CRC Crit. Rev. Plant Sci. 2:131–169.

Altieri, M. A., and L. L. Schmidt. 1986. Population trends and feeding preferences of flea beetles (*Phyllotreta cruciferae* Goeze) in collard-wild mustard mixtures. Crop Prot. 5:170–175.

Altieri, M. A., and W. H. Whitcomb. 1979. The potential use of weeds in the manipulation of beneficial insects. Hortic. Sci. 14:12–18.

Altieri, M. A., A. Schoonhoven, and J. D. Doll. 1977. The ecological role of weeds in insect pest management systems: a review illustrated with bean (*Phaseolus vulgaris* L.) cropping systems. PANS 23:185–206.

Altieri, M. A., W. J. Lewis, D. A. Nordlund, R. G. Gueldner, and J. W. Todd. 1981. Chemical interactions between plants and *Trichogramma* spp. wasps in Georgia soybean fields. Prot. Ecol. 3:259–263.

Altieri, M. A., S. Annamalai, K. P. Katiyar, and R. A. Flath. 1982. Effects of plant extracts on the rates of parasitization of *Anagasta kuehniella* (Lep.: *Pyralidae*) eggs by *Trichogramma pretiosum* (Hym.: Trichogrammatidae) under greenhouse conditions. Entomophaga 27:431–438.

Andow, D. A. 1986. Plant diversification and insect population control in agroecosystems. In D. Pimentel (ed.), Some Aspects of Integrated Pest Management. pp. 277–368. Department of Entomology, Cornell University, Ithaca, N.Y.

Arthur, A. P. 1962. Influence of host tree on abundance of *Itoplectis conquisitor* (Say) (Hymenoptera: Ichneumonidae), a polyphagous parasite of the European pine shoot moth, *Rhyacionia buoliana* (Schiff.) (Lepidoptera: Olethreutidae). Can. Ent. 94:337–347.

Atema, J., and G. D. Burd. 1975. A field study of chemotactic responses of the marine mudsnail, *Nassarius obsoletus*. J. Chem. Ecol. 1:243–251.

Atsatt, P. R. 1981. Lycaenid butterflies and ants: selection for enemy-free space. Amer. Nat. 118:636–654.

Bach, C. E. 1980. Effects of plant density and diversity on the population dynamics of a specialist herbivore, the striped cucumber beetle, *Acalymma vittatta* (Fab.). Ecology 61:1515–1530.

Baker, H. G., and I. Baker. 1975. Studies of nectar-constituents and pollinator plant coevolution. In L. E. Gilbert and P. H. Raven (eds.), Coevolution of Plants and Animals. pp. 100–140. University of Texas Press, Austin, Tex.

Baker, H. G., and I. Baker. 1983. Floral nectar sugar constituents in relation to pollinator type. In C. E. Jones and R. L. Little (eds.), Handbook of Experimental Pollination Biology. pp. 117–141. Van Nostrand Reinhold, New York.

Baldwin, I. T., and J. C. Schultz. 1983. Rapid changes in tree leaf chemistry induced by damage: evidence for communication between plants. Science 221:277–279.

Barbosa, P., and J. A. Saunders. 1985. Plant allelochemicals: linkages between herbivores and their natural enemies. In G. A. Cooper-Driver, T. Swain, and E. E. Conn (eds.), Chemically Mediated Interactions between Plants and Other Organisms, Vol. 19. Recent Advances in Phytochemistry. pp. 107–137. Plenum, New York.

Barbosa, P., J. A. Saunders, J. Kemper, R. Trumbule, J. Olechno, and P. Martinat. 1986. Plant allelochemicals and insect parasitoids: the effects of nicotine on *Cotesia congregata* (Say) and *Hyposoter annulipes* (Cresson). J. Chem. Ecol. 12:1319–1328.

Barker, R. J. 1978. Poisoning by plants. In R. A. Morse (ed.), Honey Bee Pests, Predators, and Diseases. pp. 273–296. Cornell University Press, Ithaca, N.Y.

Barth, F. G. 1985. Insects and Flowers. Princeton University Press, Princeton, N.J.

Bauer, T. 1982. Prey capture in a ground-beetle larvae. Anim. Behav. 30:203–208.

Beattie, A. J. 1985. The Evolutionary Ecology of Ant-Plant Mutualisms. Cambridge University Press, Cambridge.

Beevers, M., W. J. Lewis, H. R. Gross, Jr., and D. A. Nordlund. 1981. Kairomones and their use for management of entomophagous insects. X. Laboratory studies on manipulation of host-finding behavior of *Trichogramma pretiosum* Riley with a kairomone extracted from *Heliothis zea* (Boddie) moth scales. J. Chem. Ecol. 7:635–648.

Belcher, D. W., and R. Thurston. 1982. Inhibition of movement of larvae of the convergent lady beetle by leaf trichomes of tobacco. Environ. Ent. 11:91–94.

Ben Saad, A. A., and G. W. Bishop. 1976a. Effect of artificial honeydews on insect communities in potato fields. Environ. Ent. 5:453–457.

Ben Saad, A. A., and G. W. Bishop. 1976b. Attraction of insects to potato plants through use of artifical honeydews and aphid juice. Entomophaga 21:49–57.

Bentley, B. L. 1977. Extra-floral nectaries and protection by pugnacious body guards. Annu. Rev. Ecol. Syst. 8:407–427.

Bequaert, J. 1924. Galls that secrete honeydew. A contribution to the problem as to whether galls are altruistic adaptions. Bull. Brooklyn Ent. Soc. 19:101–124.

Bergman, J. M., and W. M. Tingey. 1979. Aspects of interaction between plant genotypes and biological control. Bull. Ent. Soc. Amer. 25:275–279.

Bergstrom, G., and J. Lofqvist. 1971. *Camponotus ligniperda* Latr. A model for the composite volatile secretions of Dufours gland in formicine ants. In A. S. Tahori (ed.), Chemical Releasers in Insects. Proceedings of the 2nd Int, IUPAC Congress, Vol. 3. pp. 195–223. Gordon and Breach, New York.

Blum, M. 1966. Chemical releasers of social behavior. VIII. Citral in the mandibular gland secretions of *Lestrimelitta limao*. Ann. Ent. Soc. Amer. 59:962–964.

Blum, M. 1970. Citral in stingless bees: isolation and function in trail-laying and robbing. J. Insect Physiol. 16:1637–1648.

Blum, M. S. 1974. Deciphering the communicative rosetta stone. Bull. Ent. Soc. Amer. 20:30–35.

Blum, M. S. 1980. Arthropods and ecomones: better fitness through ecological chemistry. In R. Gilles (ed.), Animals and Environmental Fitness. pp. 207–222. Pergamon Press, New York.

Blum, M. S. 1981. Chemical Defenses of Arthropods. Academic Press, New York.

Blum, M. S. 1983. Detoxication, deactivation, and utilization of plant compounds by insects. In P. Hedin (ed.), Plant Resistance to Insects. American Chemical Society Symposium Series 208. pp. 265–275. American Chemical Society, Washington, D.C.

Blum, M. S., and J. M. Brand. 1972. Social insect pheromones: their chemistry and function. Amer. Zool. 12:553–576.

Boecklen, W. 1984. The role of extra-floral nectaries in the herbivore defense of *Cassia fasiculata*. Ecol. Ent. 9:243–249.

Boethel, D. J., and R. D. Eikenbary (eds.). 1986. Interactions of Plant Resistance and Parasitoids and Predators of Insects. Wiley, New York.

Bombosch, S., and S. Volk. 1966. Selection of oviposition site by *Syrphus corol-*

lae F. In I. Hodek (ed.), Ecology of Aphidophagous Insects. pp. 117–119. Academia Prague.

Boppre, M., U. Seibt, and W. Wickler. 1984. Pharmacophagy in grasshoppers? Ent. Exp. Appl. 35:115–117.

Boucher, D. H, S. James, and K. Keeler. 1982. The ecology of mutualism. Annu. Rev. Ecol. Syst. 13:315–347.

Bowers, M. D., and G. M. Puttick. 1986. Fate of ingested iridoid glycosides in lepidopteran herbivores. J. Chem. Ecol. 12:169–178.

Brattsten, L. B. 1979. Biochemical defense mechanisms in herbivores against plant allelochemicals. In G. A. Rosenthal and D. H. Janzen (eds.), Herbivores: Their Interaction with Secondary Plant Metabolites. pp. 199–270. Academic Press, New York.

Brower, L. P. 1968. Ecological chemistry and the palatabilty spectrum. Science 161:1349–1351.

Brower, L. P. 1969. Ecological chemistry. Sci. Amer. 220:22–29.

Brower, L. P. 1984. Chemical defense in butterflies. In R. J. Vane-Wright and P. R. Ackery (eds.), The Biology of Butterflies. pp. 109–134. Academic Press, New York.

Brower, J. V. Z., and L. P. Brower. 1962. Experimental studies of mimicry. 6. The reaction of toads (*Bufo terrestris*) to honeybees (*Apis mellifera*) and their dronefly mimics (*Eristalis vinetorum*). Amer. Nat. 96:297–307.

Brower, J. V. Z., and L. Brower. 1965. Experimental studies of mimicry. 8. Further investigations of honeybees (*Apis mellifera*) and their dronefly mimics (*Eristalis* spp.). Amer. Nat. 99:173–187.

Brower, L. P., and J. V. Z. Brower. 1964. Birds, butterflies, and plant poisons: a study in ecological chemistry. Zoologica 49:137–159.

Brower, L. P., and W. H. Calvert. 1985. Foraging dynamics of bird predators on overwintering monarch butterflies in Mexico. Evolution 39:852–868.

Brower, L. P., and L. S. Fink. 1985. A natural toxic defense system: cardenolides in butterflies versus birds. Ann. N.Y. Acad. Sci. 443:171–186.

Brower, L. P., and S. C. Glazier. 1975. Localization of heart poisons in the monarch butterfly. Science 188:19–25.

Brower, L., and C. Moffitt. 1974. Palatability dynamics of cardenolides in the monarch butterfly. Nature 249:280–283.

Brower, L. P., P. B. McEvoy, K. L. Williamson, and M. A. Flannery. 1972. Variation in cardiac glycoside content of monarch butterflies from natural populations in eastern North America. Science 177:426–429.

Brower, L. P., J. N. Seiber, C. J. Nelson, S. P. Lynch, and P. M. Tuskes. 1982. Plant-determined variation in the cardenolide content, thin-layer chromatography profiles, and emetic potency of monarch butterflies, *Danaus plexippus* reared on the milkweed, *Asclepias eriocarpa* in California. J. Chem. Ecol. 8:579–633.

Brower, L. P., J. N. Seiber, C. J. Nelson, S. P. Lynch, and M. M. Holland. 1984a. Plant-determined variation in the cardenolide content, thin-layer chromatography profiles, and emetic potency of monarch butterflies, *Danaus plexippus* L. reared on milkweed plants in California. 2. *Asclepias speciosa*. J. Chem. Ecol. 10:601–639.

Brower, L. P., J. N. Seiber, C. J. Nelson, S. P. Lynch, M. P. Hoggard, and J. A. Cohen. 1984b. Plant-determined variation in cardenolide content and thin-layer chromatography profiles of monarch butterflies, *Danaus plexippus* reared on the milkweed. 3. *Asclepias californica*. J. Chem. Ecol. 10:1823–1857.

Brower, L. P., B. E. Horner, M. A. Marty, C. M. Moffitt, and B. Villa-R. 1985. Mice (*Peromyscus maniculatus, P. spicilegus,* and *Microtus mexicanus*) as predators of overwintering monarch butterflies (*Danaus plexippus*) in Mexico. Biotropica 17:89–99.

Bultman, T. L., and S. H. Faeth. 1986. Selective oviposition by a leaf miner in response to temporal variation in abscission. Oecologia (Berlin) 69:117–120.

Burghardt, G. M. 1970. Defining "communication." In J. W. Johnston, Jr., D. G. Moulton and A. Turk (eds.), Advances in Chemoreception, Vol. 1, Communication by Chemical Signal. pp. 5–18. Appleton-Century-Crofts, New York.

Buschinger, A., W. Ehrhardt, and V. Winter. 1980. The organization of slave raids in dulotic ants—a comparative study (Hymenoptera: Formicidae). Z. Tierpsychol. 53:245–264.

Calvert, W. H., L. E. Hedrick, and L. P. Brower. 1979. Mortality of the monarch butterfly (*Danaus plexippus* L.): avian predation at five overwintering sites in Mexico. Science 204:847–851.

Campbell, B. C., and S. S. Duffey. 1979. Tomatine and parasitic wasps: potential incompatibility of plant antibiosis with biological control. Science 205:700–702.

Carroll, C. R., and C. A. Hoffman. 1980. Chemical feeding deterrent mobilized in response to insect herbivory and counteradaptation by *Epilachna tredecimnotata*. Science 209:414–416.

Carter, W. 1939. Injuries to plants caused by insect toxins. Bot. Rev. 5:273–326.

Carter, M. C., and A. F. G. Dixon. 1984. Honeydew: an arrestant stimulus for coccinellids. Ecol. Ent. 9:383–387.

Chabot, B. F., and D. J. Hicks. 1982. The ecology of leaf life spans. Annu. Rev. Ecol. Syst. 13:229–259.

Chadab, R. 1979. Early warning cues for social wasps attacked by army ants. Psyche 86:115–123.

Chambliss, L. O., and C. M. Jones. 1966. Cucurbitacins: specific insect attractants in Cucurbitaceae. Science 153:1392–1393.

Chew, F. S. 1977. Coevolution of pierid butterflies and their cruciferous food plants. II. The distribution of eggs on potential food plants. Evolution 31:568–579.

Clark, A. H. 1927. Fragrant butterflies. Smithson. Inst. Annu. Rep. 1926:421–446.

Clausen, C. 1940. Entomophagous Insects. McGraw-Hill, New York.

Coakley, J. M., F. Maxwell, and J. Jenkins. 1969. Influence of feeding, oviposition, and egg and larval development of the boll weevil on abscission of cotton squares. J. Econ. Ent. 62:244-245.

Colburn, R., and D. Asquith, 1970. A cage used to study the finding of a host by the ladybird beetle, *Stethorus punctum*. J. Econ. Ent. 63:1376–1377.

Compton, S. G. 1987. *Aganais speciosa* and *Danaus chrysippus* (Lepidoptera) sabotage the latex defenses of their host plants. Ecol. Entomol. 12:115–118.

Coppinger, R. P. 1970. The effects of experience and novelty on avian feeding behavior with reference to the evolution of warning coloration in butterflies. II. Reactions of naive birds to novel insects. Amer. Nat. 104:323–336.

Corbet, S. A. 1971. Mandibular gland secretion of larvae of the flour moth, *Anagasta kuehniella*, contains an epideitic pheromone and elicits oviposition movement in a hymenopteran parasite. Nature 232:481–484.

Curio, E. 1976. The Ethology of Predation. Springer-Verlag, New York.

Dahms, R. 1972. Managing insects to reduce environmental problems. J. Environ. Qual. 1:211.

Daloze, D., J. Braekman, and J. Pasteels. 1986. A toxic dipeptide from the defense glands of the Colorado potato beetle. Science 233:221–223.

Detling, J. K., and M. I. Dyer. 1981. Evidence for potential plant growth regulators in grasshoppers. Ecology 62:485–488.

Dimock, M. B., and G. G. Kennedy. 1983. The role of glandular trichomes in the resistance of *Lycopersicon hirsutum* F. *glabratum* to *Heliothis zea*. Ent. Exp. Appl. 33:263–268.

Dixon, C. A., J. M. Erickson, D. N. Kellett, and M. Rothschild. 1978. Some adaptations between *Danaus plexippus* and its food plant, with notes on *Danaus chrysippus* and *Euploea core* (Insecta: Lepidoptera). J. Zool. London 185:437–467.

Duffey, S. S. 1977. Arthropod allomones: chemical effronteries and antagonists. In Proceedings of the 15th International Congress on Entomology. pp. 323–399. Entomological Society of America, College Park, Md.

Duffey, S. S. 1980. Sequestration of plant natural products by insects. Annu. Rev. Ent. 25:447–477.

Duffey, S. S., and G. G. E. Scudder. 1972. Cardiac glycosides in North American Asclepiadaceae, a basis for unpalatability in brightly colored Hemiptera and Coleoptera. J. Insect Physiol. 18:63–78.

Duffey, S. S., M. S. Blum, H. M. Fales, S. L. Evans, R. W. Roncadoni, D. L. Tiemann, and Y. Nakagawa. 1977. Benzoyl cyanide and mandelonitrile benzoate in the defensive secretions of millipedes. J. Chem. Ecol. 3:101–113.

Eberhard, W. G. 1977. Aggressive chemical mimicry by a bolas spider. Science 198:1173–1175.

Edwards, P. J., and S. Wratten. 1980. Ecology of Insect-Plant Interactions. Edward Arnold, London.

Eisenberg, J. F., and P. Leyhausen. 1972. The phylogenesis of predatory behavior in mammals. Z. Tierpsychol. 30:59–93.

Eisner, H. E., D. W. Alsop, and T. Eisner. 1967. Defense mechanisms of arthropods. XX. Quantitative assessment of hydrogen cyanide production in two species of millipedes. Psyche 74:107–117.

Eisner, T. 1960. Defense mechanisms of arthropods. II. The chemical and mechanical weapons of an earwig. Psyche 67:62–70.

Eisner, T. 1970. Chemical defense against predation in arthropods. In E. Sondheimer and J. P. Simeone (eds.), Chemical Ecology. pp. 157–217. Academic Press, New York.

Eisner, T. 1972. Chemical ecology: on arthropods and how they live as chemists. Verh. Dtsch. Zool. Ges. 65:123–137.

Eisner, T. 1980. Chemistry, defense, and survival: case studies and selected topics. In M. Locke and D. S. Smith (eds.), Insect Biology in the Future. pp. 847–878. Academic Press, New York.

Eisner, T., L. B. Hendry, D. B. Peakall, and J. Meinwald. 1971. 2,5-Dichlorophenol (from ingested herbicide?) in defensive secretion of grasshopper (Romalea microptera: Orth.: Acrididae). Science 172:277–278.

Eisner, T., S. Nowicki, M. Goetz, and J. Meinwald. 1980. Red cochineal dye (carminic acid): its role in nature. Science 208:1039–1042.

Eisner, T., M. Goetz, D. Aneshansley, G. Ferstandig-Arnold, and J. Meinwald. 1986. Defensive alkaloid in blood of mexican bean beetle (Epilachna varivestis). Experientia 42:204–207.

Ekbom, B. S. 1977. Development of a biological control program for greenhouse whiteflies (Trialurodes vaporarium Westwood) using its parasite Encarsia formosa (Gahan) in Sweden. Z. Angew. Ent. 84:145–154.

Elsey, K. D. 1974. Influence of plant host on searching speed of two predators. Entomophaga 19:3–6.

Elzen, G. W., H. J. Williams, and S. B. Vinson. 1984. Role of diet in host selection of *Heliothis virescens* by the parasitoid *Campoletis sonorensis* (Hymenoptera: Ichneumonidae). J. Chem. Ecol. 10:1535–1541.

Ernsting, G., J. C. Jager, J. van de Meer, and W. Slob. 1985. Locomotory activity of a visually hunting carabid beetle in response to non-visual prey stimuli. Ent. Exp. Appl. 38:41–47.

Evens, W. G. 1982. *Oscillatoria* sp. (Cyanophyta) mat metabolites implicated in habitat selection in *Bembidion obtusidens* (Coleoptera: Carabidae). J. Chem. Ecol. 8:671–678.

Faeth, S. 1985. Host leaf-selection by leaf miners: interactions among three trophic levels. Ecology 66:870–875.

Faeth, S. H., E. F. Connor, and D. S. Simberloff. 1981. Early leaf abscission: a neglected source of mortality for folivores. Amer. Nat. 117:409–415.

Farish, D. J., and R. C. Axtell. 1966. Sensory functions of the palps and first tarsi of *Macrocheles muscaedomesticae* (Acarina: Macrochelidae), a predator of the housefly. Ann. Ent. Soc. Amer. 59:165–170.

Felt, E. P. 1940. Plant Galls and Gall Makers. Comstock, Ithaca, N.Y.

Ferguson, J. E., and R. L. Metcalf. 1985. Cucurbitacins: plant-derived defense compounds for the diabroticites (Coleoptera: Chrysomelidae). J. Chem. Ecol. 11:311–318.

Ferguson, J. E., R. L. Metcalf, and D. C. Fischer. 1985. Disposition and fate of cucurbitacin B in five species of diabroticites. J. Chem. Ecol. 11:1307–1321.

Fink, L. S., and L. P. Brower. 1981. Birds can overcome the cardenolide defense of monarch butterflies in Mexico. Nature 291:67–70.

Fink, L. S., L. P. Brower, R. B. Waide, and P. R. Spitzer. 1983. Overwintering monarch butterflies as food for insectivorous birds in Mexico. Biotropica 15:151–153.

Fleschner, C. A. 1950. Studies on the searching capacity of the larvae of three predators of the citrus red mite. Hilgardia 20:233–265.

Flint, H. M., S. S. Salter, and S. Walters. 1979. Caryophyllene: an attractant for the green lacewing *Chrysopa carnea* Stephens. Environ. Ent. 8:1123–1125.

Foster, M., and W. Ruesink. 1984. Influence of flowering weeds associated with reduced tillage in corn on a black cutworm (Lepidoptera: Noctuidae) parasitoid, *Meteorus rubens* (Nees von Esenbeck). Environ. Ent. 13:664–668.

Freidberg, A. L., and W. N. Mathis. 1985. On the feeding habits of *Rhynchopsilopa* (Dip: Ephydridae). Entomophaga 30:13–21.

Fritz, R. 1982. An ant-treehopper mutualism: effects of *Formica subsericea* on the survival of *Vanduzea arquata*. Ecol. Ent. 7:267–276.

Galil, J., and D. Eisikowitch. 1969. Further studies on the pollination ecology of *Ficus sycomorus* L. (Hymenoptera, Chalcidoidea, Agaonidae). Tijdschr. Ent. 112:1–13.

Garnett, W. B., R. D. Akre, and G. Sehlke. 1985. Cocoon mimicry and predation by myrmecophilous Diptera (Diptera: Syrphidae). Fla. Ent. 68:615–621.

Gibson, R. W., and J. A. Pickett. 1983. Wild potato repels aphids by release of aphid alarm pheromone. Nature 302:608–609.

Goh, S. H., C. H. Chuah, Y. P. Tho, and G. D. Prestwich. 1984. Extreme intraspecific chemical variability in soldier defense secretions of allopatric and sympatric colonies of *Longipeditermes longipes*. J. Chem. Ecol. 10:929–944.

Gould, F. 1984. Mixed function oxidases and herbivore polyphagy: the devil's advocate position. Ecol. Ent. 9:29–34.

Greany, P. D., and K. S. Hagen. 1981. Prey selection. In D. A. Nordlund, R. L. Jones, and W. J. Lewis (eds.), Semiochemicals: Their Role in Pest Control. pp. 121–135. Wiley, New York.

Greany, P. D., S. B. Vinson, and W. J. Lewis. 1984. Insect parasitoids: finding new opportunities for biological control. BioScience 34:690–696.

Greenblatt, J. A., and W. J. Lewis. 1983. Chemical environment manipulation for pest insect control. Environ. Manag. 7:35–41.

Gressitt, J. L. 1966a. Epizoic symbiosis: the Papuan weevil genus *Gymnopholus* (Leptopiinae) symbiotic with cryptogamic plants, orbatid mites, rotifers and nematodes. Pac. Insects 8:221–280.

Gressitt, J. L. 1966b. The weevil *Pantorhytes* (Coleoptera), involving cacao pests and epizoic symbiosis with cryptogamic plants and microfauna. Pac. Insects 8:915–965.

Gueldner, R. C., D. A. Nordlund, W. J. Lewis, J. E. Thean, and D. M. Wilson. 1984. Kairomones and their use for management of entomophagous insects. XV. Identification of several acids in scales of *Heliothis zea* moths and comments on their possible role as kairomones for *Trichograma pretiosum*. J. Chem. Ecol. 10:245–251.

Guerrant, E. O., Jr., and P. L. Fiedler. 1981. Flower defenses against nectar pilferage by ants. Biotropica 13:25–33.

Gurney, B., and N. W. Hussey. 1970. Evaluation of some coccinellid species for the biological control of aphids in protected cropping. Ann. Appl. Biol. 65:451–458.

Haase, E. 1893. Untersuchungen über die Mimikry auf Grundlage eines natürlichen Systems der Papilionidea. 2. Untersuchungen über die Mimikry. Nagel, Stuttgart, West Germany.

Hagen, K. S. 1962. Biology and ecology of predaceous Coccinellidae. Annu. Rev. Ent. 7:289–326.

Hagen, K. S. 1986. Ecosystem analysis: plant cultivars (HPR), entomophagous species and food supplements. In D. J. Boethel and R. D. Eikenbary (eds.), Interactions of Plant Resistance and Parasitoids and Predators of Insects. pp. 151–197. Wiley, New York.

Hagen, K. S., and G. W. Bishop. 1979. Use of supplemental foods and behavioral chemicals to increase the effectiveness of natural enemies. In D. W. Davis, J. A. McMurtry, and S. C. Hoyt (eds.), Biological Control and Insect Management. California Agricultural Experiment Station, Publication 4096. pp. 49–60. California Agricultural Experiment Station, Calif.

Hagen, K. S., and R. van den Bosch. 1968. Impact of pathogens, parasites and predators on aphids. Annu. Rev. Ent. 13:325–384.

Hagen, K. S., E. F. Sawall, Jr., and R. L. Tassan. 1971. The use of food sprays to increase effectiveness of entomophagous insects. Proc. Tall Timbers Conf. Ecol. Anim. Control Habitat Manag. 2:59–81.

Hagen, K. S., P. Greany, E. F. Sewall, Jr., and R. L. Tassan. 1976. Tryptophan in artificial honeydews as a source of an attractant for adult *Chrysopa carnea*. Environ. Ent. 5:458–463.

Harborne, J. B. 1982. Introduction to Ecological Biochemistry. Academic Press, London.

Haukioja, E., and S. Neuvonen. 1985. Induced long-term resistance of birch foliage against defoliators: defensive or incidental? Ecology 66:1303–1308.

Haukioja, E., and P. Niemela. 1977. Retarded growth of geometrid larvae after mechanical damage to leaves of its host tree. Ann. Zool. Fenn. 14:48–52.

Heads, P. A., and J. H. Lawton. 1985. Bracken, ants and extrafloral nectaries. III. How insect herbivores avoid ant predation. Ecol. Ent. 10:29–42.

Heinrich, B., and S. L. Collins. 1983. Caterpillar leaf damage, and the game of hide-and-seek with birds. Ecology 64:592–602.

Hendry, L. B., J. K. Wichmann, D. M. Hindenlang, K. M. Weaver, and S. H. Korzeniowski. 1976. Plants—the origin of kairomones utilized by parasitoids of phytophagous insects. J. Chem. Ecol. 2:271–283.

Hespenheide, H. A. 1985. Insect visitors to extrafloral nectaries of *Byttneria aculeata* (Sterculiaceae): relative importance and roles. Ecol. Ent. 10:191–204.

Hislop, R. G., and R. Prokopy. 1981. Mite predator responses to prey and predator-emitted stimuli. J. Chem. Ecol. 7:895–904.

Hoffman, H. 1930. Über den Fluchtreflex bie *Nassa*. Z. Vgl. Physiol. 2:662–688.

Holldobler, B. 1969. Host finding by odor in the myrmecophilous beetle *Atemeles pubicollis* (Coleoptera: Staphylinidae). Science 166:757–758.

Holldobler, B. 1971. Communication between ants and their guests. Sci. Amer. 224:86–93.

Holling, C. S. 1961. Principles of insect predation. Annu. Rev. Ent. 6:163–182.

Hori, K. 1976. Plant growth-regulating factor in the salivary gland of several heteropterous insects. Comp. Biochem. Physiol. 53B:435–438.

Horn, D. J. 1981. Effect of weedy backgrounds on colonization of collards by green peach aphid. *Myzus persicae,* and its major predators. Environ. Ent. 10:285–289.

House, H. L. 1976. Nutrition of natural enemies. In R. L. Ridgway and S. B. Vinson (eds.), Biological Control by Augmentation of Natural Enemies, pp. 151–182. Plenum, New York.

Howard, R. W., C. A. McDaniel, and G. J. Blomquist. 1980. Chemical mimicry as an integrating mechanism: cuticular hydrocarbons of a termitophile and its host. Science 210:431–433.

Howse, P. E. 1984. Sociochemicals of termites. In W. Bell and R. Cardé (eds.), Chemical Ecology of Insects. pp. 475–519. Chapman & Hall, London.

Huber, I. 1979. Prey attraction and immobilization by allomone from nymphs of *Womersia strandtmanni* (Acarina: Trombiculidae). Acarologia 20:112–115.

Hunter, B. T. 1964. Gardening without Poisons. Houghton Mifflin, Boston.

Jacobson, E. 1911. Biological notes on the hemipteran. *Ptilocerus ochraceus.* Tijdschr. Ent. 54:175–179.

Jalil, M., and J. G. Rodriguez. 1970. Studies of the behavior of *Macrocheles muscaedomesticae* (Acarina: Macrochelidae) with emphasis on its attraction to the house fly. Ann. Ent. Soc. Amer. 63:738–744.

Janzen, D. H. 1966. Coevolution of mutualism between ants and acacias in Central America. Evolution 20:249–275.

Janzen, D. 1975. The Ecology of Plants in the Tropics. Studies in Biology 58. Edward Arnold, London.

Janzen, D. H. 1977. Why ants don't visit flowers. Biotropica 9:252.

Janzen, D. H. 1978. Cicada (*Diceroprocta apache* (Davis)) mortality by feeding on *Nerium oleander*. Pan-Pac. Ent. 5:69–70.

Janzen, D. H. 1979. New horizons in the biology of plant defenses. In G. Rosenthal and D. H. Janzen (eds.), Herbivores: Their Interaction with Secondary Plant Metabolites. pp. 331–350. Academic Press, New York.

Johnson, J. B., and K. S. Hagen. 1981. A neuropterous larva uses an allomone to attack termites. Nature 289:506–507.

Jones, C. G., T. A. Hess, D. W. Whitman, P. J. Silk, and M. S. Blum. 1986. Idiosyncratic variation in chemical defenses among individual generalist grasshoppers. J. Chem. Ecol. 12:749–761.

Jones, C. G., T. A. Hess, D. W. Whitman, P. J. Silk, and M. S. Blum. 1987. Effects of diet breadth on autogenous chemical defense of a generalist grasshopper. J. Chem. Ecol. 13:283–297.

Jones, C. G., D. W. Whitman, S. J. Compton, P. J. Silk, and M. S. Blum. 1988. Diet breadth determines the type of chemical defense in a grasshopper (submitted).

Jones, D., G. A. Jones, T. Hagen, and E. Creech. 1985. Wild species of *Nicotiana* as a new source of tobacco resistance to the tobacco hornworm *Manduca sexta*. Ent. Exp. Appl. 38:157–164.

Jones, R. L., W. J. Lewis, M. Beroza, B. A. Bierl, and A. N. Sparks. 1973. Host-seeking stimulants (kairomones) for the egg parasite *Trichogramma evanescens*. Environ. Ent. 2:593–596.

Kahn, D. M., and H. V. Cornell. 1983. Early leaf abscision and folivores: comments and consideration. Amer. Nat. 122:428–432.

Kammer, A. E. 1970. Thoracic temperature, shivering and flight in the monarch butterfly, *Danaus plexippus* (L.). Z. Vgl. Physiol. 68:334–344.

Karban, R., and J. Carey. 1984. Induced resistance of cotton seedlings to mites. Science 225:53–54.

Kartohardjono, A., and E. A. Heinrichs. 1984. Populations of the brown planthopper, *Nilaparvata lugens* (Stal) (Homoptera: Delphacidae), and its predators on rice varieties with different levels of resistance. Environ. Ent. 13:359–365.

Kennedy, B. 1984. Effect of multilure and its components on parasites of *Scolytus multistriatus* (Coleoptera: Scolytidae). J. Chem. Ecol. 10:373–385.

Kesten, U. 1969. Zur Morphologie und Biologie von *Anatis ocellata* (L.) (Coleoptera: Coccinellidae). Z. Angew. Ent. 63:412–445.

Kistner, D. 1982. The Social Insects' Bestiary. In H. R. Hermann (ed.), Social Insects, Vol. III. pp. 1–224. Academic Press, New York.

Kogan, M. 1977. The role of chemical factors in insect/plant relationships. In Proceedings of the 15th International Congress on Entomology. pp. 211–227. Entomological Society of America, College Park, Md.

Kogan, M. 1986. Natural chemicals in plant resistance to insects. Iowa State J. Res. 60:501–528.

LaMon, B., and H. Topoff. 1981. Avoiding predation by army ants: defensive behavior of three ant species of the genus *Camponotus*. Anim. Behav. 29:1070–1081.

Lane, C. 1957. Preliminary note on insects eaten and rejected by a tame shama (*Kittacincla malabarica* Gm.), with the suggestion that in certain species of butterflies and moths females are less palatable than males. Ent. Mon. Mag. 93:172–179.

Lecomte, C., and E. Thibout. 1984. Étude olfactometrique de l'action de diverses substances allelochimiques végétales dans la recherche de l'hôte par *Diadromus pulchellus* (Hymenoptera, Ichneumonidae). Ent. Exp. Appl. 35:295–303.

Leius, K. 1967. Influence of wild flowers on parasitism of tent caterpillar and codling moth. Can. Ent. 99:444–446.

Letourneau, D. K. 1986. Associational resistance in squash monocultures and polycultures in tropical Mexico. Environ. Ent. 15:285–292.

Letourneau, D. K., and M. A. Altieri. 1983. Abundance patterns of a predator, *Orius tristicolor* (Hemiptera: Anthocoridae), and its prey, *Frankliniella occidentalis* (Thysanoptera: Thripidea): habitat attraction in polycultures versus monocultures. Environ. Ent. 12:1464–1469.

Levin, D. A. 1973. The role of trichomes in plant defense. Rev. Biol. 48:3–15.

Lewis, W. J., and R. L. Jones. 1971. Substance that stimulates host-seeking by *Microplitis croceipes* (Hymenoptera: Braconidae), a parasite of *Heliothis* species. Ann. Ent. Soc. Amer. 64:471–473.

Lewis, W. J., and D. A. Nordlund. 1985. Behavior-modifying chemicals to enhance natural enemy effectiveness. In M. A. Hoy and D. C. Herzog (eds.), Biological Control in Agriculture IPM Systems. pp. 89–101. Academic Press, New York.

Lewis, W. J., R. L. Jones, D. A. Nordlund, and A. N. Sparks. 1975. Kairomones and their use for the management of entomophagous insects. I. Evaluation for increasing the rate of parasitization by *Trichogramma* spp. in the field. J. Chem. Ecol. 1:343–347.

Lewis, W. J., D. A. Nordlund, H. R. Gross, Jr., R. L. Jones, and S. L. Jones. 1977. Kairomones and their use for the management of entomophagous insects. V. Moth scales as a stimulus for predation of *Heliothis zea* (Boddie) eggs by *Chrysopa carnea* Stephens larvae. J. Chem. Ecol. 3:483–487.

Lincoln, C., G. Dean, B. A. Waddle, W. C. Yearian, J. R. Philips, and L. Roberts. 1971. Resistance of frego-type cotton to boll weevil and bollworm. J. Econ. Ent. 64:1326–1327.

Loke, W. H., and T. R. Ashley. 1984. Sources of fall armyworm, *Spodoptera frugiperda* (Lepidoptera: Noctuidae), kairomones eliciting host-finding behavior in *Cotesia* (= *Apanteles*) *marginiventris* (Hymenoptera: Braconidae). J. Chem. Ecol. 10:1019–1027.

Longhurst, C., and P. E. Howse. 1978. The use of kairomones by *Megaponera foetens* (Hymenoptera: Formicidae) in the detection of its termite prey. Anim. Behav. 26:1213–1218.

Mani, M. S. 1964. Ecology of Plant Galls. Dr. W. Junk, The Hague.

Maschwitz, U., and W. Kloft. 1971. Morphology and function of the venom apparatus in insects—bees, wasps, ants, and caterpillars. In W. Bücherl and E. Buckley (eds.), Venomous Animals and Their Venoms, Vol. 3. pp. 1–60. Academic Press, New York.

Maxwell, F. G. 1972. Morphological and chemical changes that evolve in the development of host plant resistance to insects. J. Environ. Qual. 1:265–270.

McLain, K. 1979. Terrestrial trail following by three species of predatory stink bugs. Fla. Ent. 62:152–154.

McMahan, E. A. 1983. Adaptations, feeding preferences, and biometrics of a termite-baiting assassin bug (Hemiptera: Reduviidae). Ann. Ent. Soc. Amer. 76:483–486.

McNaughton, S. J. 1983a. Compensatory plant growth as a response to herbivory. Oikos 40:329–336.

McNaughton, S. J. 1983b. Physiological and ecological implications of herbivory. In O. L. Lange, P. S. Nobel, C. B. Osmond, and H. Ziegler (eds.), Physiological Plant Ecology III. pp. 657–677. Springer-Verlag, Berlin.

Meinwald, J., K. Erickson, M. Hartshorn, Y. C. Meinwald, and T. Eisner. 1968. Defensive mechanisms of arthroods. XXIII. An allenic sesquiterpenoid from the grasshopper *Romalea microptera*. Tetrahedron Lett. 25:2959–2962.

Metcalf, R. L., W. C. Mitchell, T. R. Fukuto, and E. R. Metcalf. 1975. Attraction of the oriental fruit fly, *Dacus dorsalis,* to methyl eugenol and related olfactory stimulants. Proc Natl. Acad. Sci. USA 72:2501–2505.

Miller, J. R., and K. L. Strickler. 1984. Finding and accepting host plants. In W. Bell and R. Cardé (eds.), Chemical Ecology of Insects. pp. 127–158. Chapman & Hall, London.

Mitchell, E. R. (ed.). 1981. Management of Insect Pests with Semiochemicals. Plenum, London.

Mitchell, E. R. 1986. Pheromones: as the glamour and glitter fade—the real work begins. Fla. Ent. 69:132–139.

Mizell, R. F., III, J. L. Frazier, and T. E. Nebeker. 1984. Response of the clerid predator *Thanasimus dubius* (F.) to bark beetle pheromones and tree volatiles in a wind tunnel. J. Chem. Ecol. 10:177–187.

Monteith, L. G. 1955. Host preferences of *Drino bohemica* Mesn. (Diptera: Tachinidae) with particular reference to olfactory responses. Can. Ent. 87:509–530.

Monteith, L. G. 1958. Influence of food plant of host on attractiveness of the host to tachinid parasites with notes on preimaginal conditioning. Can. Ent. 90:478–482.

Monteith, L. G. 1960. Influence of plants other than the food plants of their host on host-finding by tachinid parasites. Can. Ent. 92:641–652.

Morse, R. A. 1978. Honey Bee Pests, Predators, and Diseases. Cornell University Press, Ithaca, N.Y.

Mudd, A., J. H. H. Walters, and S. A. Corbet. 1984. Relative kairomonal activites of 2-acylcyclohexane-1,3-diones in eliciting oviposition behavior from parasite *Nemeritis canescens* (Grav.). J. Chem. Ecol. 10:1597–1601.

Mueller, T. F. 1983. The effect of plants on the host relations of a specialist parasitoid of *Heliothis* larvae. Ent. Exp. Appl. 34:78–84.

Nault, L. R., M. E. Montgomery, and W. S. Bowers. 1976. Ant aphid association: role of aphid alarm pheromone. Science 192:1349–1351.

Nelson, C. J., J. N. Seiber, and L. P. Brower. 1981. Seasonal and intraplant variation of the cardenolide content in the California milkweed, *Asclepias eriocarpa,* and implications for plant defense. J. Chem. Ecol. 7:981–1010.

Nettles, W. C. 1980. Adult *Eucelatoria* sp.: response to volatiles from cotton and okra plants and from larvae of *Heliothis virescens, Spodoptera eridania* and *Estigmene acrea.* Environ. Ent. 9:759–763.

Nichols, P. R., and W. W. Neel. 1977. The use of food wheast as a supplemental food for *Coleomegilla maculata* (DeGeer) (Coleoptera: Coccinellidae) in the field. Southwest. Ent. 2:102–106.

Nishio, S., M. S. Blum, and S. Takahashi. 1983. Intraplant distribution of cardenolides in *Asclepias humistrata* (Asclepiadaceae), with additional notes on their fates in *Tetraopes melanurus* (Coleoptera: Cerambycidae) and *Rhyssomatus lineaticollis* (Coleoptera: Curculionidae). Mem. Coll. Agric. Kyoto Univ. 122:43–52.

Noldus, L. P. J. J., and J. C. van Lenteren, 1985a. Kairomones for the egg parasite *Trichogramma evanescens* Westwood. I. Effect of volatile substances released by two of its hosts, *Pieris brassicae* L. and *Mamestra brassicae* L. J. Chem. Ecol. 11:781–791.

Noldus, L. P. J. J., and J. C. van Lenteren. 1985b. Kairomones for the egg parasite *Trichogramma evanescens* Westwood. II. Effect of contact chemicals produced by two of its hosts, *Pieris brassicae* L. and *Pieris rapae* L. J. Chem. Ecol. 11:793–800.

Nordlund, D. A. 1981. Semiochemicals: a review of the terminology. In D. A. Nordlund, R. L. Jones, and W. J. Lewis (eds.), Semiochemicals: Their Role in Pest Control. pp. 13–28. Wiley, New York.

Nordlund, D. A., and W. J. Lewis. 1976. Terminology of chemical-releasing stimuli in intraspecific and interspecific interactions. J. Chem. Ecol. 2:211–220.

Nordlund, D. A., and W. J. Lewis. 1985. Response of females of the braconid parasitoid *Microplitis demolitor* to frass of larvae of the noctuids, *Heliothis zea* and *Trichoplusia ni* and to 13-methylhentriacontane. Ent. Exp. Appl. 38:109–112.

Nordlund, D. A., W. J. Lewis, R. L. Jones, H. R. Gross, Jr., and K. S. Hagen. 1977. Kairomones and their use for management of entomophagous insects. VI. An examination of the kairomones of the predator *Chrysopa carnea*

Stephens at the oviposition sites of *Heliothis zea* (Boddie). J. Chem. Ecol. 3:507–511.

Nordlund, D. A., W. J. Lewis, and R. C. Gueldner. 1983. Kairomones and their use for management of entomophagous insects. XIV. Response of *Telenomus remus* to abdominal tips of *Spodoptera frugiperda,* (Z)-9-tetradecene-1-ol acetate and (Z)-9-dodecene-1-ol acetate. J. Chem. Ecol. 9:695–701.

Nordlund, D. A., R. B. Chalfont, and W. J. Lewis. 1985. Response of *Trichogramma pretiosum* females to extracts of two plants attacked by *Heliothis zea.* Agric. Ecosyst. Environ. 12:127–133.

Obrycki, J. J. 1986. The influence of foliar pubescence on entomophagous species. In D. J. Boethel and R. D. Eikenbary (eds.), Interactions of Plant Resistance and Parasitoids and Predators of Insects. pp. 61–83. Wiley, New York.

Obrycki, J., and M. Tauber. 1984. Natural enemy activity on glandular pubescent potato plants in the greenhouse: an unreliable predictor of effects in the field. Environ. Ent. 13:679–683.

Obrycki, J. J., M. J. Tauber, C. A. Tauber, and B. Gollands. 1985. *Edovum puttleri* (Hymenoptera: Eulophidae), an exotic egg parasitoid of the Colorado potato beetle (Coleoptera: Chrysomelidae): response to temperate zone conditions and resistant potato plants. Environ. Ent. 14:48–54.

O'Dowd, D. J. 1980. Pearl bodies of a neotropical tree, *Ochroma pyramidale:* ecological implications. Amer. J. Bot. 67:543–549.

Orr, D. B., and D. J. Boethel. 1986. Influence of plant antibiosis through four trophic levels. Oecologia (Berlin) 70:242–249.

Otte, D., and W. Cade. 1976. On the role of olfaction in sexual and interspecies recognition in crickets (*Acheta* and *Gryllus*). Behaviour 24:1–6.

Owen, M. D. 1978. Venom replenishment as indicated by histamines, in honeybee (*Apis mellifera*) venom. J. Insect Physiol. 24:433–437.

Pasteels, J. M. 1982. Is kairomone a valid and useful term? J. Chem. Ecol. 8:1079–1081.

Pasteels, J. M., and J. Gregoire. 1984. Selective predation on chemically defended chrysomelid larvae: a conditioning process. J. Chem. Ecol. 10:1693–1700.

Pasteels, J. M., J. Gregoire, and M. Rowell-Rahier. 1983. The chemical ecology of defense in arthropods. Annu. Rev. Ent. 28:263–289.

Payne, T. L., J. C. Dickens, and J. V. Richerson. 1984. Insect predator-prey coevolution via enantiomeric specificity in a kairomone-pheromone system. J. Chem. Ecol. 10:487–492.

Pellmyr, O., and L. B. Thien. 1986. Insect reproduction and floral fragrances: keys to the evolution of the angiosperms? Taxon 35:76–85.

Peterson, B. 1964. Monarch butterflies are eaten by birds. J. Lepid. Soc. 18:165–169.

Phillips, D. W. 1977. Avoidance and escape responses of the gastropod mollusc *Olivella biplicata* (Sowerby) to predatory asteroids. J. Exp. Mar. Biol. Ecol. 28:77–86.

Pierce, N. E. 1984. Amplified species diversity: a case history of an Australian lycaenid butterfly and its attendant ants. In R. I. Vane-Wright and P. R. Ackery (eds.), the Biology of Butterflies. pp. 197–200. Academic Press, London.

Pierce, N. E., and P. S. Mead. 1981. Parasitoids as selective agents in the symbiosis between lycaenid butterfly larvae and ants. Science 211:1185–1187.

Pough, F., I., Brower, H. Meck, and S. Kessell. 1973. Theoretical investigations and automimicry: multiple trial learning and the palatability spectrum. Proc. Natl. Acad. Sci. USA 70:2261–2265.

Poulton, E. B. 1932. Attacks by birds upon butterflies, especially the Danainae. Proc. Roy. Ent. Soc. London (A) 7:54–55.

Powell, J. A., and R. A. Mackie. 1966. Biological interrelationships of moths and *Yucca whipplei*. Univ. Calif. Publ. Ent. 42:1–59.

Prestwich, G. D. 1983. Chemical systematics of termite exocrine secretions. Annu. Rev. Ecol. Syst. 14:287–311.

Price, P. W. 1980. Semiochemicals in evolutionary time. In D. A. Nordlund, R. L. Jones, and W. J. Lewis (eds.), Semiochemicals: Their Role in Pest Control. pp. 251–279. Wiley, New York.

Price, P. W., C. E. Bouton, P. Gross, B. A. McPheron, J. N. Thompson, and A. E. Weis. 1980. Interactions among three tropic levels: influence of plants on interactions between insect herbivore and natural enemies. Annu. Rev. Ecol. Syst. 11:41–65.

Pritchard, I. M., and R. James. 1984. Leaf-fall as a source of leaf miner mortality. Oecologia (Berlin) 64:140–142.

Prokopy, R. J., and R. P. Webster. 1978. Oviposition deterring pheromone of *Rhagoletis pomonella:* a kairomone for its parasitoid *Opius lectus*. J. Chem. Ecol. 4:481–494.

Ramirez, B. W. 1969. Fig wasps: mechanism of pollen transfer. Science 163:580–581.

Raros, R. S. 1973. Prospects and Problems of Integrated Pest Control in Multiple Cropping. IRRI Saturday Seminar. IRRI, Los Banos, Phillippines.

Read, D. P., P. P. Feeny, and R. B. Root. 1970. Habitat selection by the aphid parasite *Diaeretiella rapae* (Hymenoptera: Braconidae) and hyperparasite *Charips brassicae* (Hymenoptera: Cynipidae). Can. Ent. 102:1567–1578.

Regnier, F. E., and E. O. Wilson. 1971. Chemical communication and "propaganda" in slave-maker ants. Science 172:267–269.

Reichstein, T. 1967. Cardenolide (herzwirksame Glycoside) als Abwehrstoffe bei Insekten. Naturwiss. Rundsch. 20:499–511.

Reichstein, T., J. von Euw, J. A. Parsons, and M. Rothschild. 1968. Heart poisons in the monarch butterfly. Science 161:861–868.

Reissig, W. H., B. H. Stanley, and W. L. Roelofs. 1985. Tests of synthetic apple volatiles in traps as attractants for apple maggot flies (Diptera: Tephritidae) in commercial apple orchards. Environ. Ent. 14:55–59.

Rence, B., and W. Loher. 1977. Contact chemoreceptive sex recognition in the male cricket, *Teleogryllus commodus*. Physiol. Ent. 2:225–236.

Rhoades, D. F. 1983. Responses of alder and willow to attack by tent caterpillars: evidence for pheromonal sensitivity of willow. In P. Hedin (ed.), Plant Resistance to Insects American Chemical Society Symposium Series 208. pp. 55–68. American Chemical Society, Washington, D.C.

Rhoades, D. F. 1985. Offensive–defensive interactions between herbivores and plants: their relevance in herbivore population dynamics and ecological theory. Amer. Nat. 125:205–238.

Rhodes, A. M., R. L. Metcalf, and E. R. Metcalf. 1980. Diabroticite beetle responses to cucurbitacin kairomones in *Cucurbita* hybrids. J. Amer. Soc. Hortic. Sci. 105:838–842.

Risch, S. J. 1981. Insect herbivore abundance in tropical monocultures and polycultures: an experimental test of two hypotheses. Ecology 62: 1325–1340.

Risch, S. J., and F. R. Rickson. 1981. Mutualism in which ants must be present before plants produce food bodies. Nature 291:149–150.

Risch, S. J., D. Andow, and M. A. Altieri. 1983. Agro-ecosystem diversity and pest control: data, tentative conclusions, and new research directions. Environ. Ent. 12:625–629.

Robinson, F. A., and E. Oertel. 1975. Sources of nectar and pollen. In Dadant and Sons (eds.), The Hive and the Honeybee. pp. 283–302. Dadant and Sons, Hamilton, Ill.

Rodriquez, E., P. L. Healey, and I. Mehta (eds.). 1984. Biology and Chemistry of Plant Trichomes. Plenum, New York.

Roeske, C. N., J. N. Seiber, L. P. Brower, and C. M. Moffitt. 1976. Milkweed cardenolides and their comparative processing by monarch butterflies (*Danaus plexippus* L.). In J. W. Wallace and R. L. Mansell (eds.), Biochemical Interaction between Plants and Insects, Vol. 10, Recent Advances in Phytochemistry. pp. 93–167. Plenum, New York.

Rogers, C. E. 1985. Extrafloral nectar: entomological implications. Bull. Ent. Soc. Amer. 31:15–20.

Rogers, D. J., and M. S. Sullivan. 1986. Nymphal performance of *Geocoris punctipes* (Hemiptera:Lygaeidae) on pest-resistant soybeans. Environ. Ent. 15:1032–1036.

Rosenthal, G. A., and D. H. Janzen (eds.). 1979. Herbivores: Their Interaction with Secondary Plant Metabolites. Academic Press, New York.

Rothschild, M. 1966. Experiments with captive predators and the poisonous grasshopper *Poekilocerus bufonius* (Klug). Proc. Roy. Ent. Soc. London (C) 31:32, 40–41.

Rothschild, M. 1973. Secondary plant substances and warning colouration in insects. Symp. Roy. Ent. Soc. London 6:59–83.

Rothschild, M. 1985. British aposematic Lepidoptera. In J. H. Heath and A. M. Emmet (eds.), The Moths and Butterflies of Great Britain and Ireland, pp. 9–62. Harley Books, Essex, England.

Rothschild, M., and J. A. Edgar. 1978. Pyrrolizidine alkaloids from *Senecio vulgaris* sequestered and stored by *Danaus plexippus*. J. Zool. London 186:347–349.

Rothschild, M., and D. Kellet. 1972. Reactions of various predators to insects storing heart poisons (cardiac glycosides) in their body tissues. J. Ent. (A) 46:103–110.

Rothschild, M., and N. Marsh. 1978. Some peculiar aspects of danaid/plant relationships. Ent. Exp. Appl. 24:437–450.

Rothschild, M., B. P. Moore, and W. V. Brown. 1984. Pyrazines as warning odor components in the monarch butterfly, *Danaus plexippus*, and in moths of the genera *Zygaena* and *Amata* (Lepidoptera). Biol. J. Linn. Soc. 23:375–380.

Rowell-Rahier, M., and J. M. Pasteels. 1986. Economics of chemical defense in Chrysomelinae. J. Chem. Ecol. 12:1189–1203.

Ryan, C. A. 1983. Insect-induced chemical signals regulating natural plant protection responses. In R. F. Denno and M. S. McClure (eds.), Variable Plants and Herbivores in Natural and Managed Systems. pp. 43–60. Academic Press, New York.

Ryan, C. A., and T. R. Green. 1974. Proteinase inhibitors in natural plant protection. In V. C. Runeckles and E. E. Conn (eds.), Metabolism and Regulation of Secondary Plant Products, Vol. 8, Recent Advances in Phytochemistry. pp. 123–140. Plenum, New York.

Sabelis, M. W., and H. E. van de Baan. 1983. Location of distant spider mite colonies by phytoseiid predators: demonstration of specific kairomones emitted by *Tetranychus urticae* and *Panonychus ulmi*. Ent. Exp. Appl. 33:303–314.

Sauls, C. E., D. A. Nordlund, and W. J. Lewis. 1979. Kairomones and their use for management of entomophagous insects. VIII. Effect of diet on the kairomonal activity of frass from *Heliothis zea* (Boddie) larvae for *Microplitis croceipes* (Creson). J. Chem. Ecol. 5:363–369.

Schmidt, J. O. 1982. Biochemistry of insect venoms. Annu. Rev. Ent. 27:339–368.

Schmutterer, H., K. R. S. Ascher, and H. Rembold (eds.). 1982. Natural Pesticides from the Neem Tree. G. T. Z., Eschborn, Germany.

Schneirla, T. C. 1971. Army Ants: A Study in Social Organization. W. H. Freeman, San Francisco.

Schoonhoven, L. 1981. Chemical mediators between plants and phytophagous insects. In D. Nordlund, R. Jones, and W. J. Lewis (eds.), Semiochemicals: Their Role in Pest Control. pp. 31–50. Wiley, New York.

Schoonhoven, L. M. 1982. Biological aspects of antifeedants. Ent. Exp. Appl. 31:57–69.

Schroth, M., and U. Maschwitz. 1984. Zur Larvalbiologie und Wirtsfindung von *Maculinea teleius* (Lepidoptera: Lycaenidae), eines Parasiten von *Myrmica laevinodis* (Hymenoptera: Fromicidae). Ent. Gen. 9:225–230.

Seiber, J. N., P. M. Tuskes, L. P. Brower, and C. J. Nelson. 1980. Pharmacodynamics of some individual milkweed cardenolides fed to larvae of the monarch butterfly (*Danaus plexippus* L.). J. Chem. Ecol. 6:321–339.

Shah, M. A. 1982. The influence of plant surfaces on the searching behavior of coccinellid larvae. Ent. Exp. Appl. 31:377–380.

Shahjahan, M. 1974. *Erigeron* flowers as a food and attractive odor source for *Peristenus pseudopallipes,* a braconid parasitoid of the tarnished plant bug. Environ. Ent. 3:69–72.

Shahjahan, M., and F. A. Streams. 1973. Plant effects on host-finding by *Leiophron pseudopallipes* (Hymenoptera: Braconidae), a parasitoid of the tarnished plant bug. Environ. Ent. 2:921–925.

Shapiro, A. M. 1977. Avian predation on butterflies—again. Ent. Rec. J. Var. London 89:293–295.

Sheehan, W. 1986. Response by specialist and generalist natural enemies to agroecosystem deversification: a selective review. Environ. Ent. 15:456–461.

Shmida, A., and M. Auerbach. 1983. The strange mustard smell of the crucifers. Isr. Land Nat. 9:61–66.

Simpson, B. B., and J. L. Neff. 1983. Evolutionary and diversity of floral rewards. In C. E. Jones and R. J. Little (eds.), Handbook of Experimental Pollination Biology. pp. 142–159. Van Nostrand Reinhold, New York.

Sivinski, J. M., and C. Calkins. 1986. Pheromones and parapheromones in the control of tephritids. Fla. Ent. 69:157–168.

Slater, J. W. 1877. On the food of gaily-colored caterpillars. Trans. Ent. Soc. London 25:205–209.

Smiley, J. T., J. M. Horn, and N. E. Rank. 1985. Ecological effect of salicin at three trophics levels: new problems from old adaptations. Science 229:649–651.

Smith, J. M. 1957. Effects of the food plant of California red scale, *Aonidiella aurantii* (Mask), on reproduction of its hymenopterous parasites. Can. Ent. 89:219–230.

Smith, R. H. 1963. Toxicity of pine resin vapors to three species of *Dendroctonus* bark beetles. J. Econ. Ent. 56:827–831.

Smithers, C. W. 1973. A note on the natural enemies of *Danaus plexippus* (L.) (Lepidoptera: Nymphalidae) in Australia. Aust. Ent. Mag. 1:37–40.

Spangler, H. G., and S. Taber III. 1970. Defensive behavior of honey bees toward ants. Psyche 77:184–189.

Städler, E. 1977. Sensory aspects of insect plant interactions. In Proceedings of the 15th International Congress on Entomology. pp. 228–248. Entomological Society of America, College Park, Md.

Städler, E., and H. R. Buser. 1984. Defense chemicals in leaf surface wax synergistically stimulate oviposition by a phytophagous insect. Experientia 40:1157–1159.

Stafford, K. C., III, C. W. Pitts, and T. L. Webb. 1984. Olfactometer studies of host seeking by the parasitoid *Spalangia endius* Walker (Acari: Macrochelidae). Environ. Ent. 13:228–231.

Starks, K. J., R. Muniappan, and R. D. Eikenbary. 1972. Interaction between plant resistance and parasitism against greenbug on barley and sorghum. Ann. Ent. Soc. Amer. 65:650–655.

Stephenson, A. G. 1981. Flower and fruit abortion: proximate causes and ultimate functions. Annu. Rev. Ecol. Syst. 12:253–279.

Stephenson, A. G. 1982a. The role of the extrafloral nectaries of *Catalpa speciosa* in limiting herbivory and increasing fruit production. Ecology 63:663–669.

Stephenson, A. G. 1982b. Iridoid glycosides in the nectar of *Catalpa speciosa* are unpalatable to nectar thieves. J. Chem. Ecol. 8:1025–1034.

Sternlicht, M. 1973. Parasitic wasps attracted by the sex pheromone of their coccid host. Entomophaga 18:339–342.

Stowe, M. K., J. H. Tumlinson, and R. R. Heath. 1987. Chemical mimicry: Bolas spiders emit components of moth prey species sex pheromones. Science 236:964–967.

Strand, M. R., and S. B. Vinson. 1982. Behavioral response of the parasitoid *Cardiochiles nigriceps* to a kairomone. Ent. Exp. Appl. 31:308–315.

Strand, M. R., and S. R. Vinson. 1983. Analysis of an egg recognition kairomone of *Telenomus heliothidis* (Hymenoptera: Scelionidae) isolation and function. J. Chem. Ecol. 9:423–432.

Streams, F. A., M. Shahjahan, and H. H. LeMasurier. 1968. Influence of plants on the parasitization of the tarnished plant bug by *Leiophron pallipes*. J. Econ. Ent. 61:996–999.

Strickler, K., and B. A. Croft. 1985. Comparative rotenone toxicity in the predator, *Amblyseius fallacis* (Acari: Phytoseidae), and the herbivore, *Tetranychus urticae* (Acari: Tetranychidae), grown on lima beans and cucumbers. Environ. Ent. 14:243–246.

Sundby, R. A. 1967. Influence of food on the fecundity of *Chrysopa carnea* Stephens (Neuroptera, Chrysopidae). Entomophaga 12:475–479.

Tahvanainen, J. O., and R. B. Root. 1972. The influence of vegetational diversity on the population ecology of a specialized herbivore, *Phyllotreta cruciferae* (Coleoptera: Chrysomelidae). Oecologia (Berlin) 10:321–346.

Tallamy, D. W. 1985. Squash beetle feeding behavior: an adaptation against induced cucurbit defenses. Ecology 66:1574–1579.

Thiery, D., and J. H. Visser. 1986. Masking of host plant odour in the olfactory orientation of the Colorado potato beetle. Ent. Exp. Appl. 41:165–172.

Thurston, R., and P. M. Fox. 1972. Inhibition by nicotine of emergence of *Apanteles congregatus* from its host, the tobacco hornworm. Ann. Ent. Soc. Amer. 65:547–550.

Tilman, D. 1978. Cherries, ants and tent caterpillars: timing of nectar production in relation to susceptibility of caterpillars to ant predation. Ecology 59:686–692.

Tinbergen, N. 1951. The Study of Instinct. Oxford University Press, New York.

Tinbergen, N. 1958. Curious Naturalists. Doubleday, New York.

Tinbergen, N. 1972. The Animal and Its World, Vol. I, Field Studies. Harvard University Press, Cambridge, Mass.

Tschinkel, W. R. 1975. A comparative study of the chemical defensive system of tenebrionid beetles: chemistry of the secretions. J. Insect Physiol. 21:753–783.

Tuskes, P. M., and L. P. Brower. 1978. Overwintering ecology of the monarch butterfly. Ecol. Ent. 3:141–153.

Ullyett, G. C. 1953. Biomathematic and insect population problems. Mem. Ent. Soc. South Afr. 2:1–89.

Urquhart, F. A. 1960. The Monarch Butterfly. University of Toronto Press, Toronto.

Urquhart, F. A. 1976. Found at last: the monarch's winter home. Natl. Geogr. Mag. 150:160–173.

Vander Meer, R. K., and D. P. Wojcik. 1982. Chemical mimicry in the myrmecophilous beetle, *Myrmecaphodius excavaticollis*. Science 218:806–807.

van Emden, H. F., and K. S. Hagen. 1976. Olfactory reactions of the green lacewing, *Chrysopa carnea*, to tryptophan and certain breakdown products. Environ. Ent. 5:469–473.

van Emden, H. F., and G. F. Williams. 1974. Insect stability and diversity in agro-ecosystems. Annu. Rev. Ent. 19:455–475.

Vince, M. A. 1964. Use of the feet in feeding by the great tit *Parus major*. Ibis 106:508–529.

Vinson, S. B. 1968. Source of a substance in *Heliothis virescens* (Lepidoptera: Noctuidae) that elicits a searching response in its habitual parasite *Cardiochiles nigriceps* (Hymenoptera: Braconidae). Ann. Ent. Soc. Amer. 61:8–10.

Vinson, S. B. 1976. Host selection by insect parasitoids. Annu. Rev. Ent. 21:109–133.

Vinson, S. B. 1984. Parasite-host relationships. In W. J. Bell and R. T. Cardé (eds.), Chemical Ecology of Insects. pp. 205–233. Chapman & Hall, London.

von Uexkull, J. 1921. Umwelt and Innenwelt der Tiere. Springer-Verlag, Berlin.

Wall, C. 1984. The exploitation of insect communication by man—fact or fantasy? In T. Lewis (ed.), Insect Communication. pp. 379–400. Academic Press, London.

Wallbank, B. E., and Waterhouse, D. F. 1970. The defensive secretions of *Polyzosteria* and related cockroaches. J. Insect Physiol. 16:2081–2096.

Watkins, J. F., II, F. R. Gehlbach, and J. C. Kroll. 1969. Attractant-repellent secretions of blind snakes (*Leptotyphlops dulcis*) and their army ant prey (*Neivamyrmex nigrescens*). Ecology 50:1098–1102.

Way, M. 1963. Mutualism between ants and honeydew-producing Homoptera. Annu. Rev. Ent. 8:307–344.

Weatherston, J., and J. E. Percy. 1970. Arthropod defensive secretions. In M. Beroza (ed.), Chemicals Controlling Insect Behavior. pp. 95–114. Academic Press, New York.

Weaver, E. C., E. T. Clarke, and N. Weaver. 1975. Attractiveness of an assassin bug to stingless bees. J. Kans. Ent. Soc. 48:17–18.

Weis, A. E., and A. Kapelinski. 1984. Manipulation of host-plant development by the gall-midge *Rhabdophaga strobiloides*. Ecol. Ent. 9:457–465.

Weldon, P. J. 1980. In defense of "kairomone" as a class of chemical releasing stimuli. J. Chem. Ecol. 6:719–725.

Went, F. W. 1970. Plants and their chemical environment. In E. Sondheimer and J. Simeone (eds.), Chemical Ecology. pp. 71–82. Academic Press, New York.

Wesloh, R. M. 1976. Behavioral responses to the parasite, *Apanteles melanoscelus*, to gypsy moth silk. Environ. Ent. 5:1128–1132.

Whitman, D. W., M. S. Blum, and C. Jones. 1985. Chemical defense in *Taeniopoda eques* (Orthoptera: Acrididae): role of the metathoracic secretion. Ann. Ent. Soc. Amer. 78:451–455.

Whitman, D., M. Blum, and C. Jones. 1986. Olfactorily mediated attack suppression in the southern grasshopper mouse toward an unpalatable prey. Behav. Processes 13:77–83.

Wiklund, C. 1981. Generalist vs. specialist oviposition behavior in *Papilio machaon* (Lepidoptera) and functional aspects of the hierarchy of oviposition preferences. Oikos 36:163–170.

Wilbert, H. 1974. Die Wahrnehmung von Beute durch die Eilarven von *Aphidoletes aphidimyza* (Cecidomyiidae). Entomophaga 19:173–181.

Williams, W. G., G. G. Kennedy, R. T. Yamamota, J. D. Thacker, and J. Bordner. 1980. 2-Tridecanone: a naturally occurring insecticide from the wild tomato *Lycopersicon hirsutum* f. *glabratum*. Science 207:888–889.

Williamson, D. L. 1971. Olfactory discernment of prey by *Medetera bistriata* (Diptera: Dolichopodidae). Ann. Ent. Soc. Amer. 64:586–589.

Wilson, E. O. 1971. The Insect Societies. Harvard University Press, Cambridge, Mass.

Wilson, E. O. 1975. Enemy specifications in the alarm-recruitment system of an ant. Science 190:798–800.

Wood, D. L. 1982. The role of pheromones, kairomones, and allomones in the host selection and colonization behavior of bark beetles. Annu. Rev. Ent. 27:411–446.

Wyatt, I. J. 1965. The distribution of *Myzus persicae* (Sulz.) on year-round chrysanthemums. I. Summer season. Ann. Appl. Biol. 56:439–459.

Wyatt, I. J. 1970. The distribution of *Myzus persicae* (Sulz.) on year-round chrysanthemums. II. Winter season. The effect of parasitism by *Aphidius matricariae* Hal. Ann. Appl. Biol. 65:31–41.

Zoebelein, G. 1956. Der Honigtau als Nahrung der Insekten. Tel. I. I. Z. Angew. Ent. 38:369–416.

CHAPTER 2

Influences of Plant-Produced Allelochemicals on the Host/Prey Selection Behavior of Entomophagous Insects

D. A. Nordlund
USDA Agricultural Research Service
College Station, Texas

W. J. Lewis
USDA Agricultural Research Service
Tifton, Georgia

M. A. Altieri
University of California
Berkeley, California

CONTENTS

1 Introduction
2 Plant-produced allelochemicals and host/prey habitat location
   2.1 General plant influences
   2.2 The role of allelochemicals: synomones
   2.3 The role of allelochemicals: associational resistance
3 Effects of plants on herbivore kairomones
4 Implications for the use of entomophagous insects in applied biological control
5 Conclusion
   References

1   INTRODUCTION

DeBach (1964) defined biological control as "the action of parasites, predators, and pathogens in maintaining another organism's density at an average lower than would occur in their absence." Under this definition, biological control includes both natural biological control, which occurs without human involvement, and applied biological control, which requires purposeful introduction or manipulation of parasites, predators, or pathogens to reduce populations of undesirable organisms. Entomophagous insects (parasitoids and predators) are major factors in natural biological control, acting as regulators of host/prey populations and as agents of selection (Price 1984). In addition, entomophagous insects are cornerstones of applied biological control of insect pests, our major competitors for food and fiber. Thus, knowledge of the interactions of entomophagous insects with their hosts or prey is of great theoretical and practical value.

Successful reproduction by entomophagous insects is dependent on the location of suitable hosts or prey on which to oviposit or feed. Successful parasitization requires several steps: host habitat location, host location, host acceptance, host suitability, and host regulation (Laing 1937, Flanders 1953, Doutt 1964, Vinson 1975a). The first three of these steps make up the host selection sequence. Predators, at least those that actively hunt for their prey, appear to go through essentially the same basic selection process (Vinson 1981), giving us a host/prey selection behavior sequence of: host/prey habitat location, host/prey location, and host/prey acceptance. There is growing evidence that plants influence these behaviors. The influences of plant-produced allelochemicals on the host/prey-habitat location and host/prey location by entomophagous insects and their implications for biological control are discussed in the following pages.

2   PLANT-PRODUCED ALLELOCHEMICALS AND HOST/PREY HABITAT LOCATION

2.1   General Plant Influences

The choice of habitat in which to search has long been recognized as an important first step in the host/prey selection process. For those entomophagous insects with many potential host or prey species, host/prey specificity may be largely an effect of host/prey habitat location. The latter may place a greater limitation on the number of species actually attacked than does host or prey suitability (Townes 1960, Flanders 1962). Picard and Rabaud (1914), probably the first to recognize this, found that many parasitic Hymenoptera would attack larvae from different families, and even different orders, if they were found on the same food plant and suggested that the food plant of the host influenced host selection.

Many entomophagous insects exhibit a high degree of specificity for microhabitats in which to search. Vet et al. (1984), for example, reported that *Asobara tabida* (Nees), a parasitoid of Drosophilidae, preferred decaying fruit, where yeast was the predominant agent of decay, whereas *Asobara rufescens* (Forester) preferred to search in decaying plant leaves, where bacteria were the predominant agent of decay. The specificity was so strong that *Drosophila* larvae in a fruit–yeast mixture, placed in a beet field, were parasitized only by *A. tabida,* while *A. rufescens* females parasitized only larvae found in the adjacent decaying beet leaves. There are numerous examples of entomophagous insects attacking hosts or prey on some plant species or in some microhabitats, but at much lower rates or not at all in others (Taylor 1932, Walker 1940, Zwolfer and Kraus 1957, Salt 1958, Sekhar 1960, Arthur 1962, Streams et al. 1968, Young and Price 1975, Martin et al. 1976, Harrington and Barbosa 1978, Zehnder and Trumble 1984, Vet and van Alphen 1985). Entomophagous insects also have been observed searching host-free plants (Vinson 1975b). Nordlund et al. (1984), for example, found that *Geocoris,* Nabidae, Reduviidae, and Coccinellidae were more numerous (Table 2.1), and predation of *Heliothis zea* (Boddie) eggs was greater, in plots of

TABLE 2.1 Mean Number of Insects per 3.0 Meters of Row Sampled with a D-Vac in Plots of Corn or Tomato[a]

Insect Group	Corn	Tomato
Geocoris	1.12	0.02
Nabidae	0.10	0.00
Reduviidae	0.12	0.02
Coccinellidae	1.12	0.00

Source: Data from Nordlund et al. (1984).
[a] Means for each insect group are significantly different ($P < 0.05$) as determined by Duncan's multiple-range test.

TABLE 2.2   Predation and Parasitism of _Heliothis zea_ Eggs After About 24 Hours of Exposure in Plots of Corn or Tomato[a]

Corn	Tomato
Percent of Eggs Missing	
82.9	46.0
Percent Parasitism by **Trichogramma** *spp.*	
1.5	42.9

Source: Data from Nordlund et al. (1984).
[a] Means are significantly different ($P < 0.05$) as determined by Duncan's multiple-range test.

corn than in nearby plots of tomato (Table 2.2). Parasitism of *H. zea* eggs by *Trichogramma* spp., however, was significantly higher in tomato plots than it was in corn plots (Table 2.2). These findings indicate that plants do influence the host/prey habitat location behavior of at least some entomophagous insects.

2.2   The Role of Allelochemicals: Synomones

Allelochemical-mediated interactions between plants and herbivorous insects have received considerable attention (Wallace and Mansell 1976, Kogan 1977, Rosenthal and Janzen 1979, Schoonhoven 1981), as have interactions between herbivores and entomophagous insects (Arthur 1981, Greany and Hagen 1981, Weseloh 1981). Although many characteristics of the environment and plants upon which herbivores are found may influence the host/prey selection behavior of entomophagous insects, allelochemicals appear to be extensively involved in this behavior (Nordlund et al. 1981; see also Chapter 1). Despite the recognized importance of plants to the host/prey selection behavior of entomophagous insects, little research has been done on allelochemical-mediated interactions among plants, herbivores, and entomophagous insects (Price et al. 1980, Vinson 1981, Barbosa and Saunders 1985).

   Allelochemicals produced by plants that stimulate host/prey selection behavior of entomophagous insects are referred to as synomones. A synomone is a substance produced or acquired by an organism which when it contacts an individual of another species in the natural context, evokes in the receiver a behavioral or physiological reaction adaptively favorable to both emitter and receiver (Nordlund and Lewis 1976). Several examples of synomones influencing entomophagous insect behavior have been reported. Monteith (1955, 1956, 1958) reported that *Drino bohemica* Mesnil and *Bessa harveyi* Townsend were attracted, in an olfactometer, to white spruce, the food plant of their host. *Itoplectis conquisitor* (Say), which attacks hosts on Scotch pine but not red pine, were attracted to Scotch pine in an olfactometer, but not to red pine (Arthur 1962). Kesten (1969) reported that *Anatis ocellata* (L.) was attracted by pine needles, the food source of their aphid prey. The aphid parasitoid, *Diaeretiella rapae* M'Intosh, was attracted, in an olfactometer, to collard leaves and to allyl isothiocyanate (mustard oil), which is found in collards (Read et al. 1970). Caryophyllene, from cotton plants, similarly attracted adult *Chrysopa* (= *Chrysoperla*) (Flint et al. 1979). *Eucelatoria* sp., a tachinid parasitoid of *Heliothis* spp., was shown to be attracted to volatile chemicals from cotton, corn, and okra (Nettles 1979, 1980). Altieri (unpublished data) demonstrated that *Microplitis croceipes* (Cresson) females were attracted to *Geranium carolinianum* and *Amaranthus* spp. Water extract of *Amaranthus* increased parasitism by *Trichogramma* spp. on soybean, cowpea, and tomato (Table 2.3). Elzen et al. (1983) reported that *Campoletis sonorensis* Cameron females were attracted to and stimulated to antennate and probe parts of some plant species but not others. Six terpenoids from cotton that attract *C. sonorensis* were isolated and identified (Elzen et al. 1984; see also Chapter 5). Finally, host selection behavior of *Trichogramma pretiosum* (Riley) females was found to be stimulated by synomones from tomato. In addition, tomato extract, when applied to corn, increased rates of parasitism in the field (Table 2.4).

TABLE 2.3   Percent Parasitism of *Heliothis zea* Eggs by Naturally Occurring *Trichogramma* spp. in Crop Systems Sprayed with a Water Extract of *Amaranthus* or Water[a]

	Parasitism in Crop			
Treatment	Soybean	Cowpeas	Tomato	Cotton
Amaranthus	21.4a	45.4a	24.3a	13.6a
Water	12.6b	31.6b	17.6b	4.2b

Source: Data from Altieri et al. (1981).
[a] Means followed by different letters in each column are significantly different ($P < 0.05$) as determined by Duncan's multiple-range test.

TABLE 2.4   Percent Parasitism of *Heliothis zea*
Eggs by *Trichogramma pretiosum* on Corn Plants
Treated with a Hexane Extract of Tomato[a]

Treated	Control
37.7	28.5

Source: Data from Nordlund et al. (1985).
[a] Means significantly different ($P < 0.05$) as determined by ANOVA.

Responses to allelochemicals, however, are not immutable and can be influenced by a variety of factors, including the physiological state of the entomophagous insect. The state of ovarian development in female parasitoids with long preoviposition periods, for example, influences the response of females to plant-produced allelochemicals. Thorpe and Caudle (1938) reported that *Pimpla ruficollis* Gravenhorst, a parasitoid of the European pine shoot moth (*Rhyacionia buoliana* Schiff), which emerges sometime before host larvae are available for oviposition, leaves the pine environment to feed on other plants, only to return after 3 or 4 weeks to search for hosts. During the first 3 or 4 weeks of adult life, *P. ruficollis* females are repelled by Scotch pine. During this period, the ovaries are small and probably not able to produce mature eggs. When the ovaries mature (after 3 to 4 weeks), females are attracted by materials in the oil of Scotch pine. Nishida (1956) also found that ovarian development in *Opius fletcheri* Silvestri must be completed before the female responds to the plants on which its host feeds. *Eucarcelia rutilla* Vill. females preferred oak to pine during the preoviposition period, but at the end of that period were repelled by oak and attracted by pine (Herrebout and van der Veer 1969). Similarly, Shahjahan (1974) found that the response of *Peristenus pseudopallipes* (Loan) females to synomones from *Erigeron* flowers was influenced by the number of eggs in females' ovaries.

2.3   The Role of Allelochemicals: Associational Resistance

Tahvanainen and Root (1972) coined the term "associational resistance" to describe the phenomenon of lower herbivore densities on a particular plant host in multispecies associations. This influence on pest abundance may have contributed to the development of polycultures, such as those in agroforestry systems (Fig. 2.1) and traditional agricultural systems (Ba-

Figure 2.1 Agroforestry system of western Java. (Photo by M. A. Altieri.)

tra 1962; Guevara 1962; DeLoach 1970; Fye 1972; IRRI 1972, 1975; Baker and Norman 1975; Crookston and Kent 1976; Francis et al. 1976; Bach 1980; Ruthenburg 1980; Cromartie 1981; Zandstra et al. 1981; Matteson et al. 1984). These systems have benefited from centuries of empirical knowledge as well as trial and error and thus are exceptionally well adapted to local conditions (Johnson 1972, Harwood 1979). Most exhibit minimal pest impact (Altieri et al. 1978; Risch 1979, 1980a,b, 1981; Gagné 1982; Letourneau and Altieri 1983; Letourneau 1986).

Research on polyculture systems clearly shows that crops grown in diverse, structurally complex crop communities tend to have low numbers of pests (Elton 1958; Pimentel 1961; van Emden 1965; Tahvanainen and Root 1972; Root 1973, 1975; Dempster and Coaker 1974; Feeny 1975; Litsinger and Moody 1976; Altieri et al. 1977, 1978; Perrin 1977; Perrin and Phillips 1978; Risch 1979, 1983; Litsinger et al. 1980; Altieri and Gliessman 1983; Risch et al. 1983; Andow 1986; but see Sheehan 1986). The presence of various weed species in or around crop areas also lowers the population densities of herbivorous insects (Allen and Smith 1958, Pimentel 1961, Adams and Drew 1965, Dempster 1969, Flaherty 1969,

Potts and Vickerman 1974, Smith 1976, Altieri et al. 1977). Root (1973) suggested two hypotheses to explain the reduced number of pests commonly reported for polycultures. One hypothesis, the "resource concentration hypothesis," focuses on the movement and reproductive behavior of the pest insects. The other, the "enemies hypothesis," predicts greater numbers of entomophagous insects in diverse habitats than in simple habitats. Based on a review of available evidence, Andow (1983) and Risch et al. (1983) argued [as did Root (1973)] that the resource concentration hypothesis may be more important than the enemies hypothesis in accounting for decreased herbivore abundance in diverse habitats.

In support of the enemies hypothesis, however, a number of studies have shown that plants found in association with crops influenced the presence and/or effectiveness of entomophagous insects (Allen and Smith 1958; Read et al. 1970; Robinson 1972; Burleigh et al. 1973; Zandstra and Motooka 1978; Altieri and Whitcomb 1979, 1980; Altieri and Todd 1981; Fye 1983; Shelton and Edwards 1983). Altieri et al. (1981), for example, reported that parasitism of *H. zea* eggs by *Trichogramma* spp. was significantly higher on soybean plants associated with *Desmodium, Croton,* or *Cassia* than on weed-free soybean plots. These associated plants may have provided food such as nectar or pollen for entomophagous insects (van Emden 1963, Shahjahan 1974, Gilbert 1975, Simmons et al. 1975, Bentley 1977, Smiley 1978) and thus increased their longevity and fecundity (Leius 1963, 1967; Syme 1975, 1977). In addition, synomones from plant associates may influence the behavior and thus the effectiveness of entomophagous insects. Alternate hosts or prey are also found on associated plants (Allen 1932, van Emden 1965, Doutt and Nakata 1973, Kido et al. 1983). Not all combinations of plants increase the activities of entomophagous insects, however. Monteith (1960), for example, found that parasitism of the larch sawfly (*Pristiphora erichsonii* Hartig) was much higher in pure stands of eastern larch than in mixed stands. Burleigh et al. (1973) found that there were fewer entomophagous insects in cotton grown adjacent to corn than in cotton grown in association with sorghum. The presence of some other herbaceous plants in patches of *Erigeron* appeared to lower parasitism of *Lygus lineolaris* (Palisot de Beauvoris) by *P. pseudopallipes* (Shahjahan and Streams 1973), despite the fact that the parasitoid is attracted to *Erigeron* by olfactory cues (Shahjahan 1974).

For the most part, studies of pest insects in polyculture versus monoculture systems have failed to examine the ecological mechanisms underlying observations that pest numbers are generally lower in more complex systems. In addition, a minimum number of data are available on the role of predators, fewer still are available for parasitoids, and the enemies hypothesis has never been tested with parasitoids. It would seem safe, however, to assume that the enemies hypothesis explains at least part of this reduction. On the basis of studies cited above, it would not be surprising to find that synomones are partly responsible for the increased abundance or effectiveness of entomophagous insects observed in many polyculture systems.

3 EFFECTS OF PLANTS ON HERBIVORE KAIROMONES

Once an entomophage is in an appropriate habitat, it still has to assess whether host or prey are present, and if so, it has to locate suitable hosts or prey. Allelochemicals released by hosts or prey and used by entomophagous insects in host/prey location are referred to as *kairomones*. A kairomone is a substance, produced, acquired by, or released as a result of the activities of an organism which, when it contacts an individual of another species in the natural context, evokes in the receiver a behavioral or physiological reaction adaptively favorable to the receiver but not to the emitter (Brown et al. 1970, Nordlund and Lewis 1976). Chemicals originating from plants that stimulate host/prey selection behavior of entomophagous insects should also be considered kairomones if they are released only as a result of the activities of a herbivore or are active only in combination with materials from the herbivore (Nordlund and Lewis 1976).

For example, volatile host tree terpenes, in combination with the aggregation pheromone of bark beetles, attract several bark beetle predators and parasitoids (Wood et al. 1968; Vité and Williamson 1970; Camors and Payne 1971, 1973; Borden 1974, 1977). Similarly, Vinson (1975b) showed that *Cardiochiles nigriceps* Vierick females were attracted to damaged tobacco plants and were stimulated into an intensive search of the surrounding tissue if the damage was caused by the feeding of *Heliothis virescens* (Fabricius) larvae (indicating that a combination of plant and herbivore-released materials was involved). *Lixophaga diatraeae* (Townsend) has been shown to differentiate between adjacent infested and uninfested stalks of sugarcane (Roth et al. 1982). The host and its frass alone evoked little olfactory response. These data indicate that a volatile(s) released by larval feeding activity within the stalk, and probably produced by the plant, was important. Loke et al. (1983) reported that *Cotesia* (= *Apanteles*) *marginiventris* (Cresson) females spent more time searching leaves damaged by fall armyworm [*Spodoptera frugiperda* (J. E. Smith)] larvae (Table 2.5). This indicates that a combination of plant-produced kairomones may be involved in this interaction as well.

Most of the work on the effects of herbivore feeding on kairomone release has involved the response of parasitoids to the frass of lepidopterous larvae. Hsiao et al. (1966) found no difference in the response of *Lydella grisescens* Robineau-Desvoidy females to ethanol extracts of frass of *Ostrinia nubilalis* (Hübner) larvae fed on an artificial diet or corn. Roth et al. (1978), however, reported that *L. diatraeae* females were stimulated to larviposit by frass from *Diatraea saccharalis* (Fabricius) larvae fed on sugarcane, less stimulated by frass of larvae fed on sorghum or corn, and unaffected by frass of larvae fed on a soybean–flour–wheat germ diet. Thompson et al. (1983) reported that a methanol extract of sugarcane stalks served as a larviposition stimulant for *L. diatraeae;* the latter extract was just as active as a methanol extract of frass of sugarcane-fed *D. saccharalis* larvae. Both of these extracts had similar chromatographic properties. Mohyuddin et al. (1981) found that a Pakistan strain of *Apanteles flavipes* (Cameron) responded strongly to frass from *Chilo partellus* (Swinh.) larvae that had fed on corn, but weakly to frass from larvae fed sugarcane. *Microplitis croceipes* females responded more strongly to frass of *H. zea* larvae fed cowpea than to frass from larvae fed on laboratory diet (Sauls et al. 1979). Nordlund and Sauls (1981) reported that *M. croceipes* females responded more strongly to frass of soybean-reared larvae than to frass of corn-fed larvae. There was no difference in the response of *M. croceipes* females to frass of larvae fed corn or a CSM

TABLE 2.5 Mean Percent Time Spent Searching by
***Cotesia* (=*Apanteles*) *marginiventris* Females During a**
5-minute Observation Period on Differently Damaged
Corn Plants[a]

Damage	Percent Time
None	2.5a
Artificial damage	17.4b
Damage by *Spodoptera frugiperda* larvae	83.2c

Source: Data from Loke et al. (1983).
[a] Means followed by different letters are significantly different ($P <$ 0.05) as determined by Duncan's multiple-range test.

TABLE 2.6   Scored Host Selection Response of
Microplitis croceipes **Females to Extracts of Frass from**
Heliothis zea **Larvae Fed on Different Plants or CSM**
Laboratory Diet[a]

Soybean	Cotton	CSM Diet	Corn
1.6a	1.0b	0.3c	0.0c

Source: Data from Nordlund and Sauls (1981).
[a] Means followed by different letters are significantly different ($P <$ 0.05) as determined by Duncan's multiple-range test.

laboratory diet (Table 2.6). *Microplitis demolitor* females were shown to respond to frass of *H. zea* and *Trichoplusia ni* (Hübner) larvae fed cowpea much more strongly than to frass of larvae reared on laboratory diet (Nordlund and Lewis 1985). We do not know whether materials from plants are merely accumulated and concentrated or if the plants supply necessary precursers that are altered as they pass through the digestive tract of the larvae. Read et al. (1970), for example, found that the aphid parasitoid, *D. rapae,* was attracted to aphids that had been removed from collard leaves for only 15 minutes but not to aphids that had been removed from collard leaves for 24 hours. Finally, the specific plant species on which a herbivorous insect feeds can also affect the herbivore's chemistry and thus the kairomones released.

4   IMPLICATIONS FOR THE USE OF ENTOMOPHAGOUS INSECTS IN APPLIED BIOLOGICAL CONTROL

Entomophagous insects are important biological resources which serve as a cornerstone of biological control programs directed against insect pests. In the classical biological control approach (= importation) (Table 2.7), introduction and establishment of biological control agents have often led to such control of the target organism that it is simply no longer a pest. Unfortunately, the majority of our important agricultural insect pests are native insects and many are pests of annual row crops. Importation cannot generally be expected to work in annual row crops or against native pests (Nordlund 1984); thus biological control in these situations will involve the more interventionist approaches of periodic release and environmental manipulation. Success with these approaches will require a more thorough understanding of the interactions of entomophagous in-

TABLE 2.7   Techniques for Use of Entomophagous Insects in Biological Control Programs

I. Importation
II. Periodic releases
   A. Inundative releases
   B. Inoculative releases
III. Environmental manipulation
   A. Provision of alternate, factitious, or nonviable hosts
   B. Use of semiochemicals to improve the performance of entomophages
   C. Provision of various environmental requisites, such as food or nesting places
   D. Modification of cropping practices to favor the entomophage

Source: Reprinted with permission from Nordlund (1984).

sects with other components of the ecosystem, including plants. These interactions form a complex web, however, and interactions between entomophagous insects and plants are only a part, albeit an important part, of that web (Price 1981). An understanding of the various interactions that make up the web can lead to the development of approaches for encouraging the abundance and effectiveness of entomophagous insects to enhance the control of herbivorous pest species (Doutt and Smith 1971, Pradham 1971, Pimentel and Goodman 1978, van den Bosch 1978, Lewis 1981, Altieri et al. 1983, Lewis and Nordlund 1985).

The use of allelochemicals to manipulate entomophagous insects and improve their performance in the field holds great promise for biological control programs, particularly those directed against insect pests in annual row crop agroecosystems (Lewis and Nordlund 1984, 1985; Nordlund et al. 1986). Application of synomones could simulate the presence of a preferred plant or the diverse habitat of a polyculture (Altieri et al. 1981); for example, synomones from tomato may be applied to corn plants to improve the performance of *Trichogramma* (Nordlund et al. 1985; see also Chapter 1). With a sufficient data base we may be able to develop polyculture systems that are much less subject to pest damage than the monoculture systems currently used in developed countries (Altieri 1983, 1985), possibly leading to a change from an extensive to an intensive agricultural system. We may even be able to breed plants that produce allelochemicals that cause increases in entomophage activity and thus increase their resistance to pest insects (Lewis and Nordlund 1984). Plant breeders have, for the most part, ignored the third trophic level in their studies of host plant resistance. Plant breeding, however, can be used to improve the performance of entomophagous insects. The interaction of *Encarsia formosa* Gahan, with the host plant of its herbivore host, although unfortunately not an example involving allelochemicals, is an example of how specific changes in plants can result in greater effectiveness of entomophages. The parasitoid walks 3.5 times faster on the leaves of a hairless cucumber variety than on a hairy variety; and this results in a 20% increase in parasitism of greenhouse whitefly (*Trialeuroides vaporariorum* (Westwood) (van Lenteren et al. 1977, 1980). We would predict that comparable changes in plant chemistry would have similar results.

Selection of the best species or strain of entomophagous insect for use in a program can be critical to the program's success. A great many biological data are used in these decisions, but microhabitat preferences generally have not been considered. With a minimum amount of additional testing we should be able to determine if a particular strain of entomophage is attracted to or stimulated by the target crop plant. Considering the previously mentioned microhabitat specificity of some entomophagous insects, this procedure could result in a very significant improvement in the effectiveness of any type of release effort.

The potential for very rapid evolutionary change is a characteristic of the Insecta. The number of different species of insects is a testament to that potential. Therefore, as discussed by Price (1981), should any one control technique (be it insecticidal or biological) become so successful as to cause high mortality over a period of time, we should expect the target population to develop "resistance" to that technique. Resistance to entomophagous insects may take the form of encapsulation of parasitoid eggs or larvae, defensive secretions that deter feeding by predators or oviposition by parasitoids, shifts in microhabitat preference, and so on. In view of this potential for change, we would be wise to adopt a holistic approach to the study of our ecosystems and remain vigilant for reductions in the effectiveness of the control techniques that may be adopted.

CONCLUSION

Herbivory is a major selection pressure on plant populations. Chemistry (including nutrition), morphology, and escape in time or space are the primary means of defense for plants (Atsatt and O'Dowd 1976). Chemical defenses include the production and release of synomones that stimulate the host/prey selection behavior of entomophagous insects. As we have seen, this means of defense can result in increased mortality for herbivorous insects.

The question of whether or not plant-produced allelochemicals are important to the associational resistance of many multispecies plant associations has not been resolved. In their discussion of plant defense guilds, Atsatt and O'Dowd (1976) listed three ways in which guild members decrease herbivore abundance:

1. As "insectary plants" that aid in the maintenance of herbivore predators and parasitoids
   a. With floral and extrafloral nectaries providing an important alternative energy or by
   b. Supporting alternative hosts or prey
2. As "repellent plants," either directly or indirectly causing the herbivore to fail to locate or to reject its normal food plant
3. As "attractant-decoy plants," causing the herbivore to feed on alternative and possibly lower-quality plants

The possibility that allelochemicals might also be an important influence on entomophagous insects was not mentioned directly. In the summary of their paper, however, Atsatt and O'Dowd state that "insectary plants lower the numerical response of herbivores by increasing the efficiency of their predators and parasitoids." With additional study we should be able to demonstrate that the increased efficiency is a result, at least in part, of plant-produced allelochemicals.

We are just beginning to understand the important interactions between plants and entomophagous insects, and further research is necessary before significant advances can be expected. Lewis and Nordlund (1985) recommended a number of research approaches and objectives for the development of techniques for using semiochemicals to manipulate entomophagous insects in biological control programs (Table 2.8). Plants and plant-produced semiochemicals are mentioned a number of times in these recommendations.

Entomophagous insects have, or at least may have, a great impact on the populations of insects on which they feed. For this reason, interactions between them, their hosts or prey, and plants or other media on

**TABLE 2.8   Prioritized Listing of Research Approaches
and Objectives for the Development of Techniques for
Using Semiochemicals to Manipulate Entomophagous
Insects in Biological Control Programs**

1. Develop bioassay techniques for elucidating behavioral responses to volatile semiochemicals; and develop monitoring procedures for measuring behavioral responses to plant and host insect volatiles, during foraging among and within habitats.
2. Elucidate the roles and behavioral mechanisms governing insect responses to semiochemicals; determine their relative importance; and isolate and identify the major semiochemicals necessary for retention and efficient foraging of entomophages within target areas.
3. Determine the primary genetic and nongenetic (learning) characteristics influencing entomophage responses to the major semiochemicals.
4. Develop techniques and procedures for integrating knowledge of semiochemicals and their influences of entomophage foraging behavior into agronomic practices, such as plant breeding, cropping patterns, weed control, and irrigation, that will maximize the foraging efficiency of natural and released (imported and native) natural enemies.
5. Develop techniques and procedures for using behavior-modifying chemicals for monitoring and forecasting the density and performance of entomophage populations.

Source: Lewis and Nordlund (1985).

which the hosts or prey feed are of great academic interest and practical importance. Studies of these interactions should continue and will yield additional interesting and practical information in the next few years.

REFERENCES

Adams, J, B., and M. E. Drew. 1965. Grain aphids in New Brunswick. III. Aphid populations in herbicide-treated oatfields. Can. J. Zool. 43:789–794.

Allen, H, W, 1932. Present status of oriental fruit moth parasite investigations. J. Econ. Ent. 25:360–367.

Allen, W. W., and R. F. Smith. 1958. Some factors influencing the efficiency of *Apanteles medicaginis* Muesebeck (Hymenoptera: Braconidae) as a parasite of the alfalfa caterpillar *Colias philodice eurytheme* Boisduval. Hilgardia 23:1–42.

Altieri, M. A. 1983. Vegetational designs for insect-habitat management. Environ. Manag. 7:3–7.

Altieri, M. A. 1985. Agroecology, The Scientific Basis of Alternative Agriculture, 2nd ed. Division of Biological Control, University of California, Berkeley, Calif.

Altieri, M. A., and S. R. Gliessman. 1983. Effects of plant diversity on the density and herbivory of the flea beetle, *Phyllotreta cruciferae* (Goeze), in California collard (*Brassica oleracea*) cropping systems. Crop. Prot. 2:497–501.

Altieri, M. A., and J. W. Todd. 1981. Some influences of vegetational diversity on insect communities of Georgia soybean fields. Prot. Ecol. 3:333-338.

Altieri, M. A., and W. H. Whitcomb. 1979. The potential use of weeds in the manipulation of beneficial insects. Hortic. Sci. 14:12–18.

Altieri, M. A., and W. H. Whitcomb. 1980. Weed manipulation for insect pest management in corn. Environ. Manag. 4:483–489.

Altieri, M. A., A. van Schoonhoven, and J. D. Doll. 1977. The ecological role of weeds in insect pest management systems: a review illustrated with bean (*Phaseolus vulgaris*) cropping systems. PANS 23:195–205.

Altieri, M. A., C. A. Francis, A. van Schoonhoven, and J. D. Doll. 1978. A review of insect prevalence in maize (*Zea mays* L.) and bean (*Phaseolus vulgaris* L.) polyculture systems. Field Crops Res. 1:33–49.

Altieri, M. A., W. J. Lewis, D. A. Nordlund, R. C. Gueldner, and J. W. Todd. 1981. Chemical interactions between plants and *Trichogramma* wasps in Georgia soybean fields. Prot. Ecol. 3:259–263.

Altieri, M. A., P. B. Martin, and W. J. Lewis. 1983. A quest for ecologically based pest management systems. Environ. Manag. 7:91–100.

Andow, D. A. 1983. Plant diversity and insect populations: interactions among dry beans, insects, and weeds. Ph.D. dissertation. Cornell University, Ithaca, N. Y.

Andow, D. A. 1986. Plant diversification and insect population control in agroecosystems. In D. Pimentel (ed.), Some Aspects of Integrated Pest Management. pp. 277–368. Department of Entomology, Cornell University, Ithaca, N. Y.

Arthur, A. P. 1962. Influence of host tree on abundance of *Itoplectis conquistor* (Say) (Hymenoptera: Ichneumonidae), a polyphagous parasite of the European pine shoot moth, *Rhyacionia buoliana (Schiff.) (Lepidoptera:Oleuthreutidae)*. Can. Ent. 94:337–347.

Arthur, A. P. 1981. Host acceptance by parasitoids. In D. A. Nordlund, R. L.

Jones, and W. J. Lewis (eds.), Semiochemicals: Their Role in Pest Control. pp. 97–120. Wiley, New York.

Atsatt, P. R., and D. J. O'Dowd. 1976. Plant defense guilds. Science 193:24–29.

Bach, C. E. 1980. Effects of plant density and diversity on the population dynamics of a specialist herbivore, the striped cucumber beetle, *Acalymma vittatta* (Fab.). Ecology 61:1515–1530.

Baker, E. F. I., and D. W. Norman. 1975. Cropping systems in Northern Nigeria. In IRRI Proceedings of the Cropping Systems Workshop. pp. 334–361. IRRI, Los Banos, Philippines.

Barbosa, P., and J. Saunders. 1985. Plant allelochemicals: linkages between herbivores and their natural enemies. In G. A. Cooper-Driver, T. Swain, and E. E. Conn (eds.), Chemically Mediated Interactions between Plants and Other Organisms. pp. 107–138. Plenum, New York.

Batra, H. N. 1962. Mixed cropping and pest attack. Ind. Farm. 11:23–25.

Bentley, B. L. 1977. Extrafloral nectaries and protection by pugnacious bodyguards. Annu. Rev. Ecol. Syst. 8:407–427.

Borden, J. H. 1974. Aggregation in the scolytidae. In M. C. Birch (ed.), Pheromones. pp. 135–140. North-Holland, Amsterdam.

Borden, J. H. 1977. Behavioral responses of Coleoptera to pheromones, allomones, and kairomones. In H. H. Shorey and J. J. McKelvey, Jr. (eds.), Chemical Control of Insect Behavior. pp. 169–198. Wiley, New York.

Brown, W. L., Jr., T. Eisner, and R. H. Whittaker. 1970. Allomones and kairomones: transpecific chemical messengers. BioScience 20:21–22.

Burleigh, J. H., J. H. Young, and R. D. Morrison. 1973. Strip cropping's effect on beneficial insects and spiders associated with cotton in Oklahoma. Environ. Ent. 2:281–285.

Camors, F. B., Jr., and T. L. Payne. 1971. Response in *Heydenia unica* to *Dendroctonus frontalis* pheromones and a host-tree terpene. Ann. Ent. Soc. Amer. 65:31–33.

Camors, F. B., Jr., and T. L. Payne. 1973. Sequence of arrival of entomophagous insects to trees infected with the southern pine beetle. Environ. Ent. 2:267–270.

Cromartie, W. J., Jr. 1981. The environmental control of insects using crop diversity. In D. Pimentel (ed.), CRC Handbook of Pest Management in Agriculture, Vol. 1. pp. 223–250. CRC Press, Boca Raton, Fl.

Crookston, R. K., and R. Kent. 1976. Intercropping—a new version of an old idea. Crops Soils 28:7–9.

DeBach, P. (ed.). 1964. Biological Control of Insect Pests and Weeds. Reinhold, New York.

DeLoach, C. H. 1970. The effect of habitat diversity on predation. Proc. Tall Timbers Conf. Ecol. Anim. Control Habitat Manag. 2:223–241.

Dempster, J. P. 1969. Some effects of weed control on the numbers of the small cabbage white (*Pieris rapae* L.) on brussels sprouts. J. Appl. Ecol. 6:339–405.

Dempster, J. P., and T. H. Coaker. 1974. Diversification of crop ecosystems as a means of controlling pests. In D. Price-Jones and M. E. Solomon (eds.), Biology in Pest and Disease Control. pp. 106–114. Blackwell Scientific, Oxford.

Doutt, R. L. 1964. Biological characteristics of entomophagous adults. In P. DeBach (ed.), Biological Control of Insect Pests and Weeds. pp. 145–167. Reinhold, New York.

Doutt, R. L., and J. Nakata. 1973. The *Rubus* leafhopper and its egg parasitoid: an endemic biotic system useful in grape-pest management. Environ. Ent. 2:381–386.

Doutt, R. L., and R. F. Smith. 1971. The pesticide syndrome-diagnosis and suggested prophylaxis. In C. B. Huffaker (ed.), Biological Control. pp. 3–15. Plenum, New York.

Elton, C. S. 1958. The Ecology of Invasions by Animals and Plants. Methuen, London.

Elzen, G. W., H. J. Williams, and S. B. Vinson. 1983. Response by the parasitoid *Campoletis sonorensis* (Hymenoptera: Ichneumonidae) to chemicals (synomones) in plants: implications for host habitat location. Environ. Ent. 12:1872–1876.

Elzen, C. W., H. J. Williams, and S. B. Vinson. 1984. Isolation and identification of cotton synomones mediating searching behavior by the parasitoid *Campoletis sonorensis*. J. Chem. Ecol. 10:1251–1264.

Feeny, P. 1975. Biochemical coevolution between plants and their insect herbivores. In L. E. Gilbert and P. H. Raven (eds.), Coevolution of Animals and Plants. pp. 3–19. University of Texas Press. Austin, Tex.

Flaherty, D. 1969. Ecosystem trophic complexity and Willamette mite *Ecotetranychus willamete* (Acarina: Tetranychidae) densities. Ecology 50:911–916.

Flanders, S. E. 1953. Variation in susceptibility of citrus-infesting coccids to parasitization. J. Econ. Ent. 46:266–269.

Flanders, S. E. 1962. The parasitic Hymenoptera: specialists in population regulation. Can. Ent. 94:1133–1147.

Flint, H. M., S. S. Salter, and S. Walters. 1979. Caryophyllene: an attractant for the green lacewing *Chrysopa carnea* Stephens. Environ. Ent. 8:1123–1125.

Francis, C. A., C. A. Flor, and S. R. Temple. 1976. Adapting varieties for intercropped systems in the tropics. In M. Stelly (ed.), Multiple Cropping. American Society of Agronomy Special Publication 27. pp. 235–253. American Society of Agronomy, Madison, Wis.

Fye, R. E. 1972. The interchange of insect parasites and predators between crops. PANS 18:143–146.

Fye, R. E. 1983. Cover crop manipulation for building pear psylla (Homoptera: Psyllidae) predator populations in pear orchards. J. Econ. Ent. 76:306–310.

Gagné, W. C. 1982. Staple crops in subsistence agriculture: their major insect pests, with emphasis on biogeographical and ecological aspects. In J. L. Gressitt (ed.), Biogeography and Ecology of New Guinea. Monographs in Biology. pp. 229–259. Junk, The Hague.

Gilbert, L. E. 1975. Ecological consequences of a coevolved mutualism between butterflies and plants. In L. E. Gilbert and P. H. Raven (eds.), Coevolution of Animals and Plants. pp. 210–240. University of Texas Press, Austin, Tex.

Greany, P. D., and K. S. Hagen. 1981. Prey selection. In D. A. Nordlund, R. L. Jones, and W. J. Lewis (eds.), Semiochemicals: Their Role in Pest Control. pp. 121–135. Wiley, New York.

Guevara, J. C. 1962. Efecto de las prácticas de siembra y de cultivos sobre plagas en maíz y frijol. Fitotec. Latinoam. 1:15–26.

Harrington, E. A., and P. Barbosa. 1978. Host habitat influences on oviposition by *Parasetigena silvestris* (R-D), a larval parasite of the gypsy moth. Environ. Ent. 7:466–468.

Harwood, R. R. 1979. Small Farm Development—Understanding and Improving Farming Systems in the Humid Tropics. Westview Press, Boulder, Col.

Herrebout, W. M., and J. van der Veer. 1969. *Euarcelia rutilla* Vill. III. Preliminary results of olfactometer experiments with females of known age. Z. angew. Ent. 64:55–61.

Hsiao, T. H., F. G. Holdaway, and H. C. Chiang. 1966. Ecological and physiological adaptations in insect parasitism. Ent. Exp. Appl. 9:113–123.

IRRI. 1972. Multiple cropping. In IRRI Annual Report for 1971. pp. 21–34. IRRI, Los Banos, Philippines.

IRRI. 1975. Cropping Systems Program Report for 1975. IRRI, Los Banos, Philippines.

Johnson, A. W. 1972. Individuality and experimentation in traditional agriculture. Hum. Ecol. 1:149–159.

Kesten, V. 1969. Zur Morphologie und Biologie von *Anatis ocellata* (L.) (Coleoptera: Coccinelidae). Z. angew. Ent. 63:412–445.

Kido, H., D. L. Flaherty, D. F. Bosch, and K. A. Valero. 1983. Biological control of grape leafhopper. Calif. Agric. 37:4–6.

Kogan, M. 1977. The role of chemical factors in insect/plant relationships. Proceedings of the 15th International Congress of Entomology. pp. 211–227. Entomological Society of America, College Park, Md.

Laing, J. 1937. Host-finding by insect parasites. I. Observations on the finding of hosts by *Alysia manducator, Mormoniella vitripennis* and *Trichogramma evanescens*. J. Anim. Ecol. 6:298–317.

Leius, K. 1963. Effects of pollen on fecundity and longevity of adult *Scambus buotianae* (Htg.) (Hymenoptera: Ichneumonidae). Can. Ent. 95:202–207.

Leius, K. 1967. Influence of wild flowers on parasitism of tent caterpillar and codling moth. Can. Ent. 99:444–446.

Letourneau, D. K. 1986. Associational resistance in squash monoculture and polycultures in tropical Mexico. Environ. Ent. 15:285–292.

Letourneau, D. K., and M. A. Altieri. 1983. Abundance patterns of a predator, *Orius tristicolor* (Hemiptera: Anthocoridae), and its prey, *Frankliniella occidentalis* (Thysanoptera: Thripidae): habitat attraction in polycultures versus monocultures. Environ. Ent. 12:1464–1469.

Lewis, W. J. 1981. Semiochemicals: their role with changing approaches to pest control. In D. A. Nordlund, R. L. Jones, and W. J. Lewis (eds.), Semiochemicals: Their Role in Pest Control. pp. 3–12. Wiley, New York.

Lewis, W. J., and D. A. Nordlund. 1984. Semiochemicals influencing fall armyworm parasitoid behavior: implications for behavioral manipulation. Fla. Ent. 67:343–349.

Lewis, W. J., and D. A. Nordlund. 1985. Behavior-modifying chemicals to enhance natural enemy effectiveness. In M. A. Hoy and D. C. Herzog (eds.), Biological Control in Agricultural Integrated Pest Management Systems. pp. 89–101. Academic Press. New York.

Litsinger, J. A., and K. Moody. 1976. Integrated pest management in multiple cropping systems. In P. A. Sanchez (ed.), Multiple Cropping. American Society of Agronomy Special Publication 27. pp. 293–316. American Society of Agronomy, Madison, Wis.

Litsinger, J. A., M. O. Lumaden, J. P. Bandary, A. T. Barrion, P. C. Pantua, R. F. Apostol, and R. Jhendi. 1980. A methodology for determining insect control recommendations. IRRI Research Paper Series 46. IRRI, Los Banos, Philippines.

Loke, W. H., T. R. Ashley, and R. I. Sailer. 1983. Influence of fall armyworm, *Spodoptera frugiperda,* (Lepidoptera: Noctuidae) larvae and corn plant damage on host finding in *Apanteles marginiventris* (Hymenoptera: Braconidae). Environ. Ent. 12:911–915.

Martin, P. B., P. D. Lingren, G. L. Green, and R. L. Ridgway. 1976. Parasitization of two species of *Plusiinae* and *Heliothis* spp. after releases of *Trichogramma pretiosum* in seven crops. Environ. Ent. 5:991–995.

Matteson, P. C., M. A. Altieri, and W. C. Gagné. 1984. Modification of small farm practices for better pest management. Annu. Rev. Ent. 29:338–402.

Mohyuddin, A. I., C. Inayatulla, and E. G. King. 1981. Host selection and strain occurrence in *Apanteles flavipes* (Cameron) (Hymenoptera: Braconidae) and its bearing on biological control of graminaceous borers (Lepidoptera: Pyralidae). Bull. Ent. Res. 71:575–581.

Monteith, L. G. 1955. Host preferences of *Drino bohemica* Mesn. with particular reference to olfactory responses. Can. Ent. 87:509–530.

Monteith, L. G. 1956. Influence of host movement on selection of host by *Drino bohemica* Mesn. as determined in an olfactometer. Can. Ent. 88:583–586.

Monteith, L. G. 1958. Influence of host and its food plant on host-finding by *Drino bohemica* Mesn. and interaction of other factors. Proc. X Int. Congr. Ent. 2:603–606.

Monteith, L. G. 1960. Influence of plants other than the food plants of their host on host-finding by tachinid parasites. Can. Ent. 92:641–652.

Nettles, W. C., Jr. 1979. *Eucelatoria* sp. females: factors influencing response to cotton and okra plants. Environ. Ent. 8:619–623.

Nettles, W. C., Jr. 1980. Adult *Eucelatoria* sp.: response to volatiles from cotton and okra plants and from larvae of *Heliothis virescens, Spodoptera eridania,* and *Estigmene acrea*. Environ. Ent. 9:759–763.

Nishida, T. 1956. An experimental study of the ovipositional behavior of *Opius fletcheri* Silvestri (Hymenoptera: Braconidae) a parasite of the melon fly. Proc. Hawaii. Ent. Soc. 16:126–134.

Nordlund, D. A. 1984. Biological control with entomophagous insects. J. Ga. Ent. Soc. 19:15–27.

Nordlund, D. A., and W. J. Lewis. 1976. Terminology of chemical releasing stimuli in intraspecific and interspecific interactions. J. Chem. Ecol. 2:211–220.

Nordlund, D. A., and W. J. Lewis. 1985. Response of females of the braconid parasitoid, *Microplitis demolitor* to frass of larvae of the noctuids, *Heliothis zea* and *Trichoplusia ni* and to 13-methylhentriacontane. Ent. Exp. Appl. 38:109–112.

Nordlund, D. A., and C. E. Sauls. 1981. Kairomones and their use for management of entomophagous insects. XI. Effect of host plants on kairomonal activity of frass from *Heliothis zea* larvae for the parasitoid *Microplitis croceipes*. J. Chem. Ecol. 7:1057–1061.

Nordlund, D. A., R. L. Jones, and W. J. Lewis (eds.). 1981. Semiochemicals: Their Role in Pest Control. Wiley, New York.

Nordlund, D. A., R. B. Chalfant, and W. J. Lewis. 1984. Arthropoid populations, yield, and damage in monocultures and polycultures of corn, beans, and tomatoes. Agric. Ecosyst. Environ. 11:353–367.

Nordlund, D. A., R. B. Chalfant, and W. J. Lewis. 1985. Response of *Tricho-gramma pretiosum* females to extracts from two plants attacked by *Heliothis zea*. Agric. Ecosyst. Environ. 12:127–133.

Nordlund, D. A., W. J. Lewis. S. B. Vinson, and H. R. Gross, Jr. 1986. Behavioral manipulation of entomophagous insects. In S. Johnson, E. G. King, and J. R. Bradley (eds.), Theory and Tactics of *Heliothis* Population Management, Vol. I, Cultural and Biological Control. Southern Cooperative Series Bulletin 316. pp. 104–115. Agric. Exp. Stn., Div. Agric., Oklahoma St. Univ., Stillwater, OK.

Perrin, R. M. 1977. Pest management in multiple cropping systems. Agro-Ecol. Syst. 3:98–118.

Perrin, R. M., and M. L. Phillips. 1978. Some effects of mixed cropping on the population dynamics of insect pests. Ent. Exp. Appl. 24:385–393.

Picard, F., and E. Rabaud. 1914. Sur le parasitisme externé des braconides (Hym.). Bull. Soc. Ent. Fr. 1914:266–269.

Pimentel, D. 1961. Species diversity and insect population outbreaks. Ann. Ent. Soc. Amer. 54:76–86

Pimentel, D., and N. Goodman. 1978. Ecological basis for the management of insect populations. Oikos 30:422–437.

Potts, G. R., and G. P. Vickerman. 1974. Studies of the cereal ecosystem. Adv. Ecol. Res. 8:107–147.

Pradham, S. 1971. Revolution in pest control. World Sci. News 8:410–447.

Price, P. W. 1981. Semiochemicals in evolutionary time. In D. A. Nordlund, R. L. Jones and W. J. Lewis (eds.), Semiochemicals: Their Role in Pest Control. pp. 251–279. Wiley, New York.

Price, P. W. 1984. Insect Ecology, 2nd ed. Wiley, New York.

Price, P. W., C. E. Bouton, P. Gross, B. A. McPheron, J. N. Thompson, and A. E. Weis. 1980. Interactions among three trophic levels: influence of plants on interactions between insect herbivores and natural enemies. Annu. Rev. Ecol. Syst. 11:41–65.

Read, D. P., P. P. Feeny, and R. B. Root. 1970. Habitat selection by the aphid parasite *Diaeretiella rapae* and hyperparasite, *Charips brassicae*. Can. Ent. 102:1567–1578.

Risch, S. J. 1979. A comparison by sweep sampling of the insect fauna from corn and sweet potato monoculture and dicultures in Costa Rica. Oecologia (Berlin) 42:195–211.

Risch, S. J. 1980a. Fewer beetle pests on beans and cowpeas interplanted with banana in Costa Rica. Turrialba 30:229–30.

Risch, S. J. 1980b. The population dynamics of several herbivorous beetles in a tropical agroecosystem: the effect of intercropping corn, beans, and squash in Costa Rica. J. Appl. Ecol. 17:593–612.

Risch, S. J. 1981. Insect herbivore abundance in tropical monocultures and polycultures: an experimental test of two hypotheses. Ecology 62:1325–1340.

Risch, S. J. 1983. Intercropping as a cultural pest control: prospects and limitations. Environ. Manag. 7:9–14.

Risch, S. J., D. Andow, and M. A. Altieri. 1983. Agroecosystem diversity and pest control: data, tentative conclusions and new research directions. Environ. Ent. 12:625–629.

Robinson, R. R. 1972. Strip-cropping effects on the abundance of predatory and harmful cotton insects in Oklahoma. Environ. Ent. 1:145–149.

Root, R. B. 1973. Organization of a plant-arthropod association in simple and diverse habitats: the fauna of collards (*Brassica oleracea*). Ecol. Monogr. 43:95–124.

Root, R. B. 1975. Some consequences of ecosystem texture. In S. A. Levin (ed.), Ecosystem Analysis and Prediction. pp. 83–97. Society for Industrial and Applied Mathematics, Philadelphia.

Rosenthal, G. A., and D. H. Janzen (eds.). 1979. Herbivores: Their Interaction with Secondary Plant Metabolites. Academic Press, New York.

Roth, J. P., E. G. King, and E. Thompson. 1978. Host location behavior by the tachinid, *Lixophaga diatraeae*. Environ. Ent. 7:794–798.

Roth, J. P., E. G. King, and S. D. Hensley, 1982. Plant, host, and parasite interactions in the host selection sequence of the tachinid, *Lixophaga diatraeae*. Environ. Ent. 11:273–277.

Ruthenberg, H. 1980. Farming Systems in the Tropics, 3rd ed. Oxford University Press, Oxford.

Salt, G. 1958. Parasite behavior and the control of insect pests. Endeavor 65: 145–148.

Sauls, C. E., D. A. Nordlund, and W. J. Lewis. 1979. Kairomones and their use for management of entomophagous insects. VIII. Effect of diet on the kairomonal activity of frass from *Heliothis zea* (Boddie) larvae for *Microplitis croceipes* (Cresson). J. Chem. Ecol. 5:363–369.

Schoonhoven, L. M. 1981. Chemical mediators between plants and phytophagous insects. In D. A. Nordlund, R. L. Jones, and W. J. Lewis (eds.), Semiochemicals: Their Role in Pest Control. pp. 31–50. Wiley, New York.

Sekhar, P. S. 1960. Host relationships of *Aphidius testaceipes* (Cresson) and *Praon aguti* (Smith), primary parasites of aphids. Can. J. Zool. 38:593–603.

Shahjahan, M. 1974. *Erigeron* flowers as a food and attractive odor source for *Peristenus pseudopallipes*, a braconid parasitoid of the tarnished plant bug. Environ. Ent. 3:69–72.

Shahjahan, M. M., and F. A. Streams. 1973. Plant effects on host-finding by *Leiophron pseudopallipes* (Hymenoptera: Braconidae), a parasitoid of the tarnished plant bug. Environ. Ent. 3:911–925.

Sheehan, W. 1986. Response by specialist and generalist natural enemies to agroecosystem diversification: a selective review. Environ. Ent. 15:450–461.

Shelton, M. D., and C. R. Edwards. 1983. Effects of weeds on the diversity and abundance of insects in soybeans. Environ. Ent. 12:296–298.

Simmons, G. A., D. E. Leonard, and C. W. Chen. 1975. Influence of tree species density and composition on parasitism of the spruce budworm, *Choristoneura fumiferana* (Clem.). Environ. Ent. 4:832–836.

Smiley, J. 1978. Plant chemistry and the evolution of host specificity: evidence from *Heliconius* and *Passiflora*. Science 201:745–747.

Smith, J. G. 1976. Influence of crop background on aphids and other phytophagous insects on brussels sprouts. Ann. Appl. Biol. 83:1–13.

Streams, F. A., M. Shahjahan, and H. G. Le Measurier. 1968. Influence of plants on the parasitization of the tarnished plant bug by *Leiophron pallipes*. J. Econ. Ent. 61:996–998.

Syme, P. D. 1975. The effects of flowers on the longevity and fecundity of two native parasites of the European pine shoot moth in Ontario. Environ. Ent. 4:337–346.

Syme, P. D. 1977. Observations on the longevity and fecundity of *Orgilus obscurator* (Hymenoptera: Braconidae) and the effects of certain foods on longevity. Can. Ent. 109:995–1000.

Tahvanainen, J. O., and R. B. Root. 1972. The influence of vegetational diversity on the population ecology of a specialized herbivore *Phyllotreta crucifera* (Coleoptera: Chrysomelidae). Oecologia (Berlin) 10:321–346.

Taylor, J. S. 1932. Report on cotton insect and disease investigation. II. Notes on the American bollworm (*Heliothis obsoleta* F.) on cotton and its parasite (*Microbracon brevicornis* Wesm.). Sci. Bull. Rep. Agric. For. Union S. Afr. 113.

Thompson, A. C., J. P. Roth, and E. G. King. 1983. Larviposition kairomone of the tachinid *Lixophaga diatraeae*. Environ. Ent. 12:1312–1314.

Thorpe, W. H., and H. B. Caudle. 1938. A study of the olfactory responses of insect parasites to the food plant of their host. Parasitology 30:523–528.

Townes, H. 1960. Host selection patterns in some nearctic ichneumonids (Hymenoptera). Verh. XI Int. Kongr. Ent. (Wien), 2:738–741.

van den Bosch, R. 1978. The Pesticide Conspiracy. Doubleday, New York.

van Emden, H. F. 1963. Observations on the effect of flowers on the activity of parasitic Hymenoptera. Ent. Mon. Mag. 98:265–270.

van Emden, H. F. 1965. The role of uncultivated land in the role of crop pests and beneficial insects. Sci. Hort. 17:121–136.

van Lenteren, J. C., J. Woets, N. van der Poel, W. van Boxtel, S. van de Merendonk, R. van der Kamp, H. Nell, and L. Sevenster-van der Lelie. 1977. Biological control of the greenhouse whitefly *Trialeurodes vaporariorum* (Westwood) (Homoptera: Aleyrodidae) by *Encarsia formosa* Gahan (Hymenoptera: Aphelinidae) in Holland, an example of successful applied ecological research. Meded. Fac. Landbouwwet. Rijksuniv. Gent. 42:1333–1342.

van Lenteren, J. C., P. M. J. Ramakers, and J. Woets. 1980. Integrated control of vegetable pests in greenhouses. In A. K. Minks and P. Gruys (eds.), Integrated Control of Insect Pests in The Netherlands. pp. 109–118. Pudoc, Wageningen, The Netherlands.

Vet, L. E. M., and J. J. M. van Alphen. 1985. A comparative functional approach to the host detection behavior of parasitic wasps. I. A qualitative study on Eucoilidae and Alysiinae. Oikos 44:478–486.

Vet, L. E. M., C. Janse, C. van Achterberg, and J. J. M. van Alphen. 1984. Microhabitat location and niche segregation in two sibling species of drosophilid parasitoids: *Asobara tabida* (Nees) and *A. rufescens* (Forester) (Braconidae: Alysiinae). Oecologia (Berlin) 61:182–188.

Vinson, S. B. 1975a. Biochemical coevolution between parasitoids and their hosts. In P. Price (ed.), Evolutionary Strategies of Parasitic Insects and Mites. pp. 14–48. Plenum, New York.

Vinson, S. B. 1975b. Source of material in the tobacco budworm which initiates host-searching by the egg-larval parasitoid, *Chelonus texanus*. Ann. Ent. Soc. Amer. 68:381–384.

Vinson, S. B. 1981. Habitat Location. In D. A. Nordlund, R. L. Jones, and W. J. Lewis (eds.), Semiochemicals: Their Role in Pest Control. pp. 51–77. Wiley, New York.

Vité, J. P., and D. L. Williamson. 1970. *Thanasimus dubius:* prey perception. J. Insect Physiol. 16:233–237.

Walker, M. G. 1940. Notes on the distribution of *Cephus pygmaeus* Linn. and its parasite *Collyria calcitrator* Grav. Bull. Ent. Res. 30:551–573.

Wallace, J. W., and R. L. Mansell (eds.). 1976. Biochemical Interactions Between Plants and Insects. Plenum, New York.

Weseloh, R. M. 1981. Host location by parasitoids. In D. A. Nordlund, R. L. Jones, and W. J. Lewis (eds.), Semiochemicals: Their Role in Pest Control. pp. 79–95. Wiley, New York.

Wood, D. L., L. E. Brown, W. D. Bedard, P. E. Tilden, R. M. Silverstein, and J. O. Rodin. 1968. Response of *Ips confusus* to synthetic sex pheromones in nature. Science 59:1373–1374.

Young, J. H., and R. G. Price. 1975. Incidence, parasitism, and distribution patterns of *Heliothis zea* on sorghum, cotton, and alfalfa of southwestern Oklahoma. Environ. Ent. 4:777–779.

Zandstra, B. H., and B. S. Motooka. 1978. Beneficial effect of weeds in pest management—a review. PANS 24:333–338.

Zandstra, H. G., E. C. Price, J. A. Litsinger, and R. A. Morris. 1981. A Methodology for On-Farm Cropping Systems Research. IRRI, Los Banos, Philippines.

Zehnder, G. W., and J. T. Trumble. 1984. Host selection of *Liriomyza* species (Diptera: Agromyzidae) and associated plants of tomato and celery. Environ. Ent. 13:492–496.

Zwolfer, H., and M. Kraus. 1957. Biocoenotic studies on the parasites of two fir- and two oak-tortricids. Entomophaga 2:169–173.

PART II

MICROORGANISMS AS MEDIATORS OF INTERTROPHIC AND INTRATROPHIC INTERACTIONS

A rich body of literature has been emerging over the past several years on the role of microorganisms, especially bacteria and fungi, as mediators of interactions between insects and between insects and plants. This role is vitally important. There are numerous examples of failures to describe and interpret accurately interactions between macroscopic organisms because of a tendency for ecologists to overlook the role of microorganisms in the system. A classic example was the eventual realization, in studies of competition between two species of *Tribolium* flour beetles, that *Adelina tribolii,* a pathogenic coccidian, was playing a pivotal role in determining the outcome of interspecific competition (Park 1954). More recent examples presented by Berenbaum and by Dicke in Part II provide (1) evidence of continued misinterpretation of cause and effect in ecological relationships, (2) ways of integrating competing hypotheses on insect–plant interactions by incorporating microbes, and (3) a critique of our present language to describe semiochemically based interactions.

For example, variation in the response of herbivores to stresses imposed by food plants have been, for the most part, described as "noise" in the system. Variable plant quality can explain some of the variation some of the time. Only recently has the importance of secondary impacts such as those of plant properties on herbivore–symbiont (pathogen or mutualist) interactions been realized. Berenbaum points out, for example, that attack by many microsporidia and granulosis viruses of a host's fat body can lead to fat body destruction, which is likely to make the herbivore more susceptible to plant allelochemical toxicity. Herbivore response, then, to both the plant and to the pathogen is dependent upon whether or not the stresses coincide.

It is imaginable that a similar scenario could affect the outcome of an applied program of plant resistance. Laboratory experiments to screen plants for aphid resistance could result in the same average number of progeny per female for two generations. The line is discarded since it is clearly not resistant. If the same experiment had been run in the field, however, it might have shown heavy mortality of the first group of progeny in the third instar. The explanation? A microorganism. Under lab conditions, the fact that the experimental cultivar contains an allelochemical which interferes with the formation of cuticle is not manifested in the mortality rate of the aphid. Only when the conditions allow for pathogens to interact in the system in a normal way, and invade through areas of disrupted cuticle, does the subtle but powerful effect of the allelochemical manifest itself. Indeed, Berenbaum invokes this type of logic among other insightful arguments in her analyses of Feeny's (1970) plant apparency and the role of tannin-like compounds in current hypotheses to explain population cycles within forest communities.

It is certain that the exclusion of microorganisms as mediators of arthropod–plant interactions has been an impediment to our understanding of complex and variable systems. Even the language that we use to describe the interactions may need reconsideration. Dicke (Chapter 4) points out that a lack of awareness of the potential role of microbial factors is reflected in the language of semiochemical ecology. As one example, the term "apneumone" (Nordlund and Lewis 1976) was designated for chemical signals derived from nonliving material, without consideration of the possibility that these chemicals may indeed originate from microorganisms that occupy the nonliving material. In the general sense, as well, the following chapters challenge the terminology presented in Chapters 1 and 2. Terms such as "kairomone" and "allomone" are quite generally accepted despite a good deal of controversy (e.g., Blum 1980, Weldon 1980, Pasteels 1982), but that controversy has only recently included the roles of microorganisms. For example, Dicke describes recent work on allelochemicals that mediate interactions between *Drosophila* and their parasitoids. The allelochemicals attract parasitoids to the host, but they are not produced by the host nor by the food plant of the host.

Rather, a fungus associated with the food plant is the source. Decisions on terminology, based upon source–receiver and cost–benefit, are no longer technically valid.

It is clear, then, that microorganisms play a wide variety of important roles by chemically mediating interactions between plants, herbivores, and their parasitoids and predators. More and more often, for example, it is apparent that it is not the plant, nor is it the insect, that is constituting what has been called chemically mediated herbivore–plant interaction. It may, in fact, involve a series of intertwining interactions with microbes where, for example, alkaloids produced by an endophytic fungus are taken up by a grazing insect (e.g., Hardy et al. 1985) and where subsequent detoxification is achieved by a bacterial symbiont acting on these compounds within the insect.

A central role of microorganisms in insect–plant interactions is significant for several reasons. First, a broad consideration of the frequency of symbiotic mutualisms between insects and microorganisms may alter our perception of the role of mutualistic interactions within the ecology of resource exploitation. The role of mutualism has been underestimated routinely in the study of interactive systems (Boucher et al. 1982). However, approximately 80% of 73 examples of insect use of microorganisms listed by Jones (1984) are thought to involve obligate or facultative, symbiotic or nonsymbiotic mutualism. Insects provide nutrients, habitat (often very specialized compartments within the body), phoresy (simple-to-elaborate dispersal mechanisms), and a habitat for other microbial associates. Microorganisms peform many functions that promote efficiency, growth, development, survival, and reproduction of insects. They may synthesize amino acids not adequately supplied by the insect's food resource or detoxify potentially detrimental allelochemicals in plants or in prey/hosts (Jones 1984).

Second, the consideration of microorganisms expands the range of cause-and-effect relationships involving plant allelochemicals. The production of toxic substances in plants, in some cases, may be modified by, or entirely a reponse to, microorganisms rather than insects (see the discussion of induction in Chapter 6). The complexity of interactions that make up a plant's "experience" becomes staggering when considering the phylloplane community, the inhabitants of the rhizosphere, and invasive organisms living within plant tissues. It is possible that allelochemicals affecting herbivores as selective factors, then, can be primarily a response to microorganisms associated with the plant. Extreme cases are those in which the microorganism actually produces the active compounds (e.g., Hardy et al. 1985). For example, microorganisms often use the same resource as herbivores. The decaying processes initiated by microorganisms can be viewed as a competitive interaction. Janzen (1977) notes that microbial infections render plant tissues less suitable as food for insect herbivores by producing toxicants and antibiotics. On the other hand, some plant-produced substances ingested by herbivores have been shown to reduce mortality to herbivores through their antibiotic properties for insect pathogens (Frings et al. 1948; Chapter 6) or to affect herbivores negatively by harming their gut symbionts (Chapter 3).

Third, turnover rates (generation times) for single-celled organisms are orders of magnitude more rapid than in either insects or plants. This should reflect the rapidity with which selection pressures can produce significant shifts in gene frequencies and thus the rate of adaptation. The generation of biotypes in herbivores is a possible example. Biotypes of *Schizaphis graminum,* the greenbug (aphid), are morphologically indistinguishable. Eisenbach and Mittler (1987) suggest extranuclear inheritance in greenbug as a mechanism for their successive ability to overcome resistant varieties of sorghum as they are developed. It is reasonable to ask if the ability is due to adaptations in symbionts rather than in the herbivore itself (Eisenbach and Mittler 1987). The role of microorganisms, then, in the ecology of herbivore–plant interactions is emerging in processes of host switching, specialism versus generalism, population cycling, trophic interactions, and community dynamics. The chapters in this section provide a wide range of examples and a synthetic treatment that demonstrate its tremendous importance in insect–plant interactions.

D. K. LETOURNEAU

REFERENCES

Blum, M. S. 1980. Arthropods and ecomones: better fitness through ecological chemistry. In R. Gilles (ed.), Animals and Environmental Fitness. pp. 207–222. Pergamon Press, Oxford.

Boucher, D. H., S. James, and K. H. Keeler. 1982. The ecology of mutualism. Annu. Rev. Ecol. Syst. 13:315–347.

Eisenbach, J., and T. E. Mittler. 1987. Extra-nuclear inheritance in a sexually produced aphid: the ability to overcome host plant resistance by biotype hybrids of the greenbug, *Schizaphis graminum.* Experientia 49:332–334.

Feeny, P. 1970. Seasonal changes in oak leaf tannins and nutrients as a cause of spring feeding by winter moth caterpillars. Ecology 51:565–581.

Frings, H., E. Goldberg, and J. C. Arentzen. 1948. Antibacterial action of the blood of the large milkweed bug. Science 108:689–690.

Hardy, T. N., K. Clay, and A. M. Hammond, Jr. 1985. Fall armyworm (Lepidoptera: Noctuidae): a laboratory bioassay and larval study for the fungal endophyte of perennial ryegrass. J. Econ. Ent. 78:571–575.

Janzen, D. H. 1977. Why fruits rot, seeds mold and meat spoils. Amer. Nat. 111:691–713.

Jones, C. G. 1984. Microorganisms as mediators of plant resource exploitation by insect herbivores. In P. W. Price, W. S. Gaud, and C. N. Slobodchikoff (eds.), A New Ecology: Novel Approaches to Integrative Systems. pp. 53–99. Wiley, New York.

Nordlund, D. A., and W. J. Lewis. 1976. Terminology of chemical releasing stimuli in intraspecific and interspecific interactions. J. Chem. Ecol. 2:211–220.

Park, T. 1954. Experimental studies of interspecies competition. II. Temperature, humidity, and competition in two species of *Tribolium*. Physiol. Zool. 27:177–238.

Pasteels, J. M. 1982. Is kairomone a valid and useful term? J. Chem. Ecol. 8:1079–1081.

Weldon, P. J. 1980. In defense of "kairomone" as a class of chemical releasing stimuli. J. Chem. Ecol. 6:719–725.

CHAPTER 3

Allelochemicals in Insect–Microbe–Plant Interactions; Agents Provocateurs in the Coevolutionary Arms Race

M. R. Berenbaum
University of Illinois
Urbana, Illinois

CONTENTS

1 Introduction
2 Allelochemical effects on microbe–insect interactions
  2.1 Microbe–insect mutualisms: symbiosis
    2.1.2 Endosymbiosis
    2.1.2 Ectosymbiosis
  2.2 Microbe–insect antagonisms: pathogenesis
    2.2.1 Allelochemical interference with pathogens
    2.2.2 Allelochemical enhancement of pathogenesis
    2.2.3 Community-level implications
3 Allelochemical effects on microbe–plant interactions
  3.1 Microbe–plant mutualisms: symbiosis
    3.1.1 Nitrogen fixation and allelochemicals
    3.1.2 Endophyte infections and allelochemicals
  3.2 Microbe–plant antagonisms: pathogenesis
    3.2.1 Phytoalexin responses to infection and plant suitability to insects
    3.2.2 Pathogen-produced attractants
4 Evolutionary implications of microbial participation in plant–insect interactions
References

1 INTRODUCTION

Life on the planet earth, from an extraterrestrial overview, would appear to consist primarily of herbivorous insects and terrestrial plants. Both taxa number in excess of 300,000 species and occupy most conceivable habitats. Interactions between these two taxa take on bewildering complexity; indeed, as the two most abundant life forms on earth, it would be a surprise if plants and insects did *not* interact. However, what is not immediately apparent, either from an extraterrestrial vantage point or one closer to home, is that there is often a third party involved, adding another dimension to these interactions, that is, microbes: consisting in the broadest sense of fungi, bacteria, protozoa, rickettsiae, and other unicellular oganisms. Microbes may enter into plant–insect interactions in a variety of ways (Jones 1984). They may be pathogenic to either plant or insect, or they may be mutualistic to plant or insect. In either case, the three-way interactions are almost inevitably mediated by plant allelochemicals. Most, if not all, plant allelochemicals are broadly biocidal, targeting sensitive processes common to most living organisms: respiration, DNA transcription, protein synthesis, and cell membrane activity, to cite a few (Table 3.1). Furanocoumarins, for example, benz-2-pyrone derivatives common in plants in the Rutaceae and Umbelliferae, bind irreversibly to thymine in DNA; when cross-linkage occurs, DNA template formation is prevented. Furanocoumarins are thus toxic to bacteria, fungi, nematodes, insects, snails, fish, amphibians, birds, mammals, and probably a lot of other organisms that have yet to be tested (Murray et al. 1982). Microbial associates of plant-feeding insects must therefore deal with the same potentially toxic secondary metabolites as do the insect herbivores. By the same token, microbial associates of plants may produce changes in plant chemistry with which insect herbivores must contend. In this chapter, the effects of allelochemicals on interactions among plants, insects, and microbes are examined relative to the nature of the interaction. The outcome of interactions and their ecological implications in community organization depend both on the nature of the relationship among plant, microbe, and insect and on relative sensitivities of the species to allelochemicals.

This chapter is therefore organized at one level according to classes of interaction, mutualism, or antagonism, and at a finer level according to the relative sensitivities of plant, insect, and microbe to allelochemicals.

TABLE 3.1 Biochemical Target Sites for Plant Allelochemicals

I.	Electron transport inhibitors	
	a. Between NAD^+ and CoQ	Rotenone
	b. Succinate oxidation	Tropanes, quinones, quinolines, nitrocompounds, imidazoles, aldehydes
	c. Cytochrome c oxidase	Glucosinolates, nitriles, N-nitrosamines, cyanogenic thiocyanates
II.	Uncouplers of oxidative phosphorylation	Diterpenes, flavonoids, polyacetylenes, phenols, aromatic acids, coumarins, fatty acids
III.	Inhibitors of energy transfer systems	
	a. Glycolysis	Fluoroacetic acid
	b. Fatty acid oxidation	
IV.	Protein synthesis inhibitors	
	a. Amino acid activation	Nonprotein amino acids
	b. Protein function	Tannins, stilbenes, ''resins,'' quinones
	c. Ribosomal initiation complex	Toxic proteins (e.g., ricin), unusual 6- substituted purine bases
	d. Elongation	Indole alkaloids
V.	DNA synthesis inhibitors	
	a. Transitional mutations	Base analogs (e.g., 5-methyl cytosine)
	b. Frame shift	Quinine
	c. Replication	Colchicine, *Veratrum* alkaloids, diaminosteroid alkaloids, furanocoumarins, coumarins
	d. Transcription	Hydrazines
VI.	Cell function disrupters	
	a. Cell membrane integrity	Polyacetylenes, sesquiterpenes, triterpenes, steroidal alkaloids

2   ALLELOCHEMICAL EFFECTS ON MICROBE–INSECT INTERACTIONS

2.1   Microbe–Insect Mutualisms: Symbiosis

2.1.1   Endosymbiosis

Like cows and other ruminants, many herbivorous insects are dependent upon microbes for making up nutritional deficiencies inherent in the food plant. Endosymbionts may be housed either extra- or intracellularly. With extracellular localization, the symbionts usually occur in diverticula or pouches in among the food material (e.g., scarabeid fermentation chambers, tephritid midgut "blind sacs," hemipteran "crypt guts," etc.). With intracellular localization, the microbes actually reside in the cells of the host, usually gut epithelial cells often highly modified for the purpose. Intracellular symbionts are protected from the risks of involuntary expulsion with food or frass while deriving nutrition directly from the host. For their part, the microbial associates, including bacteria, yeasts, rickettsiae, and protozoan flagellates, contribute all manner of nutritional supplementation. Among the more familiar symbionts are flagellates in termite hindguts, which allow the host insect to break down the otherwise indigestible structural carbohydrate cellulose. Indeed, symbionts are almost invariably associated with a nutritionally inadequate diet for herbivores (Slansky and Rodriguez 1987). Plant proteins are often compositionally inadequate to insect nutritional needs, and many symbionts produce essential amino acids. Nitrogen-fixing bacteria also supplement protein nutritional needs in many taxa (e.g., Prestwich et al. 1980). Aphid symbionts metabolize plant urates, a necessary nutritional contribution in view of the fact that aphids lack Malpighian tubules. The latter are prerequisites for extensive nitrogen metabolism. Symbiotic bacteria in cockroach mycetomes also appear to metabolize urates (Wigglesworth 1987). In sap feeders, vitamins are a major contribution from symbionts, especially thiamine, riboflavin, folic acid, and the other B- complex vitamins. Cholesterol, a dietary requirement for all insects that is poorly met by the structurally distinct plant sterols, is produced by many symbionts as well (see chapters in Henry 1967).

Another contribution made by symbionts is direct or indirect interference with metabolic detoxication of foreign substances. Defaunated termites are more susceptible, for example, to the effects of insecticide toxicity. Normal and defaunated *Coptotermes formosanus* differ in their metabolism of the organophosphate insecticide chlorpyrifos. By topical application, the LD_{50} for normal termites is 3.18 $\mu g/g$ and for defaunated termites is 2.83 $\mu g/g$, or 12% lower. One metabolite, probably a trichloro-pyridinol conjugate, was entirely lacking in defaunated termites. The flagellate symbionts live in the hindgut posterior to the Malpighian tubules and may conjugate the termite metabolite in order to reduce toxicity to the flagellates themselves (Khoo and Sherman 1981). It has been suggested (P. Dowd, personal communication) that intracellular microbial symbionts of detritivorous nitidulid beetles are a primary source of detoxication enzymes for metabolizing plant allelochemicals, fungal toxins, and insecticides. Andrews and Spence (1980) suggested that a *Bacterium* species in the midgut of the Douglas fir tussock moth may contribute to the metabolism of dietary terpenes from the host *Pseudotsuga menziesii*.

The consequences for an insect of the loss of its symbionts vary with the taxon. With most aposymbiosis, manifestations include reduced survival, physical deformation and reduction in size, and prolonged development time (Henry 1967, Koch 1967). In the laboratory, aposymbiosis can be induced chemically with various disinfectants or antibiotics. Plant products with antibiotic properties can do essentially the same thing in nature. The health and well-being of both termites and their intestinal symbionts depend on the species of wood ingested. Mauldin et al. (1981) exposed *Reticulotermes flavipes* termites to wood from 21 sympatric trees and examined the effects on gut flora. *Sassafras albidum, Catalpa,* yellow poplar (*Liriodendron tulipifera*), post oak (*Quercus stellata*), *Magnolia grandiflora,* and black walnut (*Juglans nigra*) eliminated the protozoan population altogether and reduced termite survivorship to less than 2%. Termites on northern white cedar (*Thuja occidentalis*) and sycamore (*Platanus occidentalis*) not only lost their symbionts, but died before the 3-week experiment ended. In contrast, on eastern red cedar (*Juniperus virginiana*), the mean number of protozoa present in the gut per termite was 30,000 and 65% of the termites survived. In a second study (Carter et al. 1981), the survival of the Formosan subterranean termite, *Coptotermes formosanus,* was less than 50% on northern white cedar (*Thuja occidentalis*), mulberry (*Morus* sp.), and tulip-poplar. Mean protozoan populations per termite were, respectively, 1700, 2920, and 1520, as compared to populations on slash pine in excess of 3850.

2.1.2   Ectosymbiosis

Another mutualistic interaction between insects and microbes is ectosymbiosis, in which there is regular cohabitation of microbe and insect but the association is not internalized. Symbionts, if housed, reside only temporarily in the insect body. Many bark beetles, borers of diseased or felled trees, have such associations (see Francke-Grossman 1967). Females have bag-like organs or mycetangia specifically for storing fungal spores. These are mechanically deposited on the tunnel walls, which are prepared with wood fragment and fecal linings. In most cases the beetles feed directly on the fungus. Plant chemistry is important in both establishing and maintaining infestations of bark beetles. Plant resistance in grand fir (Raffa and Berryman 1982) is a function of the wound response; resistant trees respond to attack by increasing the flow of terpene resins (Table 3.2). Many of the terpenoid constituents of the resin are toxic not only to *Scolytis ventralis* (fir engraver beetle) but to the symbiotic fungus, *Trichosporium symbioticum,* as well (Raffa and Berryman 1985). Some components of the resin exudate can reduce the growth rate of the fungus up to threefold.

Plant allelochemicals may interfere with maintaining, as opposed to establishing, an infestation. Scolytid bark beetles have a complex phero-

TABLE 3.2   Effect of Monoterpenes in Grand Fir on
***Trichosporium symbioticum,* a Fungal Symbiont of Fir**
Engraver Beetles

	Days to Colonize 90% of Agar Plate (95% Confidence Limits)	
Monoterpene	White Mycelium	Brown Mycelium
β-Pinene	9.1 (6.4–35.0)	10.4 (9.6–14.4)
Camphene	7.6 (6.0–13.8)	10.1 (9.5–11.9)
Δ^3-Carene	8.0 (6.0–21.9)	12.5 (10.5–)[a]
Limonene	12.3 (8.0–156)	15.1 (11.4–)[a]
Myrcene	25.9 (10.5–)[a]	9.9 (9.3–11.5)
Resin	14.2 (8.7–110)	16.0 (12.1–)[a]
Control	5.6 —	9.8 (9.3–11.0)

Source: Raffa and Berryman (1985).
[a] As presented in original publication.

monal system which includes aggregation and antiaggregation pheromones as well as sex pheromones; aggregation or mass attack serves the
purpose of overcoming the wound response of trees. In *Ips paraconfusus*
the precursor of the aggregation pheromone verbenol is α-pinene, produced by the attacked plants in response to attack. It is not *I. paraconfusus,* however, that produces the actual aggregation pheromone; the gut
symbiont *Bacillus cereus* does (Brand et al. 1975; but see Conn et al.
1985). The mycangial fungi of *Dendroctonus frontalis* produce sulcatol, a
beetle pheromone (Brand et al. 1976; see references in Conn ct al. 1984).
Failure to produce certain pheromones may prevent beetle population
growth and cut short infestations since, in the absence of recruitment and
subsequent mass attack, beetles may not be able to overcome the wound
response of their host. Some boring beetles require bacterial stimulation
by symbionts to initiate oocyte maturation. *Xyleborus ferruginneus* females can reproduce parthenogenetically, without benefit of sperm, but
cannot mature ova without microbial symbionts. Without oviposition, an
infestation is necessarily limited (Henry 1967).

Host plant selection, then, is critical to woodboring beetles with ecto-
symbiotic microbes. It is equally important for attine ants (Formicidae,
Myrmecinae, Attini), leaf cutter ants, or parasol ants, which cut and
transport leaf material to provide substrate for subterranean fungus gardens. *Atta cephalotes,* for example, avoids plants that contain substances
fungicidal to their alimentary fungus. Hubbell et al. (1983) demonstrated
that the leaves of *Hymenaea courbaril,* a tree seldom touched by *Atta
cephalotes,* contain a compound highly repellent to the ants. The compound, identified as caryophyllene epoxide, is not only highly repellent to
the ants, it is also highly fungicidal (Arrhenius and Langenheim 1983). At
concentrations of 100 μ/ml the compound is toxic to the attine fungus and
to almost 15 of 45 other species of animal and plant pathogenic fungi
tested. Hubbell et al. (1983) argue that caryophyllene epoxide is repellent
because it is toxic to the alimentary fungus. Hubbell and Howard (1984)
showed further that there is seasonal variability to the chemical repellency
of 42 plant species to these ants in the dry forest of Santa Rosa National
Park, Costa Rica. Seventy-five percent of the species tested demonstrated
nonpolar extractible repellents (terpenoids, steroids, and waxes), chemicals often associated with fungicidal properties. In almost all of the species, a dramatic decline in the repellency coincided with the onset of the
dry season. Hubbell and coworkers suggested that trees stop producing
antifungal substances before the dry season since the risk of fungal attack

declines dramatically; the increased acceptibility of leaves to leaf-cutting ants is "an incidental byproduct" of the plants' coevolutionary interaction with fungal pathogens.

2.2   Microbe–Insect Antagonisms: Pathogenesis

2.2.1   Allelochemical Interference with Pathogens

Far and away the most conspicuous interaction between insect and microbes is not mutualistic but rather, antagonistic; that is, microbes are disease-causing organisms in insect herbivores. Over 100 insect diseases have been described, involving protozoa, bacteria, fungi, viruses, rickettsia, and helminths (Maddox 1982). Plant allelochemicals, as much a part of a herbivorous insect's environment as humidity, temperature, and other microclimatological considerations, can influence the outcome between pathogen and host insect. If the microbe is more sensitive than the host to the effects of the allelochemical, the insect ingesting the allelochemical can benefit in the form of reduced mortality to the disease (after the fashion of taking antibiotics for an infection). Although this is beneficial from the insect perspective, from the point of view of those involved in biocontrol programs it is at the very least counterproductive.

Cases in which plant products reduce mortality from disease in insect herbivores have recently been reviewed by Barbosa and Saunders (1985) and include a variety of insect and microbe taxa. Most examples of host plant or host plant extract effects on pathogens stem from work with *Bacillus thuringiensis,* a gram-positive rod bacillus that infects many species of Lepidoptera. One reason for its broad-spectrum efficacy is that the rods from inclusions or crystals contain an endotoxin. In caterpillar guts, the bacterial wall is autolyzed to release both spore and crystal; the crystal then dissolves, in some susceptible species, to release the toxin. In the midgut, however, the bacillus can mingle with plant material that has been ingested. If that plant material contains bactericidal substances, the efficacy of *B. thuringiensis,* will be greatly reduced. The same principle applies to other insect pathogens (Table 3.3).

2.2.2   Allelochemical Enhancement of Pathogenesis

If microbes are relatively insensitive to allelochemicals in comparison with the insect host, pathogens can actually enhance the toxic effects of allelochemicals. This is particularly the case in sublethal or chronic infectious diseases, one symptom of which is a general inactivation of organ systems responsible for detoxication. Microsporidia and granulosis viruses, in general, take up residence in and destroy fat body tissue. Since fat body can contain essential metabolic enzymes, infection can reduce the efficiency with which xenobiotics are metabolized and exported, leaving greater quantities of active toxicant circulating longer throughout the insect's body. Although this mechanism has not been demonstrated with plant allelochemicals, this type of interaction occurs with insecticides. *Anthonomus grandis* infected with the protozoan *Mattesia grandis* is five to seven times more susceptible to DDT, malathion, azinphos-methyl, and carbaryl due to fat body destruction (Jaques and Morris 1980). Tetreault (1985) demonstrated similar synergistic interaction with *Nosema pyrausta,* a microsporidian parasite of *Ostrinia nubilalis,* or European corn borer. Insecticides applied topically to third instars were up to 11 times more toxic to larvae with a sublethal infection of *N. pyrausta* than to healthy larvae (Tables 3.4 and 3.5), presumably due to fat body inactivation.

Other pathogens, which destroy membrane integrity or are localized in midguts, can act in a similar fashion by disarming the membrane-bound cytochrome P-450 complex enzymes. These enzymes are largely responsible for converting lipophilic substances into excretable nontoxic hydrophilic metabolites (Brattsten 1979). Cytoplasmic polyhedrosis virus infect the cytoplasm of midgut epithelium of caterpillars. Although they are not as pathogenic as nuclear polyhedrosis virus, they may synergize toxins (synthetic or plant derived) by eliminating the cytoplasm-soluble glutathione-*S*-transferases (L. Brattsten, personal communication; see also Bakke et al. 1981).

Malpighian tubules involved in transport and excretion of xenobiotics provide an additional sensitive site prone to incapacitation by pathogens. *Nosema otiorrhychi* infects the Malpighian tubules of *Otiorrhynchus ligustici* and slows down the excretion of insecticides to such an extent that toxic levels of insecticide are attained more rapidly (Benz 1971).

Pathogen–xenobiotic synergism is effectively a two-way street; not only can pathogens render ineffective an insect's metabolic capacity for detoxication, but the metabolic effects of allelochemicals can facilitate the entry and establishment of disease organisms. Pathogens must overcome a formidable number of barriers to establish an infection. Not the least of these is insect cuticle, a multilayer consisting of a thick endocuticle (10 to

TABLE 3.3 Studies Demonstrating Host Plant Effects on Pathogens of Insects

Pathogen	Insect	Active Fraction	Reference
Bacillus cereus B. thuringiensis	Pristiphora erichsonii, Malacosoma disstria	Leaf extracts	Kushner and Harvey (1962)
B. thuringiensis		Leaf extracts	Smirnoff and Hutchison (1965)
B. entomocidus (Agrotis), B. thuringiensis (S. littoralis), and "Biospor" (Laphygma)	Agrotis ypsilon, Laphygma (= Spodoptera) exigua, and S. littoralis	Leaf extracts	Afify and Merdan (1969)
Nuclear polyhedrosis virus	Bombyx mori	Leaf extracts	Matsubara and Hayashiya (1969)
Nuclear polyhedrosis virus	Hyphantria cunea	Leaf extracts	Kunimi and Aruga (1974)
Gut bacteria	Spodoptera littoralis, S. exigua, Pieris rapae, and Bombyx mori	Leaf extracts	Merdan et al. (1975)
Nuclear polyhedrosis virus	Bombyx mori	Leaf extract	Kinoshita and Inouye (1977)
Nuclear polyhedrosis virus	Bombyx mori	Protein	Hayashiya (1978)
Gut bacteria	Anthonomus grandis	Myrcene, β-caryophyllene, α-pinene, gossypol, catechin, tannin, gallic acid	Hedin et al. (1978)
	Bombyx mori	Caffeic acid	Koike et al. (1979)
	Bombyx mori	2-furaldehyde, 2-furoic acid	Jones et al. (1981)
	Bombyx mori	Chlorogenic acid, caffeic acid	Iizuke et al. (1974)
Fungus (Beauveria bassiana)	Blissus leucopterus	Host plant	Ramoska and Todd (1985)
Bacillus thuringiensis	Manduca sexta	Canavanine	Felton and Dahlman (1984)

TABLE 3.3 *(Continued)*

Pathogen	Insect	Active Fraction	Reference
B. cereus	—	*Viburnum cassinoides* extract	Smirnoff (1975)
B. cereus	—	*Abies balsamea*	Smirnoff (1968)
B. cereus	—	α- and β-Pinene, limonene, phellandrene, fenchone, and thujone	Smirnoff (1972)
B. thuringiensis	*Bombyx mori*	*Aralia, Rhaponthicum,* ecdysterone	Chernysh et al. (1983)
B. thuringiensis	—	*Pseudotsuga menziesii:* limonene, camphene, isobornyl acetate	Andrews et al. (1980)
Beauveria bassiana	*Leptinotarsa decemlineata*	*Solanum* spp.	Hare and Andreadis (1983)
"Fungus"	*Dryocosmus dubiosus*	Tannins	Taper et al. (1986)

Source: Based in part on Barbosa and Saunders (1985).

TABLE 3.4 Effect of *Nosema pyrausta* on Susceptibility of *Ostrinia nubilalis* to Insecticides

Compound	LD_{50} (Healthy) ($\mu g/g$)	LD_{50} (Infected) ($\mu g/g$)	Ratio
Carbaryl[a]	787	130	6.05
Carbofuran[a]	168	33	5.09
Methomyl[a]	1320	234	5.64
DDT	780	9	2.14
Permethrin	8	9	0.88
Chlorpyrifos[a]	58	5	11.60
Diazinon	193	209	0.92
Fonofos	139	146	0.95
Parathion	24	22	1.09
Terbofos	288	131	1.74

Source: Data from G. Tetreault (unpublished).

[a] Significantly different at 95% level.

TABLE 3.5 Effect of Insecticides with *Nosema pyrausta* on European Corn Borer Larvae[a]

	Percent Larval Mortality	Spores/μg Larval Tissue
Carbaryl + *N. pyrausta*	64a	326a
Carbofuran + *N. pyrausta*	58a	269a
N. pyrausta	44b	518b

Source: Data from Lublinkhof and Lewis (1980).
[a] Means within columns followed by the same letter are not significantly different at the 0.05 level according to Duncan's multiple-range test.

200 μm), an exocuticle of variable thickness, and an epicuticle 0.03 to 4 μm thick. The thin epicuticle is itself a multilayer composed of an outer cement layer, a wax layer, a cuticulin layer, and a homogeneous inner epicuticle (Chapman 1971). The cuticle is variously waxy and, in many groups, tough and impermeable due to extensive chitin-protein complexing (sclerotization). Certain plant allelochemicals interfere with cuticle synthesis and composition and thus can potentially alter the resistance of insects to pathogens. Furanocoumarins, for example, interfere with sclerotization of the cuticle of *Heliothis zea* in the presence of ultraviolet light through photosensitization (J. Neal, personal communication). Neonate caterpillars surviving their first molt on an artificial diet containing furanocoumarins show areas of depigmentation and cuticular thinning, ostensibly due to damage of the epidermal layer by furanocoumarins. Since a number of pathogenic fungi, such as *Beauveria bassiana,* enter the host through the cuticle by excreting in sequence proteases and chitinases, a thinner cuticle could make penetration by fungi easier and render the insect more susceptible to infection.

The gut wall is another barrier to colonization of insects by pathogens. In the midgut, the major portion of the alimentary tract in insects, several cell types perform specific functions. Goblet cells primarily transport potassium ions to the hindgut, regenerative cells or nidi produce new cells to replace damaged ones, and the columnar or ventricular epithelial cells are responsible for digestion, absorption, and secretion of a chitinous peritrophic membrane (which serves as a mechanical barrier between objects in the lumen and the epithelial cells). As is true of epidermal tissue in general, a noncellular basement membrane underlies the epidermal columnar cells (Chapman 1971). Any material that disrupts the integrity of the midgut has the potential for releasing normally nonpathogenic bacteria with little or no invasive power that reside in the gut lumen into the hemolymph, where they can establish a lethal septicemia. Even in the absence of plant chemicals, many pathogenic bacteria gain entry into the host via the production of toxins that break down gut wall integrity (Bucher 1960).

Gut wall ulceration with concomitant septicemia appears to be a possible mode of action of plant tannins. Tannins are a heterogeneous class of protein-binding plant constituents. Condensed tannins are polymers of flavan-3-ols and/or flavan-3,4-diols (leucoanthocyanins) that are widespread in the heartwood, bark, and leaves of many tree species. Hydrolyzable tannins are polyhydric alcohols variously esterified with gallic or hexahydroxyphenic acid. It has been assumed that, since tannins bind to proteins, they act to reduce digestibility of leaf proteins and thus reduce the nutritional quality of the leaves as food to insects, imposing slow and inefficient growth upon them (Feeny 1970). However, when tannins are fed to two species of swallowtail, one (*Papilio polyxenes*) which is totally unaccustomed to tannin in the diet, and the other (*P. glaucus*) which is frequently found feeding on many species of tanniniferous trees, no effects on digestive efficiencies could be observed. Extremely high mortality in the black swallowtails, however, was observed in the presence of tannin in the diet, along with a baggy appearance and rank odor characteristically associated with bacterial septicemia (Berenbaum 1983). Histological examination revealed that tannins have dramatic effects on the gut wall integrity of *P. polyxenes* (Steinly and Berenbaum 1985). When fed tannin-treated leaves, second and fifth instars of both species develop lesions. In *P. polyxenes,* these regions spanned gut wall up to 20 to 40 cells wide, and the basement membrane was left totally exposed. Secretogogues, normally associated with enzyme production in digestion, were observed forming cytoplasmic columnar cell extrusions. While both species showed small necrotic regions, those in *P. polyxenes* bordered areas in which columnar and goblet cells were detached from the basement membrane, and breakdown of lateral-apical cell junctions and lateral degeneration of tight junctions in columnar cells were construed. None of this sort of injury was seen in the guts of caterpillars fed food plants without added tannin (Steinly and Berenbaum 1985). Gut necrosis would ostensibly allow bacteria to move into the hemocoel and establish a lethal infection in those caterpillars fed tannin-containing diets.

The lesion-forming sequence characteristic of tannin damage may be representative of a general degenerative response to toxic substances. Similar tissue damage is documented for *Bacillus thuringiensis* endotoxin and cyanide (Brattsten et al. 1983) and is consistent with bacterial septicemia induced by sublethal DDT concentrations (Benz 1971). Some insecticides act to decrease the number of circulating hemocytes. Since phagocytosis is an important line of defense against pathogens, an insecticidal substance may prolong an infection or promote its establishment. Although there is no a priori evidence, plant secondary compounds may exert similar effects and facilitate pathogenesis.

2.2.3 Community-Level Implications
The interactions between insect pathogens and plant secondary compounds have community implications as well as physiological implications. Population cycles in forest Lepidoptera have long attracted scientific notice and, along with such notice, inspired hypothesis formation. Haukioja (1980) and others invoked inducible defense mobilization as a possible mechanism. Leaves subjected to insect herbivory can respond by producing growth-inhibiting substances; subsequent herbivory produces insects that are less fecund or viable, with attendant population consequences. Rhoades (1979) made the point that many of these inducible defenses are phenolic in nature and include both hydrolyzable and condensed tannins. In contrast, Anderson and May (1980) proposed an alternative hypothesis that attributes forest insect cycles to infections due to long-lived pathogens. The pathogens are always present in a population but transmission efficiency is a function of density; below a certain threshold density, a pathogen cannot become epizootic. Commenting on periodic caterpillar outbreaks, Schultz and Baldwin (1982) remarked that "not only may reduced food quality retard pest growth but it can also make the pests more susceptible to disease, parasites and predators. This could explain why so many different hypotheses about the initiation and decline of outbreaks appear reasonable." Such may be the case. Tannins or other phenolics, for example, by increasing the incidence of lesions in the gut wall after herbivory or by suppressing cellular defense responses, may increase the susceptibility of herbivorous insects to bacterial septicemia and other pathogenic diseases. Both factors, induced defenses and susceptibility to disease, may be involved in maintaining population cycles.

3   ALLELOCHEMICAL EFFECTS ON MICROBE-PLANT INTERACTIONS

3.1   Microbe–Plant Mutualisms: Symbiosis

3.1.1   Nitrogen Fixation and Allelochemicals

Just as insects have developed mutualistic associations with microbes, many plants rely on microbes to supply them with nutritional requirements. Several thousand species in over a dozen plant families are associates of nitrogen-fixing bacteria. Best studied among these associations is the relationship between legumes and various nodule-forming bacteria that fix atmospheric nitrogen to provide nitrogen in a form utilizable by the plant associate. Nitrogen fixation can influence the allelochemical profile of a plant in several ways. Nitrogen fixation may make available to the plant more nitrogen to channel into the production of toxic nitrogenous secondary substances such as alkaloids. Vogel and Weber (1922), for example, demonstrated that unfertilized lupines associated with nodulating *Rhizobium* produce more alkaloids than do lupines lacking rhizobial associates. Nitrogen fixation may also influence plant–insect interactions by increasing the nitrogen available to insects in various plant parts. Inasmuch as nitrogen is often limiting to insect growth, nitrogen fixation may actually enhance the suitability of plant foliage to insects. Moreover, the toxicity of many allelochemicals is influenced by the ratio of nutrient to toxicant (Reese 1979; see also references in Slansky and Rodriguez 1987). Increasing the nitrogen, amino acid, and/or protein content of foliage may reduce the toxicity of allelochemicals present in the same concentrations (e.g., Broadway and Duffey 1986). Finally, nitrogen fixation may alter the availability of nitrogen in plant parts to insects in that nitrogen fixed by bacteria can be stored by plants in different molecular forms. Tropical legumes (such as soybeans) form primarily ureides, whereas temperate legumes (such as peas) use the amino acids aspartate and glutamate. Since many herbivores cannot metabolize ureides, the storage form of the nitrogen can determine the quality of the plant tissue to insect herbivores (Wilson and Stinner 1984).

There are also nonnitrogen-fixing nonpathogenic bacteria that promote plant growth (Elad et al. 1985). The precise mechanisms are unknown and may in fact vary with the microbial taxon. Some may suppress plant pathogens; others may solubilize minerals, excrete growth substances, or enhance water uptake. Overall, by increasing plant vigor these bacterial associates may free plant resources for biosynthesis of allelochemicals.

3.1.2   Endophyte Infections and Allelochemicals

Microbial associates of plants may also contribute symbiotically to their hosts by providing an active chemical defense system to the plant. Such appears to be the case with grasses infected with the endophytic fungi in the tribe Balansiae in the family Clavicipitaceae (Clay et al. 1985a,b). The fungus has, for the most part, no adverse effects on the plant. Rather, infected plants have higher nitrogen contents, improved growth, and better drought resistance (Ahmad et al. 1985 and references within). These endophytic fungi can alter plant suitability to insect herbivores (Table 3.6), although the chemical mechanism by which alteration is mediated is not always known. In the case of tall fescue infected with *Acremonium coenophialum,* N-formyl and N-acetyl loline alkaloids are produced that confer protection against both mammalian and insect herbivores. In ryegrass, endophyte infection confers protection against insect herbivores but the chemical agent responsible has not been identified (Ahmad et al. 1985, 1986). Ergot alkaloids produced by fungi in the related genus *Claviceps* cause major physiological disruption in mammalian herbivores (Clay et al. 1985a) and substituted indole alkaloids (e.g., Lolitrem B) prolong the development time of weevil larvae (Prestidge and Gallagher 1985).

3.2   Microbe–Plant Antagonisms: Pathogenesis

3.2.1   Phytoalexin Responses to Infection and Plant Suitability to Insects

Plant pathogens can profoundly alter the allelochemical profile of a host plant and can therefore directly alter the suitability of plant tissue as food for insects. Many plant species are capable of forming phytoalexins, low-molecular-weight compounds produced *de novo* in response to microbial stimulation (Creasy 1985). Despite the rather restrictive definition, virtually all phytoalexins examined are nonspecific in terms of their mode of action and virtually all examined are broadly biocidal. Reports on the animal toxicity of phytoalexins are numerous (reviewed in Smith 1982) and their effects include inhibition of mitochondrial respiration, repression of electron transport, and uncoupling of oxidative phosphorylation on the cellular level and reproductive abnormalities and feeding inhibition on the organism level. Microbial toxicity and insect toxicity may in fact be closely correlated. Russell et al. (1978) found that feeding deterrency of

TABLE 3.6   Effects of Endophytic Fungal Infection on Plant Suitability to Insect Herbivores

Plant Species	Insect Species	Effects	References
Cyperus spp.	Fall armyworm	Reduced preference, developmental delay	Clay et al. (1985b)
Perennial ryegrass	Argentine stem weevil	Reduced survival reduced preference, oviposition reduction, developmental delay	Barker et al. (1983), Barker et al. (1984a,b), Prestidge and Gallagher (1985)
	Bluegrass billbug	Reduced preference	Ahmad et al. (1986)
	Fall armyworm	Reduced preference	Hardy et al. (1985)
	House cricket	Reduced survival	Ahmad et al. (1985)
	Sod webworm	Reduced consumption	Funk et al. (1983)
Tall fescue	Fall armyworm	Reduced preference, developmental delay	Funk et al. (1983), Clay et al. (1985a)
	Argentine stem weevil	Feeding deterrence, developmental delay	Prestidge et al. (1985)
Perennial ryegrass	Japanese beetle	No difference	Funk et al. (1983)
Tall fescue	Aphids	Increased preference	Johnson et al. (1985)
	Tobacco hornworm, tobacco budworm, southern armyworm	No differences	Johnson et al. (1985)

plant compounds to grass grubs (*Costelytra zealandica*) correlated with their bacterial toxicity. Similar observations have been made for fungicidal phytoalexins and it is highly likely that the mode of action and target sites in microbes and higher animals are similar (Tuveson et al. 1986). Insect feeding can even elicit phytoalexin-like responses in plants. Feeding damage by *Cyclas formicarum* weevils in sweet potato induces the formation of the phytoalexin ipomeamarone (Mansfield 1982), and infection of potato by tobacco mosaic virus confers protection against Colorado potato beetle (Hare and Dodd 1978).

Plant responses to fungal infections can involve chemicals other than low-molecular-weight phytoalexins (Ayers et al. 1985): these chemicals can also influence host suitability to insects. Additional responses include induction of enzymes that destroy membranes, production of enzyme inhibitors such as protease inhibitors that can act to reduce digestibility of plant proteins, and production or modification of cell wall constituents to lower digestibility. The latter include lignification, synthesis of hydroxy-proline-rich proteins, and cork accumulation. In some cases the biochemical basis for reduced suitability of fungus-infected plant tissue as food for arthropods is not known (e.g., Karban et al. 1987).

3.2.2   Pathogen-Produced Attractants

Microbial infection can render plant tissue less suitable as food for insect herbivores by virtue of the fact that many microbes produce toxicants and antibiotics. These chemicals may act to protect the microbes from inadvertent consumption by the herbivore along with host tissue (Janzen 1977). Infection of corn (*Zea mays*) by *Aspergillus flavus* results in the production by the fungus of aflatoxin G1 and related compounds which can reduce growth of European corn borer (*Ostrinia nubilalis*) at concentrations in the part per billion range (Jarvis et al. 1984). Far more frequently, however, microbial infection (by either pathogenic or saprophytic organisms) enhances the nutritional value of plant tissue for insects. One general explanation behind enhanced suitability is that microbial activity can serve to break down structural carbohydrates and other materials refractory to insect digestive enzymes. Such appears to be the reason behind the accelerated development of European corn borer on corn infected with maize anthracnose (*Colletotrichum graminicola*). A 20% reduction in development time was attributed to fungal decomposition of complex carbohydrates by macerating enzymes associated with stalk rot (Carruthers et al. 1986). A similar response was observed in field populations of European corn borer infesting corn infected with *Fusarium graminearum,* another stalk-rotting fungus (Chiang and Wilcoxson 1961). Particularly among wood-feeding insects, acquisition of digestive enzymes from microbes infecting plant tissue may greatly enhance nutritional suitability and enlarge host range (Martin 1979, 1984). Fungal enzymes in particular can be acquired by ingesting the fungus directly or by consuming the substrate into which fungal enzymes have been released. Wood-rotting Basidiomycetes produce large quantities of extracellular cellulases, hemicellulases, and pectinases, all of which would serve to break down complex carbohydrates into simpler digestible sugars. In addition, certain saprophytic fungi produce phenoloxidizing enzymes which may detoxify phenolic substances otherwise toxic to insect herbivores (Martin 1979).

Pathogenic or saprophytic microbes not only may enhance the nutritional suitability of plant tissue for insect herbivores, they may produce chemicals that facilitate host finding as well. Female onion flies (*Hylemya antiqua*), for example, distinguish between decomposing and intact onion seedlings and preferentially oviposit near decomposing onions. The principal volatile attractants, N-propyl sulfide and propane thiol, are not present in intact onion; they are released only following microbial action or tissue damage, which serves to release enzymes that degrade propenylcysteine sulfoxide (Dindonis and Miller 1981). Microbial activity in general, with its concomitant tissue damage, may serve to increase the release of low-molecular-weight volatile substances (via oxidation of membrane lipids). These lipid oxidation products may be the chief constituents of the "green leaf volatiles" (aldehydes, ketones, and alcohols) which many insect herbivores use for long-range orientation and host finding (Harborne 1982).

4   EVOLUTIONARY IMPLICATIONS OF MICROBIAL PARTICIPATION IN PLANT–INSECT INTERACTIONS

Much discussion (and vituperation) has been generated over defining precisely the phenomenon, or even the existence, of plant–insect coevolution. Conventional wisdom now has it that plants and insects respond under certain circumstances in reciprocal fashion; changes in gene frequencies in one population effect a corresponding change in gene frequencies in an interacting population, which in turn act upon the initial population to bring about another genetic change (freely adapted from Janzen 1980). Traditionally, plant response has been viewed as a genetic change which can lead to a change in plant chemistry (Ehrlich and Raven 1964). A change in the structure or quantity of an allelochemical can bring about a change in the insecticidal properties of a substance to render a plant unsuitable to extant herbivores. In turn, changes in the insect genome, affecting the structure or activity of detoxication enzymes, for example (Brattsten 1979), can bring about resistance. It is increasingly apparent, however, that other genes may be mediating these interactions. More specifically, microbial mediators of plant–insect interactions may temper or alter selection pressures and thereby make the situation very complicated (and not easily predictable). Insect detoxication ability, for example, may result not from a change in microsomal cytochrome P-450 detoxication enzymes but rather from a change in the gut flora. By the same token, plant defense responses may depend directly on the activity and availability of rhizobial symbionts. Predicting the responses of plants to selection pressures by insects may be complicated by the relationship of plants to microbial associates. Given the differences in generation time between microbes and either plants or insects, rates of evolutionary response will differ accordingly. In fact, the relatively rapid responses of microbes mutualistic with plants may allow plants to keep pace with rapid changes in genomes of herbivorous insects. Moreover, not all chemical responses may be genetically based; endophyte-based resistance of perennial ryegrass to the bluegrass billbug, for example, is maternally transmitted (Ahmad et al. 1986). Predictions of selection response will thus require considerable information on modes of inheritance of resistance factors. All in all, evolutionary responses between plants and insects may not follow their expected trajectories because microbial participants mediating the interactions of interest may be subject to conflicting selection pressures or may be constrained genetically as to their responses. How plant allelochemicals affect the outcome of an interaction depends largely on their relative toxicity to microbe, plant, and insect herbivore at a given time. There is (at least at this stage) no a priori way to determine, effectively, which side they are working for without detailed information on the combatants in question.

ACKNOWLEDGMENTS

I thank Dr. Pedro Barbosa for unbelievable patience and encouragement while this review was being prepared, Dr. Joseph Maddox for supplying references and guidance, and Drs. James Nitao and Jonathan Neal for comments on several versions of the manuscript. This work was supported in part by NSF grant BSR835-1407.

REFERENCES

Afify, M., and A. I. Merdan. 1969. Reaktionsunterschiede von drei Noctuiden-arten bei bestimmten *Bacillus* Präparaten in Abhängingkeit von der Nahrung und Art der Behandlung. Anz. Schaedlingskd. Pflanzenschutz 42:102–104.

Ahmad, S., S. Govindarajan, C. R. Funk, and J. M. Johnson-Cicalese. 1985. Fatality of house crickets on perennial ryegrasses infected with a fungal endophyte. Ent. Exp. Appl. 39:183–190.

Ahmad, S., J. M. Johnson-Cicalese, W. K. Dickson, and C. R. Funk. 1986. Endophyte-enhanced resistance in perennial ryegrass to the bluegrass billbug, *Sphenophorus parvulus*. Ent. Exp. Appl. 41:3–10.

Anderson, R. M., and R. M. May. 1980. Infectious diseases and population cycles of forest insects. Science 210:658–661.

Andrews, R. E., and K. D. Spence. 1980. Action of Douglas fir (*Pseudotsuga menziesii*) tussock moth larvae (*Orgyia pseudotsugata*) and their microflora on dietary terpenes. Appl. Environ. Microbiol. 40:959–963.

Andrews, R. E., L. W. Parks, and K. D. Spence. 1980. Some effects of Douglas fir terpenes on certain microorganisms. Appl. Environ. Microbiol. 40:301–304.

Arrhenius, S. P., and J. H. Langenheim. 1983. Inhibitory effects of *Hymenaea* and *Copaifera* leaf resins on the leaf fungus, *Pestalotia subcuticularis*. Biochem. Syst. Ecol. 11:361–366.

Ayers, A. R., J. J. Goodell, and P. L. DeAngelis. 1985. Plant detection of pathogens. In G. A. Cooper-Driver, T. Swain, and E. E. Conn (eds.), Chemically Mediated Interactions between Plants and Other Organisms. pp. 1–20. Plenum, New York.

Bakke, J. E., G. L. Larsen, P. W. Aschbacher, J. J. Rafter, J. A. Gustafsson, and B. E. Gustafsson. 1981. Role of gut microflora in metabolism of glutathione conjugates of xenobiotics. In J. D. Rosen, P. S. Magee, and J. E. Casida (eds.), Sulfur in Pesticide Action and Metabolism. American Chemical Society Symposium Series 158. pp. 165–178. American Chemical Society, Washington, D.C.

Barbosa, P., and J. Saunders. 1985. Plant allelochemicals: linkages between herbivores and their natural enemies. In G. A. Cooper-Driver, T. Swain, and E. E. Conn (eds.), Chemically Mediated Interactions between Plants and Other Organisms. pp. 107–138. Plenum, New York.

Barker, G. M., R. P. Pottinger, and P. J. Addison. 1983. Effect of tall fescue and ryegrass endophytes on Argentine stem weevil. pp. 216–219. Proceedings of the 36th New Zealand Weed and Pest Control Conference.

Barker, G. M., R. P. Pottinger, and P. J. Addison. 1984a. Effect of *Lolium* endophyte fungus infections on survival of larval Argentine stem weevil. N. Z. J. Agric. Res. 27:279–281.

Barker, G. M., R. P. Pottinger, P. J. Addison, and R. A. Prestidge. 1984b. Effect of *Lolium* endophyte fungus infections on behavior of adult Argentine stem weevil. N. Z. J. Agric. Res. 27:271–277.

Benz, G. 1971. Synergism of microorganisms and chemical insecticides. In H. D. Burges and N. W. Hussey (eds.), Microbial Control of Insects and Mites. pp. 327–355. Academic Press, New York.

Berenbaum, M. R. 1983. Effects of tannins on growth and digestion in two species of papilionids. Ent. Exp. Appl. 34:245–250.

Brand, J. M., J. W. Bracke, A. J. Markovetz, D. L. Wood, and L. E. Browne. 1975. Production of verbenol pheromone by a bacterium isolated from bark beetles. Nature 254:136–137.

Brand, J. M., J. W. Bracke, L. N. Britton, A. J. Markovetz, and S. J. Barras. 1976. Bark beetle pheromones: production of verbenone by a mycangial fungus of *Dendroctonus frontalis*. J. Chem. Ecol. 2:195–199.

Brattsten, L. B. 1979. Biochemical defense mechanisms in herbivores against plant allelochemicals. In G. A. Rosenthal and D. H. Janzen (eds.), Herbivores: Their Interaction with Secondary Plant Metabolites. pp. 200–270. Academic Press, New York.

Brattsten, L. B., J. H. Samuelian, K. Y. Long, S. A. Kincaid, and C. K. Evans. 1983. Cyanide as a feeding stimulant for the southern armyworm, *Spodoptera eridania*. Ecol. Ent. 8:125–132.

Broadway, R. M., and S. S. Duffey. 1986. Plant proteinase inhibitors: mechanism of action and effect on the growth and digestive physiology of larval *Heliothis zea* and *Spodoptera exigua*. J. Insect Physiol. 32:327–334.

Bucher, G. E. 1960. Potential bacterial pathogens of insects and their characteristics. J. Insect Pathol. 2:172–195.

Carruthers, R. I., G. C. Bergstrom, and P. A. Haynes, 1986. Accelerated development of the European corn borer, *Ostrinia nubilalis* (Lepidoptera: Pyralidae), induced by interactions with *Colletotrichum graminicola* (Melanconiales: Melanconiaceae), the causal fungus of maize anthracnose. Ann. Ent. Soc. Amer. 79:385–389.

Carter, F. L., J. K. Mauldin, and N. M. Rich. 1981. Protozoan populations of *Coptotermes formosanus* Shiraki exposed to heartwood samples of 21 American species. Mater. Org. 16:29–38.

Chapman, R. 1971. The Insects: Structure and Function. Elsevier, New York.

Chernysh, S. I., V. A. Lukhtanov, and N. P. Simonenko. 1983. Adaptation to damage in the silkworm *Bombyx mori* L. (Lepidoptera, Bombycidae). II. Effects of ecdysterone and other adaptogens on larval resistance to entobacterin. Ent. Rev. 4:1–8.

Chiang, H. C., and R. D. Wilcoxson. 1961. Interactions of the European corn borer and stalk rots in corn. J. Econ. Ent. 54:779–782.

Clay, K., T. N. Hardy, and A. M. Hammond. 1985a. Fungal endophytes of grasses and their effects on an insect herbivore. Oecologia (Berlin) 66:1–5.

Clay, K., T. N. Hardy, and A. M. Hammond. 1985b. Fungal endophytes of *Cyperus* and their effect on an insect herbivore. Amer. J. Bot. 72:1284–1289.

Conn, J. E., J. H. Borden, D. W. A. Hunt, J. Holman, H. S. Whitney, O. J. Spanier, H. D. Pierce, and A. C. Oehlschlager. 1984. Pheromone production by axenically reared *Dendroctonus ponderosae* and *Ips paraconfusus* (Coleoptera: Scolytidae). J. Chem. Ecol. 10:281–290.

Creasy, L. L. 1985. Biochemical responses of plants to fungal attack. In G. A. Cooper-Driver, T. Swain, and E. E. Conn (eds.), Chemically Mediated Interactions between Plants and Other Organisms. pp. 47–80. Plenum, New York.

Dindonis, L. L., and J. R. Miller. 1981. Onion fly and little house fly host finding selectively mediated by decomposing onion and microbial volatiles. J. Chem. Ecol. 7:421–428.

Elad, Y., and I. J. Misaghi. 1985. Biochemical aspects of plant-microbe and microbe-microbe interactions in soil. In G. A. Cooper-Driver, T. Swain, and E. E. Conn (eds.), Chemically Mediated Interactions between Plants and Other Organisms. pp. 21–46. Plenum, New York.

Ehrlich, P., and R. Raven. 1964. Butterflies and plants: a study in coevolution. Evolution 18:586–608.

Feeny, P. 1970. Seasonal changes in oak leaf tannins and nutrients as a cause of spring feeding by winter moth caterpillars. Ecology 51:565–581.

Felton, G. W., and D. L. Dahlman. 1984. Allelochemical induced stress: effects of L-canavanine on the pathogenicity of *Bacillus thuringiensis* in *Manduca sexta*. J. Invertebr. Pathol. 44:187–191.

Funk, C. R., P. M. Halisky, M. C. Johnson, M. R. Siegel, A. V. Stewart, S. Ahmad, R. H. Hurley, and I. C. Harvey. 1983. An endophytic fungus and resistance to sod webworms: association in *Lolium perenne* L. Biol. Tech. 1:189–191.

Francke-Grossman, H. 1967. Ectosymbiosis in wood-inhabiting insects. In S. Henry (ed.), Symbiosis, Vol. II. pp. 141–205. Academic Press, New York.

Harborne, J. 1982. Introduction to Ecological Biochemistry. Academic Press, New York.

Hardy, T. N., K. Clay, and A. M. Hammond. 1985. Fall armyworm (Lepidoptera: Noctuidae): a laboratory bioassay and larval preference study for the fungal endophyte of perennial ryegrass. J. Econ. Ent. 78:571–575.

Hare, J. D., and T. G. Andreadis. 1983. Variation in the susceptibility of *Leptinotarsa decemlineata* (Coleoptera: Chrysomelidae) when reared on different host plants to the fungal pathogen, *Beauveria bassiana* in the field and laboratory. Environ. Ent. 12:1892–1897.

Hare, J. D., and H. A. Dodd. 1978. Changes in food quality of an insect's marginal host species associated with a plant virus. J. N.Y. Ent. Soc. 86:292.

Haukioja, E. 1980. On the role of plant defenses in the fluctuation of herbivore populations. Oikos 35:202–213.

Hayashiya, K. 1978. Red fluorescent protein in the digestive juice of the silkworm larva fed on host plant mulberry leaves. Ent. Exp. Appl. 24:228–236.

Hedin, P., O. H. Lindig, P. P. Sikorowski, and M. Wyatt. 1978. Suppressants of gut bacteria in the boll weevil from the cotton plant. J. Econ. Ent. 71:394–396.

Henry, S. (ed.). 1967. Symbiosis, Vol. II. Academic Press, New York.

Hubbell, S. P., and J. J. Howard. 1984. Chemical leaf repellency to an attine ant: seasonal distribution among potential host plant species. Ecology 65:1067–1076.

Hubbell, S. P., D. F. Wiemer, and A. Adejare. 1983. An antifungal terpenoid defends a neotropical tree (*Hymenaea*) against attack by fungus-growing ants (*Atta*). Oecologia (Berlin) 60:321–327.

Iizuke, T., S. Koike, and J. Mizutani. 1974. Antibacterial substances in feces of silkworm larvae reared on mulberry leaves. Agric. Biol. Chem. 38:1549–1550.

Janzen, D. H. 1977. Why fruits rot, seeds mold and meat spoils. Amer. Nat. 111:691–713.

Janzen, D. H. 1980. When is it coevolution? Evolution 34:611–612.

Jaques, R. P., and O. N. Morris, 1980. Compatibility of pathogens with other methods of pest control and with different crops. In H. D. Burges (ed.), Microbial Control of Pests and Plant Diseases 1970–1980. pp. 695–715. Academic Press, New York.

Jarvis, J. L., W. D. Guthrie, and E. B. Lillenhoj. 1984. Aflatoxin and selected biosynthetic precursors: effects on the European corn borer in the laboratory. J. Agric. Ent. 1:354–359.

Johnson, M. C., D. L. Dahlman, M. R. Siegel, L. P. Bush, G. C. M. Latch, D. A. Potter, and D. R. Varney. 1985. Insect feeding deterrents in endophyte-infected tall fescue. Appl. Environ. Microbiol. 49:568–571.

Jones, C. G. 1984. Microorganisms as mediators of plant resource exploitation by insect herbivores. In P. W. Price, C. N. Slobodchikoff, and W. S. Gaud (eds.), A New Ecology: Novel Approaches to Interactive Systems. pp. 54–99. Wiley, New York.

Jones, C. G., J. R. Aldrich, and M. S. Blum. 1981. Bald cypress allelochemics and the inhibition of silkworm enteric microorganisms; some ecological considerations. J. Chem. Ecol. 7:103–114.

Karban, R., R. Adamchak, and W. C. Schnathorst. 1987. Induced resistance and interspecific competition between spider mites and a vascular wilt fungus. Science 235:678–680.

Khoo, B. K., and M. Sherman. 1981. Metabolism of chlorpyrifos by normal and defaunated Formosan subterranean termites (*Coptotermes formosanus*). J. Econ. Ent. 74:681–687.

Kinoshita, T., and K. Inouye. 1977. Bactericidal activity of the normal cell-free haemolymph of silkworms (*Bombyx mori*). Infect. Immun. 16:32–36.

Koch, A. 1967. Insects and their endosymbionts. In S. Henry (ed.), Symbiosis, Vol. II. pp. 1–106. Academic Press, New York.

Koike, S., T. Iizuka, and K. Mizutani. 1979. Determination of caffeic acid in the digestive juice of silkworm larvae and its antibacterial activity against the pathogenic *Streptococcus faecalis* AD-4. Agric. Biol. Chem. 43:1727–1731.

Kunimi, Y., and H. Aruga. 1974. Susceptibility to infection with nuclear and cytoplasmic polyhedrosis virus of the fall webworm, *Hyphantria cunea* Drury, reared on several artificial diets. J. Appl. Ent. Zool. 18:1–4.

Kushner, D. J., and G. T. Harvey. 1962. Antibacterial substances in leaves: their possible role in insect resistance to disease. J. Insect Pathol. 4:155–184.

Lublinkhof, J., and L. C. Lewis. 1980. Virulence of *Nosema pyrausta* to the European corn borer when used in combination with insecticides. Environ. Ent. 9:67–71.

Maddox, J. V. 1982. Use of insect pathogens in pest management. In R. L. Metcalf and W. H. Luckmann (eds.), Introduction to Insect Pest Management. pp. 175–216. Wiley, New York.

Mansfield, J. W. 1982. The role of phytoalexins in disease resistance. In J. A. Bailey and J. W. Mansfield (eds.), Phytoalexins. pp. 253–288. Wiley, New York.

Martin, M. M. 1979. Biochemical implications of insect mycophagy. Biol. Rev. 54:1–21.

Martin, M. M. 1984. The role of ingested enzymes in the digestive processes of insects. In J. M. Anderson, A. D. M. Rayner, and D. Walton, (eds.), Animal-Microbial Interactions. pp. 155–172. Cambridge University Press, Cambridge.

Matsubara, F., and K. Hayashiya. 1969. The susceptibility to infection with nuclear polyhedrosis virus in the silkworm reared on artificial diet. J. Seric. Sci. Jpn. 38:43–48.

Mauldin, J. K., F. L. Carter, and N. M. Rich. 1981. Protozoan populations of *Reticulitermes flavipes* (Kollar) exposed to heartwood blocks of 21 American species. Mater. Org. 16:15–28.

Merdan, A., H. Abuel-Rahman, and A. Soliman. 1975. On the influence of host-plants on insect resistance to bacterial diseases. Z. angew. Ent. 78:280–285.

Murray, R. D. H., J. Mendez, and S. A. Brown. 1982. The Natural Coumarins, Occurrence, Chemistry and Biochemistry. Wiley Chichester, West Sussex, England.

Prestidge, R. A., and R. T. Gallagher. 1985. Lolitrem B—a stem weevil toxin isolated from *Acremonium*-infected ryegrass. pp. 38–40. Proceedings of the 38th New Zealand Weed and Pest Control Conference.

Prestidge, R. A., D. R. Lauren, S. G. van der Zijpp, and M. E. DiMenna. 1985. Isolation of feeding deterrents to Argentine stem weevil in cultures of endo-phytes of perennial ryegrass and tall fescue. N. Z. J. Agric. Res. 28:87–92.

Prestwich, G. D., B. L. Bentley, and E. J. Carpenter, 1980. Nitrogen sources for neotropical nasute termites: fixation and selective foraging. Oecologia (Berlin) 46:397–401.

Raffa, R. F., and A. A. Berryman. 1982. Accumulation of monoterpenes and volatiles following inoculation of grand fir with a fungus transmitted by the fir engraver, *Scolytus ventralis* (Coleoptera:Scolytidae). Can. Ent. 114:797–810.

Raffa, K. F., and A. A. Berryman. 1985. Effects of grand fir monoterpenes on the fir engraver, *Scolytus ventralis* (Coleoptera: Scolytidae), and its symbiotic fungi. Environ. Ent. 14:552–556.

Ramoska, W. A., and T. Todd. 1985. Variation in efficacy and viability of *Beauveria bassiana* in the chinch bug (Hemiptera: Lygaeidae) as a result of feeding activity on selected host plants. Environ. Ent. 14:146–148.

Reese, J. C. 1979. Interactions of allelochemicals with nutrients in herbivore food. In G. A. Rosenthal and D. H. Janzen (eds.), Herbivores: Their Interaction with Secondary Plant Metabolites. pp. 309–330. Academic Press, New York.

Rhoades, D. R. 1979. Evolution of plant chemical defense against herbivores. In G. A. Rosenthal and D. H. Janzen (eds.), Herbivores: Their Interaction with Secondary Plant Metabolites. pp. 4–55. Academic Press, New York.

Russell, G. B., O. R. W. Sutherland, R. F. N. Hutchins, and P. E. Christmas. 1978. Vestitol: a phytoalexin with insect feeding-deterrent activity. J. Chem. Ecol. 4:571–579.

Schultz, J. C., and I. Baldwin. 1982. Oak leaf quality declines in response to defoliation by gypsy moth larvae. Science 217:149–151.

Slansky, F., and J. G. Rodriguez (eds.). 1987. Nutritional Ecology of Insects, Mites, Spiders, and Related Invertebrates. Wiley, New York.

Smirnoff, W. A. 1965. Inhibition of parasporal inclusion synthesis in crystallini-ferous sporeforming bacteria of the "*cereus*" group by an aqueous extract of

Viburnum cassinoides Linnaeus (Caprifoliaceae) leaves. J. Invertebr. Pathol. 7:71–74.

Smirnoff, W. A. 1968. Effects of volatile substances released by foliage of various plants on the entomopathogenic *Bacillus cereus* group. J. Invertebr. Pathol. 11:513–515.

Smirnoff, W. A. 1972. Effects of volatile substances released by foliage of *Abies balsamea*. J. Invertebr. Pathol. 19:32–35.

Smirnoff, W. A., and P. M. Hutchison. 1965. Bacteriostatic and bacteriocidal effects of extracts of foliage from various plant species on *Bacillus thuringiensis* var *thuringiensis* Berliner. J. Invertebr. Pathol. 7:273–280.

Smith, D. A. 1982. Toxicity of phytoalexins. In J. A. Bailey and J. W. Mansfield (eds.), Phytoalexins. pp. 218–251. Wiley, New York.

Steinly, B. A., and M. Berenbaum. 1985. Histopathological effects of tannins on the midgut epithelium of *Papilio polyxenes* and *Papilio glaucus*. Ent. Exp. Appl. 39:3–9.

Taper, M. L., E. M. Zimmerman, and T. J. Case. 1986. Sources of mortality for a cynipid gall-wasp *Dryocosmus dubiosus* (Hymenoptera: Cynipidae): the importance of the tannin/fungus interaction. Oecologia (Berlin) 68:437–445.

Tetreault, G. E. 1985. Metabolism of carbaryl, chlorpyrifos, DDT, and parathion in the European corn borer: effects of microsporidiosis on toxicity and detoxication. Ph.D. dissertation. University of Illinois at Urbana-Champaign, Urbana, Ill.

Tuveson, R. W., M. R. Berenbaum, and E. E. Heininger. 1986. Inactivation and mutagenesis by phototoxins using *Escherichia coli* strains differing in sensitivity to near—and far—ultraviolet light. J. Chem. Ecol. 12:933–948.

Vogel, Z., and E. Weber. 1922. Über den Einfluss der Stickstoffernahrung auf den Bitterstoffgehalt der Lupine. Z. Pflanzennahr. 1:85–95.

Wigglesworth, V. B. 1987. Histochemical studies of uric acid in some insects. I. Storage in the fat body of *Periplaneta americana* and the action of the symbiotic bacteria. Tissue Cell 19:83–92.

Wilson, K. G., and R. E. Stinner. 1984. A potential influence of *Rhizobium* activity on the availability of nitrogen to legume herbivores. Oecologia (Berlin) 61:337–341.

CHAPTER 4

Microbial Allelochemicals Affecting the Behavior of Insects, Mites, Nematodes, and Protozoa in Different Trophic Levels

M. Dicke
Agricultural University
Wageningen, The Netherlands

CONTENTS

1 Introduction
2 Allelochemicals of stored-product microorganisms
3 Allelochemicals of soil microorganisms
   3.1 Soil fungi–nematode interactions
      3.1.1 Interactions between fungus-feeding nematodes and fungi
      3.1.2 Interactions between nematode-feeding fungi and their nematode prey
   3.2 Soil fungi–mite interactions
4 Microbial allelochemicals attracting bacterivoral protozoa
5 Allelochemicals of plant-infesting microorganisms and their influence on the behavior of microbivores
6 Allelochemicals of aquatic microorganisms
7 Allelochemicals of insect-associated microorganisms that are used as pheromoncs by the insect
8 Microbial allelochemicals affecting natural enemies of microbe-associated insects
9 Consequences for semiochemical terminology
10 Prospects for future research
References

1 INTRODUCTION

Semiochemicals are chemicals involved in chemical interactions between organisms (Law and Regnier 1971, Nordlund and Lewis 1976). They are subdivided into pheromones [substances secreted by an organism to the outside and that cause a specific reaction in receiving organisms of the same species (Karlson and Lüscher 1959, Nordlund and Lewis 1976)] and allelochemics or allelochemicals [chemicals significant to organisms of a species different from their source, for reasons other than food as such (Whittaker and Feeny 1971, Nordlund and Lewis 1976)]. Allelochemicals can be subdivided into allomones, kairomones, and synomones, depending on which organism benefits in the interaction: the emitter (allomone), the receiver (kairomone), or both (synomone) (Nordlund and Lewis 1976). The structure of this terminology is shown in Fig. 4.1. As can be inferred from the definition of semiochemicals given above, the compounds may have many different qualities, summarized as "involved in chemical interactions." In this chapter I deal exclusively with semiochemicals that convey information.

Plant semiochemicals affect the behavior of many different organisms (see, e.g., Schoonhoven 1981, Rhoades 1985, Visser 1986). In review papers on plant semiochemicals, microorganisms (bacteria and fungi) are rarely dealt with in detail. Yet microorganisms are present in all ecosystems and are known to have an essential function in many interspecific interactions (see Chapter 3). Just as semiochemicals are known to play a role in intraspecific interactions among microbes (Strazdis and MacKay

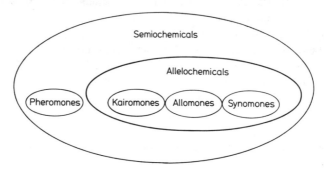

Figure 4.1 Structure of semiochemical terminology. [After Nordlund and Lewis (1976).] Reprinted, with permission, from M. Dicke, and M. W. Sabelis (1988) Infochemical terminology: based on cost-benefit analysis rather than origin of compounds. Functional Ecology 2 (in press).

1983), insects (Nordlund et al. 1981, Bell and Cardé 1984), and mammals (Albone 1984), so can semiochemicals influence interspecific relationships between microbes and other organisms.

To date, the role of microbial semiochemicals has been overlooked in many instances. For example, Nordlund and Lewis (1976), when subdividing "allelochemics", included a fourth category, apneumones, defined as substances emitted by a nonliving material that evoke a behavioral or physiological reaction adaptively favorable to a receiving organism, but detrimental to an organism of another species which may be found in or on the nonliving material. As an example of apneumones, Nordlund and Lewis (1976) mention chemicals emitted from oatmeal which attract the ichneumonid parasite, *Venturia canescens,* and aid in its search for hosts (Thorpe and Jones 1937). Another example is that of chemicals of meat to which the parasites *Alysia manducator* and *Nasonia vitripennis* are attracted (Laing 1937). [For definitions and critical discussion of the terms "attraction" and "arrestment," see Kennedy (1978).] However, in many cases it is difficult to distinguish between apneumones and allelochemicals that are produced by microorganisms. For instance, the volatile semiochemicals that are emitted from oatmeal (Thorpe and Jones 1937) might have originated from microorganisms associated with the oatmeal.

Although in some cases it is difficult to demonstrate involvement of microbial allelochemicals (see Section 7), in other instances substantial evidence has been presented for the existence of such semiochemicals. As our knowledge of the involvement of microorganisms in interactions of other organisms increases (e.g., Jones 1984), many more instances of microbial allelochemicals mediating interactions of individual organisms will be discovered. Microbial semiochemicals can be demonstrated in several ways, indicated below. These are ranked in order of increasing knowledge they provide on the existence and chemical nature of the compounds: Compared to a proper control, the behavior of an organism changes demonstrably after it has been offered (1) a substrate that obviously contains microorganisms; (2) a culture of one species of microorganisms; (3) (a) an extract of the microorganism, (b) an extract of the growth medium of the microorganism, (c) a combination of 3a and 3b, or (d) volatiles collected from the microorganism and/or its growth medium; and (4) synthetic compounds that have been shown to exist in the microorganism and/or its growth medium or in the headspace.

Microbial allelochemicals are known to affect the behavior of species in the Insecta, Arachnida, Nematoda, Protozoa, and Mammalia and in several different environments (e.g., in soil or water ecosystems and in

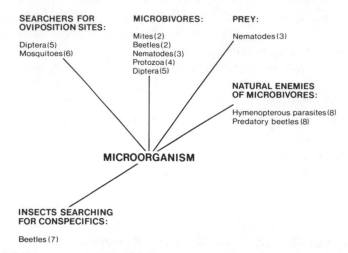

Figure 4.2   Involvement of microbial allelochemicals in several different interactions. Numbers indicate sections of this chapter that deal with these interactions.

fermenting or rotting plant material). The organisms whose behavior is influenced by microbial allelochemicals can be microbivores, prey organisms of the microbes, insects whose biology is closely associated with certain microbial species, or natural enemies of the organisms noted above (Fig. 4.2).

Knowledge of microbial allelochemicals is scattered and involves many different fields of research. In this chapter I review data on the effects of microbial allelochemicals on the behavior of animals that interact directly with microorganisms (Sections 2 to 7). Section 8 deals with microbial allelochemicals that affect the natural enemies of organisms considered in Sections 2 to 7. The involvement of microorganisms in semiochemically mediated interactions between other organisms complicates current semiochemical terminology. This aspect is considered in Section 9. Finally, in Section 10 I suggest directions for future research on microbial allelochemicals.

2   ALLELOCHEMICALS OF STORED-PRODUCT MICROORGANISMS

Stored products such as cereals or cheese may be inhabited by many microorganism and insect or mite species. The microbes live on the stored product and several insect and mite species may feed on both the stored product and the microorganisms. Sometimes it may be easier for the insects or mites to feed on grains infested with microorganisms (compared to uninfested grains) since the substrate is weakened or predigested by the microbes. That is, microorganisms infesting grains can increase the value of the grains as food source for insects (Rilett 1949). In addition, allelochemicals produced by the microorganisms may influence the behavior of insects and mites. For example, the behavior of the confused flour beetle (*Tribolium confusum*) is affected by flour compounds and by brewer's yeast (*Saccharomyces cerevisiae*) compounds (Loschiavo 1965). The beetles aggregated on a pith disk that had been dipped into extracts of flour or yeast. Compounds of yeast origin were more active than the flour compounds. Palmitic, stearic, and oleic acid from the yeast induced aggregation behavior and also acted as phagostimulants. Triglycerides of four fungus species (*Alternaria cucumerina, A. solani, Helminthosporium savitum,* and *Nigrospora sphaerica*) also elicited aggregation by *T. confusum* (Starratt and Loschiavo 1971, 1972). The intensity of the aggregation response varies with the fungal species. For high activity the triglycerides must be unsaturated (Starratt and Loschiavo 1971). Triolein was the only synthetic triglyceride of those tested that exhibited high activity. Some mono- (1-monopalmitin and 1-monostearin) or diglycerides (1,3-distearin) showed low activity. As in *S. cerevisia* (Loschiavo 1965), some fatty acids identified in *N. sphaerica* also were found to cause an aggregation response by the confused flour beetle (Starratt and Loschiavo 1971). Particularly active were stearic, oleic, elaidic, myristic, and palmitic acid and sodium oleate, identified in an extract of mycelium of *N. sphaerica*. The response to palmitic acid (Starratt and Loschiavo 1971) was much lower than that reported by Loschiavo (1965). These allelochemicals have a low volatility and thus probably affect the behavior of the beetles after contact. The beetles *Orizaephilus surinamensis* and *O. mercator* are cosmopolitan pests of stored products. Pierce et al. (1981) collected volatiles of the beetle's food [rolled oats or baker's yeast (*S. cerevisiae*)] and showed that in a pitfall olfactometer the beetles respond to volatiles of rolled oats or yeast when offered separately.

Fungi have been found to be important in the biology of *Cryptolestes ferrugineus:* mold growth, on grain used in rearing *C. ferrugineus,* reduces mortality and developmental time of the beetle (Rilett 1949), which aggregates, feeds, and develops on some species of fungi (Loschiavo and Sinha 1966). Dolinski and Loschiavo (1973) studied the distribution of the rusty grain beetle (*C. ferrugineus*) in vertical cylinders filled with wheat. Under these cylinders they placed wheat infested by one of seven fungi collected from spoiled grain. The numbers of beetles that reached the fungus-infested wheat 48 hours after their introduction on the top surface of the wheat cylinder were counted. They concluded that six of the seven fungus species attracted the beetles by volatile allelochemicals, whereas untreated wheat was inactive. However, the response of the beetles might have been caused by a nonvolatile arrestant compound.

The response of mites to volatile fungal compounds is well known to mycologists. Indeed, Dade (1960) (in Codner 1972) notes that "mites are the hated enemies of all mycologists. They are attracted by the odors of fungi and enter culture vessels bringing on their legs contaminating bacteria and moulds." Thomas and Dicke (1971) showed that food selection by the grain mite, *Acarus siro*, involves the use of volatile allelochemicals of 17 species of fungi that are associated with grain or cheese. Volatiles of eight fungus species were extracted (Thomas and Dicke 1972). The mites distinguished between the different fungus species in choice tests where two fungus species were offered (Thomas and Dicke 1971). The fungus species that were preferred the most were *Penicillium camemberti, Aspergillus flavus,* and *A. repens;* the least preferred species were *Trichoderma lignorum* and *Fusarium monoliniforme*. No active compounds were identified, but since the mites distinguished between different fungus species, specific (combinations of) compounds must be involved.

3 ALLELOCHEMICALS OF SOIL MICROORGANISMS

Soil microorganisms are involved in the decomposition of plant material, animal remains, and excretory products and thus function in nutrient cycling. The soil microflora also includes plant parasites or plant-associated fungi such as mycorrhizae or others which interact with the soil fauna. Microbial allelochemicals have been shown to mediate many of the interactions of soil fungi and soil fauna. To date, the literature on microbial allelochemicals in soil is restricted to those of fungal origin that elicit responses in nematodes and mites.

3.1 Soil Fungi–Nematode Interactions

Many nematode species respond to the allelochemicals of numerous species of soil fungi. These interactions between soil fungi and nematodes can be divided into (1) interactions between fungus-feeding nematodes and soil fungi, and (2) interactions between nematode-feeding fungi and their nematode prey. Although in both cases nematodes respond to fungal allelochemicals by attraction, the allelochemicals involved serve a totally different function in these two instances. The chemical nature of many of these allelochemicals has been elucidated. It would be interesting to know whether identical allelochemicals are involved in the two instances.

3.1.1 Interactions between Fungus-Feeding Nematodes and Fungi

Townshend (1964) showed that the mycophagous nematodes, *Aphelenchus avenae* and *Bursaphelenchus fungivorus*, can feed and reproduce on dozens of fungus species (e.g., *Fusarium oxysporum pisi, Botrytis allii, Chaetomium globosum,* and *Ophiobolis graminis*). Only a few of the fungus species tested were not suitable. He also demonstrated that these nematode species are attracted to or arrested by most of the fungus species tested. The range of attractive species is wider than the range of species on which feeding and reproduction can take place. The fungus-feeding nematodes, *Panagrellus redivivus* and *Rhabditis oxycerca*, use volatiles of baker's yeast as a kairomone (Balanova et al. 1979). Methyl, ethyl, propyl, butyl, and amyl acetate and ethyl, propyl, and amyl formate as well as ethyl propionate were identified as allelochemicals produced by *S. cerevisiae* (Balanova et al. 1979). The corresponding alcohols and acids did not elicit a response from the nematodes. The lowest concentrations of the acetates that were detected by *P. redivivus* varied from 0.0009 ppm (amyl acetate) to 1 ppm (methyl acetate) (Balan 1985). Filtrates of the culture medium of the fungi (*Gliocladium roseum, Chaetomium indicum, Rhizoctonia solani,* and *Pyrenochaeta terrestris*), which may be fed upon by the plant-parasitic nematode *Neotylenchus linfordi*, contain kairomones that function in prey location by the nematodes (Klink et al. 1970). The intensity of the response seemed to be correlated with the suitability of the particular fungus as a source of nutrition enabling nematode reproduction (Klink 1968, in Klink et al. 1970). The allelochemicals involved were thermostable molecules, soluble in methanol and unaffected by pH (Klink et al. 1970). The compounds were not ones that could serve as nutrients important to reproduction by the nematode. One of the active compounds was a yellow pigment.

3.1.2 Interactions between Nematode-Feeding Fungi and Their Nematode Prey

The response of mycophagous nematodes to fungal allelochemicals is exploited differently by various fungus species. Nematophagous fungi

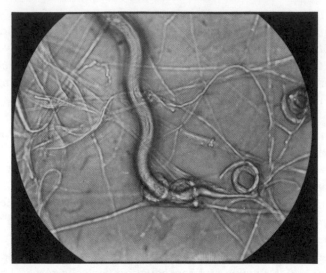

Figure 4.3   Nematode captured by the fungus, *Arthrobotrys conoides*. (Courtesy of D. Pramer.) Reprinted, with permission, from M. Alexander (1977), Introduction to Soil Microbiology. John Wiley & Sons, New York.

develop traps (Fig. 4.3) and many species have been shown to produce an allelochemical that attracts or arrests nematodes after trap formation (Field and Webster 1977). However, in some nematophagous fungi, trap formation and allelochemical production are not correlated (Jansson and Nordbring-Hertz 1979). For example, three strains of *Arthrobotrys oligospora* caused a response by the bacteria-feeding nematode, *Panagrellus redivivus,* irrespective of the presence or absence of spontaneously formed traps. Jansson and Nordbring-Hertz (1980) studied the response of 13 nematophagous fungi to five plant-parasitic nematodes. They found that the response that different fungi elicited in the nematodes tended to be stronger with increasing dependence of the fungus on nematodes for nutrients. Two of these plant-parasitic nematodes, *Ditylenchus dipsaci* and, to a lesser extent, *Pratylenchus penetrans,* showed a response to only a few of the fungus species tested. Unlike the situation in nonnematophagous fungi, the allomones of nematophagous fungi seem to be more species specific. The allelochemicals of mycelium of the fungus *Arthorbotrys musiformis,* which attract the fungus-feeding nematode *Aphelenchus avenae,* appear to be small peptides (Monoson and Ranieri 1972, Monoson et al. 1973).

The nematode *P. redivivus* was found to respond to an allomone of the predaceous fungus, *Arthrobotrys dactyloides;* one of the compounds was identified as a carbon dioxide (Balan and Gerber 1972). This general respiration product, however, cannot account for all effects or for the species specificity of the fungal allomone. Other compounds must be involved. Thin-layer chromatographic examination of culture extracts of the fungus, *Monacrosporium rutgeriensis,* suggests the presence of three different active compounds (Balan et al. 1976).

Some of the fungus-feeding nematodes (e.g., *P. redivivus*) feed on yeast, fungal mycelium, and spores of both predaceous and nonpredaceous fungi. *Panagrellus redivivus* was attracted by both types of fungi. It would be interesting to know whether the allelochemicals of these two fungal groups differ. Did predaceous fungi develop from nonpredaceous ones, and do they now benefit from the allelochemicals that attract nematodes? The allelochemicals identified to date are general compounds. However, Jansson and Nordbring-Hertz (1980) suggest that fewer nematode species respond to allelochemicals of nematophagous fungi than to compounds of nonnematophagous species. This would mean that species-specific compounds are also involved.

Predaceous fungi not only develop traps but also produce toxins that kill the nematodes that are caught. The trap-forming fungi, *A. dactyloides, Dactylaria pyriformis,* and *D. thaumasia,* also produce nematicides. This nematicidal activity is more important in killing the nematode than is strangling in the trap. The nematicidal activity of *A. dactyloides* is caused by ammonia produced by the fungus. It is not known whether the presence of ammonia affects the attractiveness of the allelochemicals, or whether ammonia production starts only after a nematode has been caught (Balan and Gerber 1972, Krizkova et al. 1976).

3.2 Soil Fungi–Mite Interactions

Soil mites also respond to microbial allelochemicals. The cheese mite, *Tyrophagus putrescentiae,* which lives in soils, uses chemicals of the cosmopolitan microfungus, *Trichothecium roseum,* as kairomones in food location (Vanhaelen et al. 1979). Several components in fungal volatiles were identified, and (*E*)-octa-1,5,-dien-3-ol and furfural elicited a positive response from the mite. However, since the concentrations of furfural needed to elicit a response were 10,000 times higher than the amounts present in 0.5 g of mycelium, Vanhaelen et al. (1979) concluded that this compound could not contribute to the attractiveness of the crude material. The response to these high furfural concentrations may imply that this compound is important to the cheese mite. Furfural in low concentrations may enhance the response to (E)-octa-1,5-diene-3-ol, but Vanhaelen et al. (1979) did not investigate this possibility. In a subsequent paper (Vanhaelen et al. 1980) the authors presented data on the more general occurrence of (E)-octa-1,5-diene-3-ol in 15 species of Homobasidiomycetes.

4 MICROBIAL ALLELOCHEMICALS ATTRACTING BACTERIVORAL PROTOZOA

Chemotaxis of the bacterial-feeding *Paramecium* sp. is mediated by a thermostable chemoattractant isolated from bacterial cultures of *Escherichia coli, E. aerogenes,* and *Bacillus subtilis* (Antipa and Martin 1980). The compound involved had a molecular weight in the range 500 to 1000. A filtrate of *E. coli* had the same effect on *Entamoeba histolytica* (Urban et al. 1983). These authors studied chemotaxis under agarose. The amebae migrated over a longer distance when a filtrate of a bacterial culture was applied at a fixed distance from the amebae compared to application of a control.

Little research has been done on the influence of microbial allelochemicals on protozoa. Although this fascinating field awaits further exploration, some interesting interactions have been reported. Cytochalasin B and colchicine have been shown to inhibit chemotaxis (Urban et al. 1983). Cytochalasin B is known to inhibit reversibly the microfilament system responsible for chemotactic migration. Colchicine affects the microtubuli system, which is important mainly for chemotactic movement.

5 ALLELOCHEMICALS OF PLANT-INFESTING MICROORGANISMS AND THEIR INFLUENCE ON THE BEHAVIOR OF MICROBIVORES

Microorganisms have an important function in the decomposition of plant material. Nutrient cycling depends for the major part on microbial activity. In degrading plant tissue, microorganisms create feeding sites for other organisms. These organisms may feed on the decomposed plant tissue, the microorganism, or its metabolic products. The behavior of many microbivores is affected by allelochemicals from their microbial food, as illustrated below. The examples of the bark beetle, *Hylastinus obscurus,* and the oribatid mite, *Pseudotrichia ardua,* illustrate feeding site selection on the basis of microbial allelochemicals. In the examples of flies of the genera *Drosophila* (fruit flies), *Delia* (onion flies), and *Fannia* and *Musca* (houseflies) microbial allelochemicals are involved in oviposition (and feeding) site selection. The leaf-cutting ants, *Atta cephalotes* and *Acromyrmex octospinosus,* represent a special relationship between insects and microorganisms. These ant species not only use microorganisms as a food source but also practice agriculture, rearing fungi in so-called fungus gardens. These ants use fungal allelochemicals in orientation toward their fungal gardens. Finally, the interrelationships between the fungi *Endothia parasitica, Ceratocystis microspora,* and *C. eucastanea,* the American chestnut, and insects attracted to fungus-infested trees illustrate the involvement of microbial allelochemicals in a system with many interdependent organisms (see the end of Section 5).

The fruit flies of the genus *Drosophila* feed, both as adults and as larvae, on yeast. Bacteria and molds have also been recovered from *Drosophila* crops, but these microorganisms seem to play a secondary role in *Drosophila* nutrition (Begon 1982). However, fruit flies have been found to feed not only on yeast cells but also on the products of yeast metabolism. These compounds stimulate larval and adult development (Parsons et al. 1979). Larvae can feed on ethanol (McKenzie and McKechnie 1979, Parsons and Spence 1981), a primary fermentation product that is toxic for many organisms. *Drosophila* species have an enzymatic detoxification system that enables them to use ethanol as a nutrient (Deltombe-Lietaert et al. 1979). The larvae also are known to use other primary alcohols (Herrewege et al. 1980) and acetic acid (Parsons 1980) as nutrients.

Adult fruit flies search for yeast either to feed on or to oviposit on. It has been known that *Drosophila* adults are attracted to yeast species by volatile fermentation products. Barrows (1907) identified fermentation products that attracted *D. ampelophila* females and Hutner et al. (1937) identified some that attracted *D. melanogaster* females. Some new microbial allelochemicals that attract females of the latter species have been identified (Fuyama 1976) (see Table 4.1). Together, these studies have elucidated 11 fermentation products that mediate oviposition site location in *Drosophila. Saccharomyces cerevisiae,* the best studied yeast species, produces many more fermentation products, 80 of which have been iden-

TABLE 4.1   Chemical Stimuli (and Concentrations) Attracting *Drosophila* Adults and/or Larval Parasites of *Drosophila*[a]

	Drosophila			Leptopilina	
				L. heterotoma	
	D. melanogaster		D. ampe-lophila	L. boulardi	
Chemical Stimulus	Hutner et al. (1937)	Fuyama (1976)	Barrows (1907)	Dicke et al. (1984)	Carton (1976)
Ethanol		+(8%)	+	+(5%)	+(0.5–10%)
Pentanol			+		
Acetic acid		+($\frac{1}{8}$%)	+	0(3%)	
Lactic acid		+(2%)	+		
Ethyl ether			+		
Diacetyl	+			$0(10^{-1}, 10^{-2}\%)$	
Acetoin	+				
Acetaldehyde	+			+(1%)	
n-Butyraldehyde		+($\frac{1}{32}$%)			
Indole	+				
Ethyl acetate		+($\frac{1}{4}$%)		$+(10^{-2}, 10^{-3}\%)$	

[a] +, Attraction; 0, neutral. Reprinted from M. Dicke et al. (1984), Chemical stimuli in host-habitat location by *Leptopilina heterotoma* (Thomson) (Hymenoptera: Eucoilidae), a parasite of *Drosophila* J. Chem. Ecol. 10:695–712, with permission of Plenum Publishing Corporation.

tified (Williams et al. 1981). Although this yeast species does not seem to be a major food source for *Drosophila* in the field, it is attractive to *Drosophila* females, and therefore screening of its identified fermentation products may yield many more compounds that mediate *Drosophila* attraction. It has been reported that one individual *Drosophila* can feed on many fungus species; however, each *Drosophila* species seems to have a special set of fungus species that is used as a food source. Thus, each species presumably prefers a different microhabitat (daCunha et al. 1957, Begon 1982). The differential use of yeast species by *Drosophila* species and the observed differential attraction (or arrestment) by these species (daCunha et al. 1957, Barker et al. 1981, Klaczko et al. 1983) may be mediated by yeast species-specific metabolites that are known for some species (e.g., Van der Walt 1970). But to date, no such species-specific microbial allelochemicals are known to mediate this species-specific attraction.

Not only does the microorganism species involved affect *Drosophila* oviposition site selection, but the substrate on which the microbe lives also has an effect. *Drosophila* females preferred to oviposit in some fruits rather than others even when the same microorganism (*S. cerevisiae*) was present in these substrates (Jaenike 1983). The latter author demonstrated that host preference may be induced or modified after contact with the same medium or with another medium (Jaenike 1982, 1983). In some instances, specific chemicals, such as ethanol (7%), were found to mediate these responses in *D. melanogaster*. This alcohol concentration deterred flies from ovipositing, but adult exposure to 7% ethanol reduced aversion to this ethanol concentration. Larval exposure did not affect adult oviposition behavior. Thus, the response to microbial allelochemicals is not fixed but is subject to modification depending on the previous experience of the flies.

Members of the Drosophilidae also inhabit other microhabitats, such as rotting beet leaves (e.g., *Drosophila fenestrarum, Scaptomyza pallida*), mushrooms (e.g., *Drosophila phalerata*), or sap streams of bleeding trees. To my knowledge, in these cases attraction by microbial allelochemicals has not been studied. However, mushrooms have been reported to produce allelochemicals that are used in oviposition site selection by another fly species, the phorid, *Megaselia halterata*. This fly is an important pest in commercial production of the mushroom, *Agaricus bisporus*. Volatiles of the mushroom have been shown to attract adult female *M. halterata* (Grove and Blight 1983). Preliminary results suggested that the microbial product oct-1-en-3-ol is one of the active compounds and that oct-1-en-3-one and octan-3-ol may be active as well.

The onion fly, *Delia antiqua* (= *Hylemya antiqua*), is a food-plant specialist that lays eggs only in a few rotting *Allium* species (e.g., onion and garlic). The larvae feed on the decaying plant material. Attempts to rear *D. antiqua* on sterile onion tissue under axenic conditions failed (Friend et al. 1959) or resulted in a 20 to 50% lower growth rate of the larvae than when reared on rotting onion tissue (Schneider et al. 1983). Recently, however, it was demonstrated that *D. antiqua* larvae can be reared on bacteria-free onion agar (Eymann and Friend 1985). These authors supplemented the agar with vitamins and amino acids and suggested that microorganisms supplemented the food components of onion with these nutrients. Thus, microorganisms that infest onion bulbs seem to be crucial to *D. antiqua* larval development. It therefore is not surprising that *D. antiqua* flies searching for oviposition sites can distinguish between decomposing and intact onion plants on the basis of allelochemicals (Dindonis and Miller 1980, 1981; Ikeshoji et al. 1980; Miller et al. 1984). Active chemicals from intact onion plants were reported to be sulfur compounds such as n-dipropyl disulfide (n-Pr$_2$S$_2$), other alkyl disulfides, and propyl mercaptan (Matsumoto and Thorsteinson 1968a,b; Pierce et al. 1978; Dindonis and Miller 1980, 1981). However, other studies have not reported such effects from these compounds (Ishikawa et al. 1981, Miller et al. 1984). Also, the host plant species involved influences the response of the flies (Hausman and Miller, in press). The flies do not respond to microorganisms isolated from onion bulbs and cultured on agar, but their metabolic products increase the attractiveness of the onion constituents (Ikeshoji et al. 1980, Dindonis and Miller 1981). A mixture of ethanol, n-butanol, 2,3-butanediol, n-butyric acid, acetyl methyl carbinol, hexanoic acid, isopropanol, and propionic acid increased the effect of n-Pr$_2$S$_2$; however, it is not clear whether all components of this mixture are needed for this (Dindonis and Miller 1981). The bacterium *Klebsiella* sp. has been isolated from decaying onions and volatiles emitted from onions inoculated with this species have been identified (Ikeshoji et al. 1980). The metabolic products ethyl acetate and tetramethyl pyrazine of this bacterium each increased the effect of n-Pr$_2$S$_2$. Ikeshoji et al. (1980) also reported that the microbial product ethyl acetate attracted (or arrested) *D. antiqua* maggots. In addition, under certain conditions the microbial products n-heptanal and tetramethyl pyrazine enhanced the moderate effect of n-Pr$_2$S$_2$. In contrast, Miller et al. (1984) also isolated a bacterium from the same genus, *Klebsiella pneumoniae,* but present data do not corroborate the data of Ikeshoji et al. (1980) on ethyl acetate and tetramethyl pyrazine.

Thus, it can be concluded that onion flies respond to allelochemicals originating from the host plant, but that they show a much stronger response to allelochemicals of a host plant infested with microorganisms. However, the data on the identity of the allelochemicals, both of the host plant and of the microbes, are ambiguous. More effort is needed to elucidate active compounds and/or their effect (singly or in combination) on onion fly behavior. In this respect it is important to note that stimuli other than semiochemicals also affect the response of the flies (Miller and Harris 1985).

Whether or not alkyldisulfides play a role in host plant location by onion flies will need further study. But even if these compounds do, microorganisms are probably also important since the production of the onion alkyldisulfides, such as n-Pr_2S_2, seems to be dependent on microorganisms. King and Coley-Smith (1969) demonstrated that several species of soil bacteria are capable of enzymatically cleaving synthetic allyl cysteine and n-propyl cysteine and their sulfoxides, the precursors of onion volatiles. If plant roots release the precursors, these soil microorganisms may produce alkyl disulfides such as n-Pr_2S_2. Thus, in this system bacteria may mediate oviposition site location of *D. antiqua* flies both directly and indirectly.

The little housefly, *Fannia canicularis,* which breeds in a variety of decaying plant and animal tissues, is attracted to microbial volatiles produced from nononion substrates, although the response of *F. canicularis* is significantly greater when n-Pr_2S_2 is offered as well (Dindonis and Miller 1981). However, traps baited with ethanol (a general fermentation product emanating from fermented carbohydrate solutions) exerted an effect on *F. canicularis* that is similar to the effect of traps baited with a fermented sucrose solution (Hwang et al. 1978). In traps baited with ethanol these authors also caught the false stable fly (*Musca stabulans*) and the housefly (*M. domestica*), but to a lesser extent.

In studies on the onion fly, the little housefly, and *Drosophila,* differential effects on males and females were observed. In traps in which females were caught, male *F. canicularis* were not caught (Dindonis and Miller 1981). Hwang et al. (1978) reported a similar sex-based differential: two-thirds of the flies caught in a stable were females. Differences between the sexes were also found in *D. antiqua,* but both sexes showed a response, although the response of males was weaker than that of females (Dindonis and Miller 1981). *Drosophila* males were also less responsive than females to microbial allelochemicals (Fuyama 1976, Barker et al. 1981). *Drosophila* adults feed on yeast and thus microbial allelochemicals may mediate food location by *Drosophila* males as well as, indirectly, mate location (by mate habitat location). In females, both location of food and location of oviposition sites are mediated by microbial allelochemicals. Presumably, in *D. antiqua* and *F. canicularis,* the response of males to microbial allelochemicals also mediates food location.

The clover root borer, *Hylastinus obscurus,* is a bark beetle that spends nearly all its life in the roots of its host plant. The beetles occur in microbially diseased clover roots, but one may question which organism is the first colonist, the microorganism or the beetle? In laboratory choice tests it was shown that adult borers preferred diseased over healthy roots. The borers did not differentiate between fungi or bacteria isolated from attractive diseased roots (that were cultured on synthetic medium) and a control of uninfested medium. However, root pieces infested with the plant-pathogenic fungi, *Colletotrichum trifolii, Fusarium roseum, F. tricinctum,* or *Rhizoctonia solani,* were preferred over healthy roots (Leath and Byers 1973). Also, the beetles preferred leachates of diseased roots over leachates of healthy roots. Thus, it would appear that host plant roots are needed to elicit a response in the clover root borer to microbial allelochemicals. These data on the beetle's response to the presence of microorganisms in clover roots and observations that root borer populations build up in the field after root rot developement (Newton and Graham 1960) suggest that the microorganisms infest the plants first. Microbial spores may enter the plants through existing cracks, but damage caused by boring may enhance inoculation.

Another organism using microbial allelochemicals to locate diseased roots is the phytophagous oribatid mite, *Pseudotritia ardua.* This mite was observed to generally occur in the soil around *Artemisia* roots but was almost absent in the root region of other plants (Führer 1961). The mite is known to feed on woody plants, although it is not a lignin feeder, and prefers diseased *Artemisia* roots over healthy ones (Führer 1961). An active compound that was shown to be of bacterial origin and an attractive *Pseudomonas* species were isolated from *Artemisia* roots (Führer 1961).

Ant behavior is characterized by the influence of many semiochemicals (see Bradshaw and Howse 1984). As mentioned earlier, some ant species rear fungi as a food source. The leaf-cutting ants, *Atta cephalotes* and *Acromyrmex octospinosus,* search for plant materials suitable for use as a substrate for the fungus culture. After chemical recognition of plant material, the ants cut pieces of leaf and transport them to their fungus garden. The importance of fungus rearing for these ant species can be demonstrated by the ability of ants to recognize leaves of host plants that are unsuitable food for the fungus. Hubbell et al. (1983) found that chemical recognition of the tree species *Hymenaea courbaril* was mediated by the plant terpenoid caryophyllene epoxide. This component, which has antifungal properties, was responsible for the rejection by *A. cephalotes* workers of this tree's foliage. Orientation toward their fungus gardens in both *A. cephalotes* and *A. octospinosus* is mediated by a volatile fungus allelochemical (Littledyke and Cherrett 1978). Unfortunately, the fungus used as an odor source had been touched by conspecific ants and thus the influence of a pheromone could not be excluded. However, since volatile semiochemicals are known to be emitted by fungi and since fungus rearing is so intricately integrated in the ant's biology, it seems very probable that fungal allelochemicals caused the response reported by Littledyke and Cherrett (1978).

An intricate interspecific interaction in which microbial allelochemicals are involved is the following, in which three fungus species, the American chestnut, and several insect species are involved (Russin et al. 1984). The fungus, *Endothia parasitica,* causes chestnut blight cankers on the American chestnut (*Castanea dentata*). This fungus is disseminated by wind or by insects. Older cankers appeared to be infested with the fungi *Ceratocystis microspora* and *C. eucastanea.* These two fungi were observed only on necrotic tissue of old blight cankers and failed to become established on living or dead tissue from 25 hardwood species tested, including the American chestnut (Russin and Shain 1984). These authors showed that growth of *Ceratocystis* was increased on agar when blighted bark extract was added. When an extract of healthy bark was used, this did not occur. They found antifungal compounds in healthy bark extract that were not detectable in blighted bark extract, and they suggested that *E. parasitica* modified inhibitory compounds in healthy bark and may have even produced metabolites that stimulate *Ceratocystis* growth. Thus, *Ceratocystis* is dependent on *Endothia* infestation of trees. In the field, traps baited with *Ceratocystis*-laden blighted chestnut bark caught more insects than did traps baited with healthy bark (Russin et al. 1984). Unfortunately, they did not use blighted bark without *Ceratocystis* species, and thus it cannot be definitely concluded whether *Ceratocystis, E. parasitica,* or both were responsible for the higher number of insects caught. Also, in a field study on the response of bark beetles toward ponderosa pines infested by *Ceratocystis wageneri,* higher numbers of beetles were trapped on infested trees compared to uninfested healthy trees (Goheen et al. 1985). However, the possibility that these results were caused by pheromones rather than by allelochemicals produced by *C. wageneri* cannot be excluded.

6 ALLELOCHEMICALS OF AQUATIC MICROORGANISMS

Aquatic microorganisms function in the decomposition of plant material. In this process they release many organic compounds in the water. Culicine and aedine mosquitoes breed in polluted water where organic nutrients are abundant because of the decomposition of plant tissue by anaerobic bacteria. These mosquitoes are attracted to, or are stimulated to oviposit in, water containing hay infusion. In many cases, involvement of microbial allelochemicals has been suggested but not demonstrated conclusively in oviposition site location by female mosquitoes (Gerhardt 1959, Gjullin et al. 1965, Ikeshoji and Mulla 1970) or semiochemicals of conspecifics (Ikeshoji and Mulla 1970). Oviposition attractant pheromones have been reported recently (e.g., Laurence et al. 1985). But in some cases microorganisms have been isolated from hay infusions, and when these microorganisms were added to sterile water it attracted (or arrested) *Culex pipiens fatigans* (Hazard et al. 1967) or stimulated oviposition by this species (Ikeshoji et al. 1967) or by *Aedes aegypti* (Roberts and Hsi 1977). The primary microorganisms were *Aerobacter aerogenes* (Hazard et al. 1967) and *Pseudomonas reptilivora* (Ikeshoji et al. 1967). Some synthetic chemicals have been found to stimulate oviposition, but these compounds have not been demonstrated to originate from the microorganisms involved (Gjullin et al. 1965, Ikeshoji et al. 1967).

7 ALLELOCHEMICALS OF INSECT-ASSOCIATED MICROORGANISMS THAT ARE USED AS PHEROMONES BY THE INSECT

Microbial allelochemicals not only function in interspecific interactions, but in some instances they are found to mediate intraspecific interactions in insects. Phenol acts as a female sex pheromone in the grass grub beetle, *Costelytra zealandica* (Henzell 1970, Henzell and Lowe 1970). This compound is known from plants but has never before been reported from insects. A bacterium was isolated from the colleterial glands (which open at the vaginal orifice) of this beetle (Hoyt et al. 1971). In a trap containing a culture of this bacterium on synthetic medium, more male *C. zealandica* were caught than in traps containing medium only or traps which were unbaited. From this bacterium a compound was isolated that had chromatographic similarities to phenol (Hoyt et al. 1971). Thus, it appears that this bacterium by itself produces a sex pheromone for the beetle. The bacterium does not seem to be obligatorily associated with *C. zealandica,* as it can be cultured on synthetic media. Presumably, the microorganism does not occur by itself in the beetle's macrohabitat, as this would mean that attracted male beetles would not benefit from responding to phenol and thus natural selection would act against phenol as a sex pheromone.

Bark beetle biology is characterized by intricate mechanisms of pheromone production (Birch 1984). Aggregation and spacing out of beetles, for instance, are mediated by volatile semiochemicals originating from both the tree and the beetles. Several microorganisms have been isolated from the hind gut of bark beetles or are known to be transmitted to the tree by the beetles. These microorganisms are found to participate in pheromone production.

The aggregation pheromones of *Ips paraconfusus* are expelled in the fecal pellets of male beetles feeding on the phloem. These substances, (Z)-verbenol, ipsenol, and ipsdienol, seem to originate from the hind gut of the beetles. Brand et al. (1975) isolated the bacterium, *Bacillus cereus,* from the beetles and showed its ability to convert α-pinene, originating from the plant into (E)- and (Z)-verbenol. Although they did not show that this conversion actually takes place *in vivo,* their data indicate that this is a distinct possibility. From *I. grandicollis* and three *Dendroctonus* species these authors also isolated a *Bacillus* species capable of converting α-pinene into the two verbenol isomers.

A symbiotic fungus (*Sporothrix* sp.) that is capable of oxidizing (E)-verbenol into verbenone was isolated from the mycangium of the southern pine beetle (*Dendroctonus frontalis*) (Brand et al. 1976). (E)-Verbenol is a component of the beetle's aggregation pheromone which is produced after infestation of the host tree. An attractive tree infested by *D. frontalis* loses its attractiveness as the number of bark beetles increases. Verbenone is thought to be involved in nullifying the effect of the beetle's aggregation pheromone (see Birch 1984).

Two other microorganism species have been isolated from *D. frontalis,* the yeasts *Hansenula holstii* and *Pichia pinus* (Brand et al. 1977). Volatile metabolites of these yeast species can greatly enhance the attractiveness of a mixture of frontalin, (E)-verbenol, and turpentine. The attractiveness of this mixture is increased when the three pheromone components are offered at concentrations with low attractiveness. The active yeast metabolites are isoamyl alcohol, isoamyl acetate, and 2-phenylethyl acetate (Brand et al. 1977).

The European elm bark beetle (*Scolytus multistriatus*) is a vector of the Dutch elm disease caused by the fungus *Ceratocystis ulmi*. The aggregation pheromone of the beetle consists of 4-methyl-3-heptanol, α-multistriatin, and α-cubebene (Gore et al. 1977). The first two of these are produced by the beetles themselves (although it has not been ascertained whether microorganisms in the gut play a role), but the third compound originates from the tree. The level of α-cubebene increases after wood is infested with *C. ulmi*. It is not known whether α-cubebene is biosynthesized by both the tree and *C. ulmi* or whether the rate at which this compound is biosynthesized by the tree increases in response to infestation with the fungus. Thus, microbial allelochemicals have been found to function as insect pheromones, but in many cases it is difficult to establish definitely the role of the microorganisms in the production of these semiochemicals.

These data on microbial allelochemicals that act as insect pheromones raise questions on the evolutionary background of this phenomenon. Microorganisms that are closely associated with insects may be favored by the fitness of the individual insect with which they are associated. Therefore it seems probable that natural selection results in the microorganism not only influencing the behavior of the individual but also that of conspecifics, if this increases the fitness of the individual insect with which the individual microorganism is associated. If, for instance, the microorganism enhances insect mate finding, this may increase microbial fitness if increased chances of mate finding for the insect improve the dissemination chances for the individual microbe or a colony of genetically related microorganisms.

8 MICROBIAL ALLELOCHEMICALS AFFECTING NATURAL ENEMIES OF MICROBE-ASSOCIATED INSECTS

In Section 5 we have seen that many insects use microbial allelochemicals to locate food or oviposition sites. Natural enemies of these microbivore species that search for food or oviposition sites need to find those sites where the microbivores or their offspring are present. It is known that natural enemies can use pheromones of their host or prey in host–prey location (Lewis et al. 1982, Wood 1982, Noldus and van Lenteren 1985). In these cases a host-prey pheromone acts as a kairomone for the natural enemy. Natural enemies may cue in on any host–prey related compound, but only the responses to those compounds that are inevitable for the host-prey are assumed to last long on an evolutionary time scale. A dual role for the host–prey pheromones as kairomone may also exist in host–

prey location by microbivore's natural enemies. In addition, these natural enemies may also use those semiochemicals that emanate from the microbivore's food, microorganism or oviposition site (i.e., compounds that have been shown to mediate the location of food or oviposition sites by the microbivores).

The woodwasp, *Sirex noctilio,* attacks dead, dying, or weakened coniferous trees. Female woodwasps drill into trees to lay eggs and they introduce a symbiotic fungus. The fungus develops in the wood, exerting a phytotoxic effect and the emerging larva feeds on the wood and fungus. Thus, the fungus may predigest wood tissue for the larva or produce complementary nutrients. Several parasite species of *S. noctilio* are known. The presence of the fungus *Amylostereum* sp. plays a role in host location by the egg-larval parasitic wasp, *Ibalia leucospoides* (Madden 1968). Odor of an *Amylostereum* sp. culture on sawdust-base medium elicited behavior in female parasites that was identical to the behavior stimulated by suitably aged host drills in logs. The parasites showed maximum response to host drills in logs or to *Amylostereum* sp. cultures when drills or cultures were 2 to 3 weeks old (temperature not mentioned). This period approximated the time needed for egg eclosion. The results suggest that volatiles of a certain fermentation stage were involved.

The parasitic wasp, *Rhyssa persuasoria,* attacks late stage larval, pupal, and rarely, adult woodwasps. Oviposition punctures of its host were not recognized by the parasite and did not stimulate drilling, but stimuli emanating from within the wood did (Spradbery 1970). Host frass elicited drilling by the parasite, but washed host larvae did not. The symbiotic fungus of the host elicited this behavior in *R. persuasoria* with maximum activity associated with 3- to 4-month-old fungal culture, a period that approximated the maturation time of host larvae. The symbiont of the woodwasp, *Sirex juvencus,* probably differs from the symbiont of other *Sirex* species and was more attractive than some common woodland fungi. Identification of the fungus species was not reported, however, but its attractiveness approximated that of an *Amylostereum* species (*A. laevigatum*). The active compound(s) could be extracted from the fungus with water or ethanol. Madden (1968) suggests that acetaldehyde might be one of the active compounds for *I. leucospoides*. However, as *R. persuasoria* females distinguished different fungus species, common fermentation products may have been involved but could not have been the only allelochemicals mediating host searching.

Many dipteran species live as juveniles in decomposing fruit. Larvae of a variety of tephritid fruit flies are attacked by the endoparasite, *Biosteres longicaudatus*. This parasite was found to use allelochemicals produced by a fungus infesting peaches (Greany et al. 1977). The fungus was identified as *Monolinia fructicola* and was also attractive when cultured on a synthetic medium. Also, an extract of a pure culture of *Penicillium digitatum* was attractive to the parasite. Greany et al. (1977) identified acetaldehyde (0.01 to 1%), ethanol (50 to 100%), and to a lesser extent, acetic acid (1 to 10%) as attractive fermentation products. Surprisingly, no attraction was found from a 10% ethanol solution, a more natural concentration in fermenting substrates, whereas solutions of as high as 50% and 100% were attractive. Apparently, the parasites were not repelled by these high ethanol concentrations. An attractive extract of *P. digitatum* contained only 0.37% ethanol. Therefore, other allelochemicals such as acetaldehyde must have been involved synergistically. Nothing is known on orientation of the tephritid flies themselves to fermentation products. These flies oviposit in fresh fruits.

Drosphila flies use several fungal allelochemicals in oviposition site location (see Section 5). These fungal allelochemicals also mediate interactions between *Drosphila* and their parasites (Carton 1976, Dicke et al. 1984, Vet 1984). In two cases some active compounds have been identified. Female *Leptopilina boulardi* (= *Cothonaspis* sp.) use ethanol (0.5 to 10%) in host searching (Carton 1976) (Table 4.1). This parasite was repelled when offered ethanol concentrations exceeding about 20% [compare these results with those of Greany et al. (1977) mentioned above]. The larval parasite, *Leptopilina heterotoma* (Figure 4.4), does not use volatile host-related allelochemicals but only volatiles orginating from yeast (in this case *S. cerevisiae*) (Dicke et al. 1984). They showed that ethanol (5%), acetaldehyde (1%), and ethyl acetate (0.01 and 0.001%) attracted the parasite, whereas acetic acid (3%) and diacetyl (0.1 and 0.01%) did not (Table 4.1). These concentrations approximated those emitted by a yeast culture. Thus, in oviposition site location these two *Leptopilina* species used some fungal allelochemicals that were also used by their hosts, ethanol and acetaldehyde (the latter only in the case of *L. heterotoma*), However, some compounds (i.e., acetic acid and diacetyl) that individually attracted its host did not attract *L. heterotoma* (see Table 4.1). Dicke et al. (1984) showed that a combination of ethanol, ethyl acetate, and acetaldehyde was less attractive than yeast to *L. heterotoma* and thus more fermentation products were probably involved in host habitat location by this parasite (Dicke et al. 1984). *Leptopilina hetero-*

Figure 4.4   Parasitic wasp, *Leptopilina heterotoma,* on *Drosophila melanogaster* Larvae in a yeast suspension on agar. (Courtesy of J. C. van Lenteren.)

toma parasitizes hosts in several different microhabitats (e.g., fermenting fruits or rotting plant material). In host habitat location, *L. heterotoma* distinguished between odors of different fermentation stages of *S. cerevisiae*. The odor of a 6-day-old *S. cerevisiae*–sugar–agar medium was preferred over the odor of a 1- or 3-day-old medium (Vet and Van Opzeeland 1985). It is unknown whether this was correlated to the presence of preferred host species or host stages in media of different ages.

The response of parasites to microbial allelochemicals is not unalterable. The response of *L. heterotoma* toward the odor of potential host habitats is subject to adult conditioning (Vet and Van Opzeeland 1985). Oviposition experience on *D. melanogaster* affected the preference to odor of decaying beet leaves when a fresh yeast suspension was the alternative. Inexperienced females preferred the odor of decaying beet leaves, whereas females that had oviposited in *D. melanogaster* (in fresh yeast suspension for 24 hours before the experiment) preferred fresh yeast odor (Vet and Van Opzeeland 1985).

The responses of parasites to odors of different host habitats are summarized in Table 4.2 (see also Vet 1984). Host habitats consisted of decaying mushrooms, decaying beet leaves, or pure cultures of the yeast *S. cerevisiae*. As with *L. heterotoma,* oviposition experience affects host habitat location by the parasite *Leptopilina clavipes* (Vet 1983). This parasite attacks *Drosophila phalerata* in decaying mushrooms but can also

TABLE 4.2 Attraction of Parasitic Hymenoptera to the Microhabitat of Dipteran Hosts

Parasite Species	Attracted by:	Dipteran Host Species	References
Leptopilina clavipes	Decaying mushroom	*Drosophila phalerata*	Vet et al. (1983), Vet (1983)
	Saccharomyces cerevisiae	Several frugivorous *Drosophila* species	Vet (1983), Vet et al. (1983)
L. fimbriata	Decaying beet leaves	*Scaptomyza pallida*	Vet et al. (1983), (1984)
L. heterotoma	*S. cerevisiae*	Several frugivorous *Drosophila* species	Dicke et al. (1984), Vet and Van Opzeeland (1984)
	Decaying beet leaves	*S. pallida*	Vet and Van Opzeeland (1985)
L. boulardi	*S. cerevisiae*	Several frugivorous *Drosophila* species	Carton (1976), Vet (1985a)
Asobara tabida	*S. cerevisiae*	Several frugivorous *Drosophila* species	Vet et al. (1983, 1984)
A. rufescens	Decaying beet leaves	*S. pallida*	Vet et al. (1984)
A. gahani	*S. cerevisiae*	Several frugivorous *Drosophila* species	Vet and Van Opzeeland (1984)
A. spec. nov. C.	*S. cerevisiae*	Several frugivorous *Drosophila* species	Vet and Van Opzeeland (1984)
Kleidotoma dolichocera	Decaying beet leaves	*S. pallida*	Vet (1985a)
Phaenocarpa canaliculata	Decaying mushroom	*Fannia canicularis*	Vet (1985a)
Aphaereta minuta	Decaying meat	*Phormia regina*	Vet (1985a)
	S. cerevisiae	*Drosophila hydei*	Vet (1985a)

attack this host species in fermenting fruits. Female parasites in a four-arm airflow olfactometer were attracted by the odor of decaying mushroom, whereas the odor of fresh mushroom was slightly repellent (Vet 1983). Inexperienced parasites were attracted by *S. cerevisiae* odor when the alternative was humidified air. However, when decaying mushroom was offered as an alternative to *S. cerevisiae*, the parasites preferred the mushroom odor. When reared on hosts in yeast, the response to yeast odor is increased. This kind of conditioning through rearing, however, did not alter the preference for mushroom odor over yeast odor. Conditioning through oviposition experience was much stronger. After females had oviposited in hosts in a yeast suspension they significantly preferred the odor of yeast over that of mushrooms. This shows that microbial allelochemicals that are used in host habitat location by *L. clavipes* act in a complex way, since their influence is subject to learning.

Of all the parasite species of Drosophilidae studied by Vet and coworkers (see Vet 1984), only two (*Asobara tabida* and *A. rufescens*) distinguished between a host habitat with host larvae and one without on the basis of volatile allelochemicals (Vet and Van Opzeeland 1984). In these two species the response to a volatile host kairomone was found to be induced by oviposition experience on the host species involved (i.e., *Drosophila melanogaster* for *A. tabida* and *Scaptomyza pallida* for *A rufescens*). *Asobara tabida* females, however, showed a response to the volatile kairomone of *S. pallida* after oviposition experience on *D. melanogaster*. In all cases the host kairomone and the microbial allelochemical were present as a mixture. It would be interesting to know whether the microbial allelochemical is needed for a response to the volatile host-related kairomone. Whether or not this is the case, microbial allelochemicals, in general, are more important than host-related kairomones in host searching by the *Drosophila* parasites studied, before the parasites contact the host microhabitat.

Host-related kairomones do play an important role in host location after the parasite has arrived in the host habitat. Nonvolatile kairomones of Drosophilidae have been reported to elicit a response from many parasitic wasps (Dicke et al. 1985, Vet 1985b). Parasites spent more time on a patch with kairomone than on one without. *Leptopilina heterotoma* only responded to the nonvolatile kairomone of *D. melanogaster* when the kairomone was presented together with yeast (Dicke et al. 1985). Here again a microbial allelochemical may be important in host searching by activating the parasites and making them perceive the nonvolatile drosophilid kairomones.

It is known that bark beetle pheromones can act as kairomones for predators, or as allomones for other bark beetles species [see Wood (1982) for a review]. Microbial allelochemicals that act as a pheromone in bark beetle biology may thus also affect the behavior of other bark beetle species or predators of these insects. For instance, in *Ips paraconfusus,* (*Z*)-verbenol may be produced by the bacterium, *Bacillus cereus* (Brand et al. 1975). This bark beetle pheromone of possible microbial origin acts as a kairomone for the predator, *Enoclerus lecontei.* As the origin of bark beetle pheromones (*sensu lato*) is more clearly elucidated, other microbial allelochemicals may be shown to mediate interspecific insect behavior as well.

9   CONSEQUENCES FOR SEMIOCHEMICAL TERMINOLOGY

Current semiochemical terminology has been based on two criteria: the origin (producer–acquirer) and cost–benefit analysis of the compounds (cf. Nordlund and Lewis 1976). In this chapter we have seen that these two criteria may conflict. For this reason Dicke and Sabelis (1988) proposed to eliminate the origin criterion from the current terminology. Because their proposal is based for a major part on involvement of microbial semiochemicals in interactions between other organisms, their reasoning is given briefly in this section.

As indicated in this chapter, there may be interactions mediated by semiochemicals in which neither of the two interacting organisms is the actual producer of the compound. From an evolutionary point of view, the mere fact that an interaction is mediated by a semiochemical urges analyzing the cost–benefit aspects for both interactants. In that context, the exact origin of the compound is not important as long as the semiochemical under consideration is pertinent to the biology of the source-related interactant. This is illustrated by two examples considered in this chapter: (1) Phenol, produced by a bacterium in the colleterial gland of *C. zealandica,* is a chemical important in the biology of the beetle (Henzell 1970, Henzell and Lowe 1970, Hoyt et al. 1971) and has been classified as a pheromone in intraspecific beetle–beetle interactions. However, according to the terminology of Nordlund and Lewis (1976), phenol should be classified as an allelochemical (producer and receiver belong to different species). (2) *Drosophila* larvae are utterly dependent on infestation of their feeding substrate with microorganisms. According to the origin criterion of Nordlund and Lewis (1976), the microbial products that attract the parasitic wasp, *L. heterotoma* (Dicke et al. 1984), can only be regarded in the interaction between the microorganism (producer) and the wasp (receiver). Subsequent use of the cost–benefit criterion may then classify the compounds as allomone, kairomone, or synomone in that interaction. However, apart from the interaction between the microorganism and the parasitic wasp, the fermentation products, which are pertinent to the biology of the fruit flies, also mediate the interaction between *Drosophila* and the wasp, even though none of these is the producer–acquirer of the compounds. In this interaction the yeast metabolites ethanol, ethyl acetate, and acetaldehyde act as kairomones.

Our knowledge of the involvement of microorganisms in interactions between other organisms is increasing (e.g., Jones 1984). Therefore, more cases will be discovered where microbial semiochemicals mediate such interactions. As a result, application of the origin criterion will increasingly complicate classification because of the role of species of the intermediate microbial trophic level. On the other hand, the cost–benefit criterion is directly related to the evolutionary context of the interaction being mediated by the semiochemical under consideration.

No cost–benefit classification criterion has existed in the pheromone definition [see Dicke and Sabelis (1988) for a discussion]. In pheromone terminology costs and benefits have been viewed on the level of the species. However, this approach violates what is known of natural selection, which does not act on this level but on the level of the individual (Alcock 1982, Dicke and Sabelis 1988). Recent research data and discussions on semiochemical terminology have led Dicke and Sabelis (1988) to advocate the elimination of the origin criterion and to suggest that terms be based solely on the costs and benefits on the level of the individual.

10   PROSPECTS FOR FUTURE RESEARCH

Microorganisms have a unique position in the research field of semiochemicals. Because of their specific characteristics, microorganisms are very well suited for studying certain aspects of semiochemicals that are difficult to study in other organisms. Semiochemicals convey information

TABLE 4.3   Microbial Allelochemicals Whose Chemical Structure Has Been Elucidated

Producing Microorganism	Responding Organism	Allelochemicals	Reference
Arthrobotrys dactyloides	*Panagrellus redivivus* (Rhabditida: Cephalobidae)	Carbon dioxide	Balan and Gerber (1972)
Hansenula holstii, Pichia pinus, P. bovis	*Dendroctonus frontalis* (Coleoptera: Scolytidae)	Isoamyl acetate, 2-phenethyl acetate	Brand et al. (1977)
Klebsiella sp.	*Delia antiqua* (Diptera: Anthomyiidae)	Ethyl acetate, tetramethyl pyrazine	Ikeshoji et al. (1980)
Monolinia fructicola	*Biosteres longicaudatus* (Hymenoptera: Braconidae)	Ethanol, acetaldehyde, acetic acid	Greany et al. (1977)
Nigrospora sphaerica	*Tribolium confusum* (Coleoptera: Tenebrionidae)	Triolein, stearic acid, oleic acid, elaidic acid, myristic acid, palmitic acid, sodium oleate	Starratt and Loschiavo (1971)
Saccharomyces cerevisiae	*Tribolium confusum* (Coleoptera: Tenebrionidae)	Palmitic acid, stearic acid, oleic acid, gluten, gliadin, glutenin	Loschiavo (1965)
S. cerevisiae	*Panagrellus redivivus* (Rhabditida: Cephalobidae) *Rhabditis oxycerca* (Rhabditida: Rhabditidae)	Methyl acetate, ethyl acetate, propyl acetate, butyl acetate, amyl acetate, ethyl formate, propyl formate, amyl formate, ethyl propionate	Balanova et al. (1979)
S. cerevisiae	*Leptopilina heterotoma* (Hymenoptera: Eucoilidae)	Ethanol, acetaldehyde, ethyl acetate	Dicke et al. (1984)
S. cerevisiae	*Fannia canicularis* (Diptera: Muscidae) *Musca stabulans* (Diptera: Muscidae) *Musca domestica* (Diptera: Muscidae)	Ethanol	Hwang et al. (1978)

TABLE 4.3 (*Continued*)

Producing Microorganism	Responding Organism	Allelochemicals	Reference
Trichothecium roseum	*Tyrophagus putrescentiae* (Acarina: Acaridae)	(*E*)-octa-1, 5-dien-3-ol	Vanhaelen et al. (1979)
?[a]	*Drosophila melanogaster* (Diptera: Drosophilidae)	Diacetyl, acetoin, acetaldehyde, indole	Hutner et al. (1937)
		Ethanol, acetic acid, lactic acid, *n*-butyraldehyde	Fuyama (1976)
?[a]	*Drosophila ampelophila* (Diptera: Drosophilidae)	Ethanol, acetic acid, lactic acid, diethyl ether	Barrows (1907)
?[a]	*Leptopilina boulardi* (Hymenoptera: Eucoilidae)	Ethanol	Carton (1976)

[a] General fermentation products studied, rather than any particular microorganism.

in many interactions. However, the production of such chemicals is usually meant for a different purpose. The possibility that other organisms take advantage of such chemicals is interesting from an evolutionary point of view. Very little is known about actual selection pressure on semiochemical production. Genetic studies of microorganisms with regard to mutation frequency, survival chances of mutants, and so on, may be fruitful because of the detailed knowledge of the genetics of many microorganisms as well as knowledge of the structure of many microbial allelochemicals (Table 4.3). The outcome of such studies will be an enormous stimulus for developing ideas on the role of natural selection on semiochemical production in general.

On the other hand, microorganisms interacting with other organisms through allelochemicals generally are poorly understood ecologically. In many instances the cost–benefit aspects of the allelochemical with respect to the microorganisms are unknown [see Nordlund (1981) and Dicke and Sabelis (1988) for discussion]. For instance, what costs and benefits for the microorganism are associated with the volatile fungal allelochemicals in the interaction between the microorganism and the larval parasites of microbivoral Diptera (Carton 1976, Greany et al. 1977, Dicke et al.

1984)? A lack of ecological knowledge of microorganisms in these interactions hinders our understanding of selection pressures. Ecological studies on the relationships of microorganisms and other organisms (e.g., Jones 1984, Starmer and Fogleman 1986) will complement the studies on the genetics of microorganisms with respect to understanding the selection forces acting on microbial production of semiochemicals.

To date, the research on microbial allelochemicals has been scattered over many research fields without much coordination (see Sections 2 to 8). The present review indicates that the knowledge that has been obtained so far may be only the tip of an immense iceberg. The field of microbial allelochemicals is likely to expand enormously, both horizontally (many more organisms) and vertically (more detailed study of the phenomena observed).

So far, it seems that data have been collected haphazardly. However, our knowledge of microbial allelochemicals will increase dramatically if future research coordinates not only issues specific to microbial allelochemicals (e.g., elucidation of origin, function served in the microorganism, evolutionary implications, etc.) but also issues related to semiochemicals in general (e.g., mutation frequency and viability of mutants that do not produce certain semiochemicals). The results of such research will help to structure the research field of semiochemicals. For research on the problems related to semiochemicals in general, an interdisciplinary model system could be established. The yeast species *S. cerevisiae* would be a good candidate organism because much is already known about this organism from different fields of research.

ACKNOWLEDGMENTS

I thank J. C. Van Lenteren, M. W. Sabelis, and L. E. M. Vet for comments on the manuscript, M. O. Harris for discussions, and J. R. Miller for sending preprints of his work. J. Burrough-Boenisch improved the English text and F. J. J. von Planta and P. J. Kostense prepared the figures.

REFERENCES

Albone, E. S. 1984. Mammalian Semiochemistry: The Investigation of Chemical Signals between Mammals. Wiley, Chichester, West Sussex, England.

Alcock, A. 1982. Natural selection and communication among bark beetles. Fla. Ent. 65:17–32.

Antipa, G. A., and K. Martin. 1980. Chemotaxis in the ciliated protozoa. Amer. Zool. 20:798.

Balan, J. 1985. Measuring minimal concentrations of attractants detected by the nematode *Panagrellus redivivus*. J. Chem. Ecol. 11:105–111.

Balan, J., and N. N. Gerber. 1972. Attraction and killing of the nematode *Panagrellus redivivus* by the predaceous fungus *Arthrobotrys dactyloides*. Nematologica 18:163–173.

Balan, J., L. Krizkova, P. Nemec, and A. Kolozsvary. 1976. A qualitative method for detection of nematode attracting substances and proof of production of three different attractants by the fungus *Monacrosporium rutgeriensis*. Nematologica 22:306–311.

Balanova, J., J. Balan, L. Krizkova, P. Nemec, and D. Bobok. 1979. Attraction of nematodes to metabolites of yeasts and fungi. J. Chem. Ecol. 5:909–918.

Barker, J. S. F., G. J. Parker, G. L. Toll, and P. R. Widders. 1981. Attraction of *Drosophila buzzatii* and *D. aldrichi* to species of yeasts isolated from their natural environment. I. Laboratory experiments. Aust. J. Biol. Sci. 34:593–612.

Barrows, W. M. 1907. The reactions of the pomace fly, *Drosophila ampelophila* Loew, to odorous substances. J. Exp. Zool. 4:515–539.

Begon, M. 1982. Yeasts and *Drosophila*. In M. Ashburner, H. L. Carson, and J. N. Thompson, Jr. (eds.), The Genetics and Biology of *Drosophila*, Vol 3B. pp. 345–384. Academic Press, New York.

Bell, W. J., and R. T. Cardé (eds.). 1984. Chemical Ecology of Insects. Chapman & Hall, London.

Birch, M. C. 1984. Aggregation in bark beetles. In W. J. Bell and R. T. Cardé (eds.), Chemical Ecology of Insects. pp. 331–353. Chapman & Hall, London.

Bradshaw, J. W. S., and P. E. Howse. 1984. Sociochemicals of ants. In W. J. Bell and R. T. Cardé (eds.), Chemical Ecology of Insects. pp. 429–473. Chapman & Hall, London.

Brand, J. M., J. W. Bracke, A. J. Markovetz, D. L. Wood, and L. E. Browne. 1975. Production of verbenol pheromone by a bacterium isolated from bark beetles. Nature 254:136–137.

Brand, J. M., J. W. Bracke, L. N. Britton, A. J. Markovetz, and S. J. Barras. 1976. Bark beetle pheromones: production of verbenone by a mycangial fungus of *Dendroctonus frontalis*. J. Chem. Ecol. 2:195–199.

Brand, J. M., J. Schultz, S. J. Barras, L. J. Edson, T. L. Payne, and R. L. Hedden. 1977. Bark beetle pheromones. Enhancement of *Dendroctonus frontalis* (Coleoptera. Scolytidae) aggregation pheromone by yeast metabolites in laboratory bioassays. J. Chem. Ecol. 3:657–666.

Carton, Y. 1976. Attraction de *Cothonaspis* sp. (Hymenoptère: Cynipide) par le milieu trophique de son hôte: *Drosophila melanogaster*. Colloq. Int. C.N.R.S. 265:285–303.

Codner, R. C. 1972. Preservation of fungal cultures and the control of mycophagous mites. In D. A. Shapton and R. G. Board (eds.), Safety in Microbiology. Society for Applied Bacteriology Technical Series 6. pp. 213–227. Academic Press, London.

daCunha, A. B., A. M. El-Tabey Shehata, and W. Oliviera. 1957. A study of the diets and nutritional preferences of tropical species of *Drosophila*. Ecology 38:98–106.

Deltombe-Lietaert, M. C., J. Delcour, N. Lenelle Montfort, and A. Elens. 1979. Ethanol metabolism in *Drosophila melanogaster*. Experienta 35:579–581.

Dicke, M., and M. W. Sabelis. 1988. Infochemical terminology: based on cost-benefit analysis rather than origin of compounds? Funct. Ecol. 2 (in press).

Dicke, M., J. C. van Lenteren, G. J. F. Boskamp, and E. van Dongen-Van Leeuwen. 1984. Chemical stimuli in host-habitat location by *Leptopilina heterotoma* (Thomson) (Hymenoptera: Eucoilidae), a parasite of *Drosophila*. J. Chem. Ecol. 10:695–712.

Dicke, M., J. C. van Lenteren, G. J. F. Boskamp, and R. van Voorst. 1985. Intensification and prolongation of host searching in *Leptopilina heterotoma* (Thomson) (Hymenoptera: Eucoilidae) through a kairomone produced by *Drosophila melanogaster*. J. Chem. Ecol. 11:125–136.

Dindonis, L. L., and J. R. Miller. 1980. Host finding responses of onion and seedcorn flies to healthy and decomposing onions and several synthetic constituents of onion. Environ. Ent. 9:467–472.

Dindonis, L. L., and J. R. Miller. 1981. Onion fly and little house fly host finding selectively mediated by decomposing onion and microbial volatiles. J. Chem. Ecol. 7:419–426.

Dolinski, M. G., and S. R. Loschiavo. 1973. The effect of fungi and moisture on the locomotory behavior of the rusty grain beetle, *Cryptolestes ferrugineus* (Coleoptera: Cucujidae). Can. Ent. 105:485–490.

Eymann, M., and W. G. Friend. 1985. Development of onion maggots (Diptera: Anthomyiidae) on bacteria-free onion agar supplemented with vitamins and amino acids. Ann. Ent. Soc. Amer. 78:182–185.

Field, J. I., and J. Webster. 1977. Traps of predacious fungi attract nematodes. Trans. Br. Mycol. Soc. 68:467–469.

Friend, W. G., E. H. Salkeld, and I. L. Stevenson. 1959. Nutrition of onion maggots, larvae of *Hylemya antiqua* (Meig.), with reference to other members of the genus *Hylemya*. Ann. N. Y. Acad. Sci. 77:384–393.

Führer, E. 1961. Der Einfluss von Pflanzenwurzeln auf die Verteilung der Kleinarthropoden im Boden, untersucht an *Pseudotritia ardua* (Oribatei). Pedobiologia 1:99–112.

Fuyama, Y. 1976. Behaviour genetics of olfactory responses in *Drosophila*. I. Olfactory and strain differences in *Drosophila melanogaster*. Behav. Genet. 6:407–420.

Gerhardt, R. W. 1959. The influence of soil fermentation on oviposition site selection by mosquitoes. Mosq. News 19:151–155.

Gjullin, C. M., J. O. Johnsen, and F. W. Plapp, Jr. 1965. The effect of odors released by various waters on the oviposition sites selected by two species of *Culex*. Mosq. News 25:268–271.

Goheen, D. J., F. W. Cobb, Jr., D. L. Wood, and D. L. Rowney. 1985. Visitation frequencies of some insect species on *Ceratocystis wageneri* infected and apparently healthy ponderosa pines. Can. Ent. 117:1535–1543.

Gore, W. E., G. T. Pearce, G. N. Lanier, J. B. Simeone, R. M. Silverstein, J. W. Peacock, and R. A. Cuthbert. 1977. Aggregation attractant of the European elm bark beetle, *Scolytus multistriatus*. Production of individual components and related aggregation behaviour. J. Chem. Ecol. 3:429–446.

Greany, P. D., J. H. Tumlinson, D. L. Chambers, and G. M. Boush. 1977. Chemically mediated host finding by *Biosteres* (*Opius*) *longicaudatus,* a parasitoid of tephritid fly larvae. J. Chem. Ecol. 3:189–195.

Grove, J. F., and M. M. Blight. 1983. The oviposition attractant for the mushroom phorid *Megaselia halterata:* the identification of volatiles present in mushroom house air. J. Sci. Food Agric. 34:181–185.

Hausman, S. M., and J. R. Miller. Production of onion fly attractants and ovipositional stimulants by bacterial isolates cultured on onion. J. Chem. Ecol. (in press).

Hazard, E. I., M. S. Mayer, and K. E. Savage. 1967. Attraction and oviposition stimulation of gravid female mosquitoes by bacteria isolated from hay infusions. Mosq. News 27:133–136.

Henzell, R. F. 1970. Phenol, an attractant for the male grass grub beetle (*Costelytra zealandica* (White)) (Scarabeidae: Coleoptera). N. Z. J. Agric. Res. 13:294–296.

Henzell, R. F., and M. D. Lowe. 1970. Sex attractant of the grass grub beetle. Science 168:1005–1006.

Hoyt, C. P., G. O. Osborne, and A. P. Mulcock. 1971. Production of an insect sex attractant by symbiotic bacteria. Nature 230:472–473.

Hubbell, S. P., D. F. Wiemer, and A. Adejare. 1983. An antifungal terpenoid defends a neotropical tree (*Hymenaea*) against attack by fungus growing ants (*Atta*). Oecologia (Berlin) 60:321–327.

Hutner, S. H., H. M. Kaplan, and E. V. Enzmann. 1937. Chemicals attracting *Drosophila*. Amer. Nat. 71:575–581.

Hwang, Y. S., M. S. Mulla, and H. Axelrod. 1978. Attractants for synanthropic flies. Ethanol as attractant for *Fannia canicularis* and other pest flies in poultry ranches. J. Chem. Ecol. 4:463–470.

Ikeshoji, T., and M. S. Mulla. 1970. Oviposition attractants for four species of mosquitoes in natural breeding waters. Ann. Ent. Soc. Amer. 63:1322–1327.

Ikeshoji, T., T. Umino, and S. Hirakoso. 1967. Studies on mosquito attractants and stimulants. IV. An agent producing stimulative effect for oviposition of *Culex pipiens fatigans* in field water and the stimulative effects of various chemicals. Jpn. J. Exp. Med. 37:61–69.

Ikeshoji, T., Y. Ishikawa, and Y. Matsumoto. 1980. Attractants against the onion maggots and flies, *Hylemya antiqua*, in onions inoculated with bacteria. J. Pestic. Sci. 5:343–350.

Ishikawa, Y., T. Ikeshoji, Y. Matsumoto, M. Tsutsumi, and Y. Mitsui. 1981. Field trapping of the onion and seed-corn fly with baits of fresh and aged onion pulp. Appl. Ent. Zool. 16:490–493.

Jaenike, J. 1982. Environmental modification of oviposition behavior in *Drosophila*. Amer. Nat. 119:784–802.

Jaenike, J. 1983. Induction of host preference in *Drosophila melanogaster*. Oecologia (Berlin) 58:320–325.

Jansson, H.-B., and B. Nordbring-Hertz. 1979. Attraction of nematodes to living mycelium of nematophagous fungi. J. Gen. Microbiol. 112:89–93.

Jansson, H.-B., and B. Nordbring-Hertz. 1980. Interactions between nematophagous fungi and plant-parasitic nematodes: attraction, induction of trap formation and capture. Nematologica 26:383-389.

Jones, C. G. 1984. Microorganisms as mediators of plant resource exploitation by insect herbivores. In P. W. Price, C. N. Slobodchikoff, and W. S. Gaud (eds.), A New Ecology: Novel Approaches to Interactive Systems. pp. 53–99. Wiley, New York.

Karlson, P., and M. Lüscher. 1959. "Pheromones" a new term for a class of biologically active substances. Nature 183:155–156.

Kennedy, J. S. 1978. The concepts of olfactory 'arrestment' and 'attraction.' Physiol. Ent. 3:91–98.

King, J. E., and J. R. Coley-Smith. 1969. Production of volatile alkyl sulphides by microbial degradation of synthetic alliin and alliin-like compounds, in relation to germination of sclerotia of *Sclerotium ceptivorum* Berk. Ann. Appl. Biol. 64:303–314.

Klaczko, L. B., J. R. Powell, and C. E. Taylor. 1983. *Drosophila* baits and yeasts: species attracted. Oecologia (Berlin) 59:411–413.

Klink, J. W., V. H. Dropkin, and J. E. Mitchell. 1970. Studies on the host-finding mechanisms of *Neotylenchus linfordi*. J. Nematol. 2:106–117.

Krizkova, L., J. Balan, P. Nemec, and A. Kolozsvary. 1976. Predacious fungi *Dactylaria pyriformis* and *Dactylaria thaumasia:* production of attractants and nematicides. Folia Microbiol. 21:493–494.

Laing, J. 1937. Host finding by insect parasites. I. Observations on finding of hosts by *Alysia manducator, Mormoniella vitripennis* and *Trichogramma evanescens*. J. Anim. Ecol. 6:298–317.

Laurence, B. R., K. Mori, T. Otsuka, J. A. Pickett, and L. J. Wadhams. 1985. Absolute configuration of mosquito oviposition attractant pheromone, 6-acetoxy-5-hexadecanolide. J. Chem. Ecol. 11:643–648.

Law, J. H., and F. E. Regnier. 1971. Pheromones. Annu. Rev. Biochem. 40:533–548.

Leath, K. T., and R. A. Byers. 1973. Attractiveness of diseased red clover roots to the clover root borer. Phytopathology 63:428–431.

Lewis, W. J., D. A. Nordlund, R. C. Gueldner, P. E. A. Teal, and J. H. Tumlinson. 1982. Kairomones and their use for management of entomophagous insects. XIII. Kairomonal activity for *Trichogramma* spp. of abdominal tips, excretion and a synthetic sex pheromone blend of *Heliothis zea* (Boddie) moths. J. Chem. Ecol. 8:1323–1331.

Littledyke, M., and J. M. Cherrett. 1978. Olfactory responses of the leaf-cutting ants *Atta cephalotes* (L.) and *Acromyrmex octospinosus* (Reich) (Hymenoptera: Formicidae) in the laboratory. Bull. Ent. Res. 68:273–282.

Loschiavo, S. R. 1965. The chemosensory influence of some extracts of brewer's yeast and cereal products on the feeding behavior of the confused flour beetle, *Tribolium confusum* (Coleoptera: Tenebrionidae). Ann. Ent. Soc. Amer. 58:576–588.

Loschiavo, S. R., and R. N. Sinha. 1966. Feeding, oviposition, and aggregation by the rusty grain beetle, *Cryptolestes ferrugineus* (Coleoptera: Cucujidae) on seed-borne fungi. Ann. Ent. Soc. Amer. 59:578–585.

Madden, J. L. 1968. Behavioural responses of parasites to the symbiotic fungus associated with *Sirex noctilio* F. Nature 218:189–190.

Matsumoto, Y., and A. J. Thorsteinson. 1968a. Effect of organic sulphur compounds on oviposition in onion maggot, *Hylemya antiqua* Meigen (Diptera: Anthomyiidae). Appl. Ent. Zool. 3:5–12.

Matsumoto, Y., and A. J. Thorsteinson. 1968b. Olfactory response of larvae of the onion maggot, *Hylemya antiqua* (Meigen) (Diptera: Anthomyiidae) to organic sulphur compounds. Appl. Ent. Zool. 3:107–111.

McKenzie, J. A., and S. W. McKechnie. 1979. A comparative study of resource utilization in natural populations of *Drosophila melanogaster* and *D. simulans*. Oecologia (Berlin) 40:299–309.

Miller, J. R., and M. O. Harris. 1985. Viewing behavior-modifying chemicals in the context of behavior: lessons from the onion fly. In T. E. Acree and D. M. Soderlund (eds.), Semiochemistry: Flavors and Pheromones. pp. 3–31. Walter de Gruyter, West Berlin.

Miller, J. R., M. O. Harris, and J. A. Breznak. 1984. Search for potent attractants of onion flies. J. Chem. Ecol. 10:1477–1488.

Monoson, H. L., and G. M. Ranieri. 1972. Nematode attraction by an extract of a predacious fungus. Mycologia 65:628–631.

Monoson, H. L., A. G. Galsky, J. A. Griffin, and E. J. Mcgrath. 1973. Evidence for and partial characterization of a nematode attraction substance. Mycologia 65:78–86.

Newton, R. C., and J. H. Graham. 1960. Incidence of root-feeding weevils, root rot internal breakdown, and virus and their effect on longevity of red clover. J. Econ. Ent. 53:865–867.

Noldus, L. P. J. J., and J. C. van Lenteren. 1985. Kairomones for the egg parasite *Trichogramma evanescens* Westwood. I. Effect of volatile substances released by two of its hosts, *Pieris brassicae* L. and *Mamestra brassicae* L. J. Chem. Ecol. 11:783–791.

Nordlund, D. A., and W. J. Lewis. 1976. Terminology of chemical releasing stimuli in intraspecific and interspecific interactions. J. Chem. Ecol. 2:211–220.

Nordlund, D. A., W. J. Lewis, and R. L. Jones. 1981. Semiochemicals: Their Role in Pest Control. Wiley, New York.

Parsons, P. A. 1980. Acetic acid vapour as resource: threshold difference among three *Drosophila* species. Experientia 36:1363.

Parsons, P. A., and G. E. Spence. 1981. Ethanol utilization threshold differences among three *Drosophila* species. Amer. Nat. 117:568–571.

Parsons, P. A., S. M. Stanley, and G. E. Spence. 1979. Environmental ethanol at low concentrations: longevity and development in the sibling species *Drosophila melanogaster* and *D. simulans*. Aust. J. Zool. 27:747–754.

Pierce, H. D., Jr., R. S. Vernon, J. H. Borden, and A. C. Oehlschlager. 1978. Host selection by *Hylemya antiqua* (Meigen). Identification of three new attractants and oviposition stimulants. J. Chem. Ecol. 4:65–72.

Pierce, A. M., J. H. Borden, and A. C. Oehlschlager. 1981. Olfactory response to beetle-produced volatiles and host-food attractants by *Orizaephilus surinamensis* and *O. mercator*. Can J. Zool. 59:1980–1990.

Rhoades, D. F. 1985. Offensive-defensive interactions between herbivores and plants: their relevance in herbivore population dynamics and ecological theory. Amer. Nat. 125:205–238.

Rilett, R. O. 1949. The biology of *Laemophloeus ferrugineus* (Steph.). Can. J. Res. 27D:112–148.

Roberts, D. R., and B. P. Hsi. 1977. A method of evaluating ovipositional attractants of *Aedes aegypti* (Diptera: Culicidae), with preliminary results. J. Med. Ent. 14:129–131.

Russin, J. S., and L. Shain. 1984. Colonization of chestnut blight cankers by *Ceratocystis microspora* and *C. eucastaneae.* Phytopathology 74:1257–1261.

Russin, J. S., L. Shain, and G. L. Nordin. 1984. Insects as carriers of virulent and cytoplasmic hypovirulent isolates of the chestnut blight fungus. J. Econ. Ent. 77:838–846.

Schneider, W. D., J. R. Miller, J. A. Breznak, and J. F. Fobes. 1983. Onion maggot, *Delia antiqua,* survival and development on onions in the presence and absence of microorganisms. Ent. Exp. Appl. 33:50–56.

Schoonhoven, L. M. 1981. Chemical mediators between plants and phytophagous insects. In D. A. Nordlund, W. J. Lewis and R. L. Jones (eds.), Semiochemicals: Their Role in Pest Control. pp. 31–50. Wiley, New York.

Spradbery, J. P. 1970. Host finding by *Rhyssa persuasoria* (L.), an ichneumonid parasite of siricid woodwasps. Anim. Behav. 18:103–114.

Starmer, W. T., and J. C. Fogleman. 1986. Coadaptation of *Drosophila* and yeasts in their natural habitat. J. Chem. Ecol. 12:1037–1055.

Starratt, A. N., and S. R. Loschiavo. 1971. Aggregation of the confused flour beetle, *Tribolium confusum,* elicited by mycelial constituents of the fungus *Nigrospora sphaerica.* J. Insect Physiol. 17:407–414.

Starratt, A. N., and S. R. Loschiavo. 1972. Aggregation of the confused flour beetle, *Tribolium confusum,* elicited by fungal triglycerides. Can. Ent. 104:757–759.

Strazdis, J. R., and V. L. MacKay. 1983. Induction of yeast mating pheromone a-factor by α cells. Nature 305:543–545.

Thomas, C. M., and R. J. Dicke. 1971. Response of the grain mite *Acarus siro* (Acarina: Acaridae), to fungi associated with stored-food commodities. Ann. Ent. Soc. Amer. 64:63–68.

Thomas, C. M., and R. J. Dicke. 1972. Attraction of the grain mite *Acarus siro* (Acarina: Acaridae), to solvent extracts of fungi associated with stored-food commodities. Ann. Ent. Soc. Amer. 65:1069–1073.

Thorpe, W. H., and F. G. W. Jones. 1937. Olfactory conditioning in a parasitic insect and its relations to the problems of host selection. Proc. Roy. Soc. London Ser. B. 124:56–81.

Townshend, J. L. 1964. Fungus hosts of *Aphelenchus avenae* Bastian, 1865, and *Bursaphelenchus fungivorous* Franklin and Hooper, 1962, and their attractiveness to these nematode species. Can. J. Microbiol. 10:727–737.

Urban, T. C. Jarstrand, and A. Aust-Kettis. 1983. Migration of *Entamoeba histolytica* under agarose. Amer. J. Trop. Med. Hyg. 32:733–737.

van der Walt, J. P. 1970. Ester production. In J. Lodder (ed.), The Yeasts. pp. 99–100. North-Holland, Amsterdam.

Vanhaelen, M., R. Vanhaelen-Fastre, J. Geeraerts, and T. Whirlin. 1979. Cis- and trans-octa-1,5-dien-3-ol, new attractants to the cheese mite *Tyrophagus putrescentiae* (Schrank) (Acarina, Acaridae) identified in *Trichothecium roseum* (Fungi Imperfecti). Microbios 23:199–212.

Vanhaelen, M., R. Vanhaelen-Fastre, and J. Geeraerts. 1980. Occurrence in mushrooms (Homobasidiomycetes) of cis- and trans-octa-1,5-dien-3-ol, attractants to the cheese mite *Tyrophagus putrescentiae* (Schrank) (Acarina, Acaridae). Experientia 36:406–407.

van Herrewege, J., J. R. David, and R. Grantham. 1980. Dietary utilization of aliphatic alcohols by *Drosophila*. Experientia 36:846–847.

Vet, L. E. M. 1983. Host-habitat location through olfactory cues by *Leptopilina clavipes* (Hartig) (Hym.: Eucoilidae), a parasitoid of fungivorous *Drosophila:* the influence of conditioning. Neth. J. Zool. 33:225–249.

Vet, L. E. M. 1984. Comparative Ecology of Hymenopterous Parasitoids. Ph.D. thesis. University of Leiden, Leiden, The Netherlands. Kanters Publishers, Alblasserdam, The Netherlands.

Vet, L. E. M. 1985a. Olfactory microhabitat location in some eucoilid and alysiine species (Hymenoptera), larval parasitoids of Diptera. Neth. J. Zool. 35:720–730.

Vet, L. E. M. 1985b. Response to kairomones by some alysiine and eucoilid parasitoid species (Hymenoptera). Neth. J. Zool. 35:486–496.

Vet, L. E. M., and K. van Opzeeland. 1984. The influence of conditioning on olfactory microhabitat and host location in *Asobara tabida* (Nees) and *A. rufescens* (Foerster) (Braconidae: Alysiinae) larval parasitoids of Drosophilidae. Oecologia (Berlin) 63:171–177.

Vet, L. E. M., and K. van Opzeeland. 1985. Olfactory microhabitat selection in *Leptopilina heterotoma* (Thomson) (Hym.: Eucoilidae), a parasitoid of Drosophilidae. Neth. J. Zool. 35:497–504.

Vet. L. E. M., J. C. van Lenteren, M. Heymans, and E. Meelis. 1983. An airflow olfactometer for measuring responses of hymenopterous parasitoids and other small insects. Physiol. Ent. 8:97–106.

Vet, L. E. M., C. Janse, C. van Achterberg, and J. J. M. van Alphen. 1984. Microhabitat location and niche segregation in two sibling species of *Drosoph-*

ila parasitoids: *Asobara tabida* (Nees) and *A. refescens* (Foerster) (Braconidae: Alysiinae). Oecologia (Berlin) 61:182–188.

Visser, J. H. 1986. Host odor perception in phytophagous insects. Annu. Rev. Ent. 31:121–144.

Whittaker, R. H., and P. P. Feeny. 1971. Allelochemics: chemical interactions between species. Science 171:757–770.

Williams, A. A., T. A. Hollands, and O. G. Tucknott. 1981. The gas chromatographic–mass spectrometric examination of the volatiles produced by the fermentation of a sucrose solution. Z. Lebensm. Unters. Forsch. 172:377–381.

Wood, D. L. 1982. The role of pheromones, kairomones and allomones in host selection and colonization behavior of bark beetles. Annu. Rev. Ent. 27:411–446.

THEORY AND MECHANISMS: PLANT EFFECTS VIA ALLELOCHEMICALS ON THE THIRD TROPHIC LEVEL

The biomass of invertebrates available as food for carnivorous insects is at least an order of magnitude less than the plant biomass available as a resource of phytophagous insects (Strong et al. 1984). However, for many reasons, including (1) physical plant defenses such as trichomes, fibrous tissues, and waxy surfaces, (2) nutritional deficiencies such as amino acid imbalance and low nitrogen levels, and (3) chemical plant defenses such as nonprotein amino acids, cardiac glycosides, and tannins, the real abundance of food for herbivores is not as great as it may seem. In addition, herbivores make up a substantial proportion of invertebrates available to carnivores, so herbivores are subject to severe constraints by the third trophic level. Indeed, Lawton and McNeill (1979) describe the ecological plight of herbivores as being caught between the devil (parasitoids and predators) and the deep blue sea (suboptimal or toxic properties of food plants).

Although in evolutionary history, host range and feeding preference (relative acceptance) of phytophagous insects have been determined by plant characteristics, by the milieu within which they search, and by their behavioral and physiological plasticity (Fox and Morrow 1981, Scriber 1983), other factors may be just as important. They include (1) selection pressures imposed by natural enemies (Lawton and McNeill 1979, Price et al. 1980, Bernays and Graham 1988) and (2) direct and indirect effects of variability in plant allelochemicals within plants and within populations (Dolinger et al. 1973, Denno 1983). It is the complexity of these combined processes which may allow for the numerous examples of herbivore–plant interactions that challenge theories of plant defense developed and widely accepted in the 1970s.

Herbivores that do depart from generalizations on resource use patterns may provide interesting systems with which to consider plant variability and aspects of natural selection by predators and parasitoids. Certainly, host plant selection by herbivores is not always predictable (Zimmerman et al. 1984). For example, Courtney (1981) documented distinct oviposition preference by pierid butterflies for a suboptimal food source. Chew (1977) showed that oviposition also occurs on plants toxic to pierid larvae. The ease of host plant location should increase with patch size and stand purity (Root 1973); but some butterflies deposit a disproportionately high number of eggs on isolated, single plants (Mackay and Singer 1983, Root and Kareiva 1984) and the squash bug, *Anasa tristis,* oviposits preferentially on plants with nonhost overgrowth rather than in pure stands of its host plant (Letourneau 1986). The facts that some of these insect–plant pairs involve recent importations (Miller and Strickler 1984), that individual variability is expected even in coevolved systems, and that complex interactive systems include random processes and evolutionary constraints warn against a blind adherence to an adaptationist interpretation, but consideration of the third trophic level may explain some of the "anomalous" behaviors.

In addition to the unpredictability of herbivore–plant interactions due to history, context, physiology/behavior, and enemies, we must add a direct consideration of the variability in plant quality (Langenheim et al. 1978, Denno 1983, Whitham 1983). In 1973, Dolinger and coworkers published a landmark paper on the role of heterogeneity of plant secondary compounds in individual plants within a population as a plant defense strategy (Dolinger et al. 1973). It is now well accepted that plants are heterogeneous resources which (1) vary in time and space in their suitability to insect herbivores, and (2) place constraints upon the ability of herbivores to evolve resistance to plant secondary compound defenses (Fox 1981). The associated hypothesis, that resource constancy favors adaptation by specialist herbivores, is based upon what might be thought of as horizontal and vertical contact with a selection pressure. With constant resource quality, it is likely that a large number of individuals and generations will experience similar selective pressures, and thus the probability of fixing in the population a trait that renders the herbivore resistant is high. The relative impact of variability for polyphagous feeders should be less than it would be for specialists.

Given the context of three trophic levels, specialist and generalist natural enemies that feed on herbivores may also be subjected to such "variability" defense, but this time in favor of the herbivore. Variable plant allelochemicals may have detrimental effects on host/prey habitat or host/prey finding. Variable kairomones, for example, may not work efficiently. Does variability in plant volatiles from plants or herbivore frass reduce their usefulness as cues to natural enemies? The effects on parasitoids and predators of plant-derived allelochemicals ingested by herbivores and used as allomones are also a recent and very important area of study. It is reasonable to assume that carnivores can concentrate nutrients so that suboptimal nutritional value is probably not a problem as it is for herbivores; but secondary compounds in the hemolymph, tissue, or gut can be harmful to parasitoids and predators of the herbivore. Again, however, we may find that predictions from two-trophic-level studies are inadequate. A pioneering study by Orr and Boethel (1986) of the effects of soybean antibiosis suggests that significant and reciprocal interaction of antibiosis and protection function through at least four trophic levels.

The impacts of variability have been discussed widely in studies of vertebrate predators of herbivores that sequester plant allelochemicals for their own defense (see Chapters 1, 7, and 8), but not so for the effects of naturally occurring variability in plant allelochemicals on invertebrate predators and parasitoids. We are only beginning to understand the roles of evolutionary history, host/prey availability (local abundance and diversity) and quality, physiological and behavioral plasticity, and variability in host/prey quality on their acceptability and suitability for predators and parasitoids.

Both the chapter by Williams, Elzen, and Vinson and that by Barbosa contribute vital new insights on the effects of plant allelochemicals as they ascend the food chain. Indeed, plant-derived allelochemicals exact a variety of physiological changes on parasitoids, some of which determine their fitness. The degree to which fitness is decreased or increased depends upon a number of factors, many of which are still to be discovered; so far, the important predictors seem to be parameters such as host range of the parasitoid, level of specificity of the herbivore, and amount of compound in the host diet, but not the category within which the plant defense compound falls. This section provides detailed discussions for a number of systems under current study, which include both mechanisms and developing hypotheses of plant effects on parasitoids, and vice versa. Given that plant defensive chemistry can have detrimental or beneficial effects on parasitoids, and that these effects are reflected in herbivory to the plant, the evolution of plant allelochemistry may be affected significantly by the third trophic level. Alternatively, the multiple levels of uncertainty may impede coevolution in anything but the most diffuse sense. A working hypothesis is proposed by Barbosa—that generalist parasitoids will suffer greater losses than will specialist parasitoids from exposure to plant-derived toxins in their hosts. We hope to spur additional research on comparative studies of generalists and specialists on at least the second and third trophic levels, to clarify the ultimate costs and benefits of plant defensive chemistry. Is the strategy of a generalist based upon spreading its progeny among a range of host types, which may be relatively easy to find but which vary in suitability, while a specialist can adapt to a more constant resource that is more difficult to find? Given the range of impacts that chemicals such as rutin, tomatine, and gossypol can have on predators and parasitoids, it may be possible to add to Lawton and McNeill's (1979) description of the herbivore's plight—that the deep blue sea can also passify, irritate, enrage, or drown the devil!

D. K. LETOURNEAU

REFERENCES

Bernays, E. A., and M. Graham. 1988. On the evolution of host range in phytophagous insects. Ecology 69 (in press).

Chew, F. S. 1977. Coevolution of pierid butterflies and their cruciferous food plants. II. The distribution of eggs on potential food plants. Evolution 31:568–579.

Courtney, S. 1981. Coevolution of pierid butterflies and their cruciferous food-plants. III. *Anthocharis cardamines* (L.) survival, development and oviposition on different plants. Oecologia (Berlin) 51:91–96.

Denno, R. F. 1983. Tracking variable plants in space and time. In R. F. Denno and M. S. McClure (eds.), Variable Plants and Herbivores in Natural and Managed Systems. pp. 291–341. Academic Press, New York.

Dolinger, P. M., P. R. Ehrlich, W. L. Fitch, and D. E. Breedlove. 1973. Alkaloid and predation patterns in Colorado lupine populations. Oecologia (Berlin) 13:191–204.

Fox, L. R. 1981. Defense and dynamics in plant-herbivore systems. Amer. Zool. 21:853–864.

Fox, L. R., and P. A. Morrow. 1981. Specialization: species property or local phenomenon? Science 211:887–897.

Langenheim, J. H., W. H. Stubblebine, D. E. Lincoln, and C. E. Foster. 1978. Implications of variation in resin composition among organs, tissues and populations in the tropical legume *Hymenaea*. Biochem. Syst. Ecol. 6:299–313.

Lawton, J. H., and S. McNeill. 1979. Between the devil and the deep blue sea: on the problem of being a herbivore. In R. M. Anderson, B. D. Turner, and L. R. Taylor (eds.), Population Dynamics. 20th Symposium of the British Ecological Society. pp. 233–244. Blackwell Scientific, Oxford.

Letourneau, D. K. 1986. Associational resistance in squash monoculture and polyculture in tropical Mexico. Environ. Ent. 15:285–292.

Mackay, D. A., and M. C. Singer. 1983. The basis of an apparent preference for isolated host plants by ovipositing *Euptychia libya* butterflies. Ecol. Ent. 7:299–303.

Miller, S. R., and K. L. Strickler. 1984. Finding and accepting host plants. In W. J. Bell and R. T. Cardé (eds.), Chemical Ecology of Insects. pp. 127–157. Chapman & Hall, London.

Orr, D. B., and D. J. Boethel. 1986. Influence of plant antibiosis through four trophic levels. Oecologia (Berlin), 70:242–249.

Price, P. W., C. E. Bouton, P. Gross, B. A. McPheron, J. N. Thompson, and A. E. Weis. 1980. Interactions among three trophic levels: influence of plants on interactions between insect herbivores and natural enemies. Annu. Rev. Ecol. Syst. 11:41–65.

Root, R. B. 1973. Organization of a plant-arthropod association in simple and diverse habitats: the fauna of collards (*Brassica oleracea*). Ecol. Monogr. 43:95–124.

Root, R. B., and P. M. Kareiva. 1984. The search for resource by cabbage butterflies (*Pieris rapae*): the ecological consequences and adaptive significance of Markovian movement in a patchy environment. Ecology 65:147–165.

Scriber, J. M. 1983. Evolution of feeding specialization, physiological efficiency, and host races in selected Papilionidae and Saturniidae. In R. F. Denno and M. S. McClure (eds.), Variable Plants and Herbivores in Natural and Managed Systems. pp. 373–412. Academic Press, New York.

Strong, D. R., J. H. Lawton, and T. R. E. Southwood. 1984. Insects on Plants. Blackwell Scientific, Oxford.

Whitham, T. G. 1983. Host manipulation of parasites: within-plant variations as a defense against rapidly evolving pests. In R. F. Denno and M. S. McClure (eds.), Variable Plants and Herbivores in Natural and Managed Systems. pp. 14–41. Academic Press, New York.

Zimmerman, M., D. A. Cibula, and B. Schulte. 1984. Oviposition behavior of *Hylemya* (*Delia*) sp. (Diptera: Anthomyiidae): suboptimal host plant choice? Environ. Ent. 13:696–700.

CHAPTER 5

Parasitoid–Host–Plant Interactions, Emphasizing Cotton (*Gossypium*)

H. J. Williams
Texas A & M University
College Station, Texas

G. W. Elzen
USDA, Agricultureal Research Service
Stoneville, Mississippi

S. B. Vinson
Texas A & M University
College Station, Texas

CONTENTS

1 Introduction
2 Attraction of parasitoids to plants: historical background
3 Chemical mediation of *Campoletis sonorensis* behavior
  3.1 Isolation and identification of biologically active cotton volatiles
  3.2 *Campoletis* bioassays and the response to plants and plant allelochemicals
4 Genetic variability in cotton volatile chemistry
5 Effects of allelochemical toxins on parasitoid and host development
6 Summary
  References

1   INTRODUCTION

Current theory of insect–plant interactions deals generally with only two trophic levels, plants and their herbivores (Gilbert and Ravens 1975, Rhoades and Cates 1976, Rosenthal and Janzen 1979, Denno and McClure 1983). Theories on the ecology of insect–plant relationships should become more rigorous when the third trophic level is also considered (Lawton and McNeill 1979, Price et al. 1980, Price 1981, Schultz 1983). Biological control processes involve interactions between plants and organisms of the third trophic level (Sweetman 1958, DeBach 1964, Hoy and Herzog 1985), and it is surprising that until recently (Boethel and Eikenbary 1986) few studies have examined the effects of plant physiology and plant allelochemicals on the biology of parasitoids and predators. These interactions are not limited to the impact of plant chemistry on herbivores and the influence of herbivore physiology and fitness on beneficials but include direct effects of plants on the third trophic level. As depicted in Fig. 5.1, attraction of herbivores and provision of herbivore refuge sites would appear to be maladaptive to the plant, while attraction of beneficials and provision of beneficial insect feeding and refuge sites would be advantageous. One might speculate that plants would evolve mechanisms to increase beneficial insect efficiency in reducing herbivore damage. In the

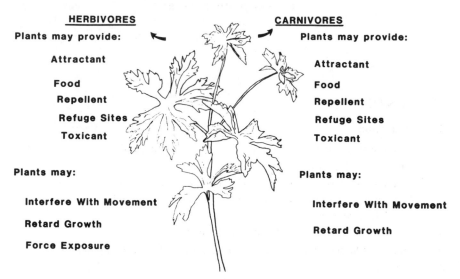

Figure 5.1   Adaptive benefits of attraction and defense.

case of hymenopteran parasitoids, there is circumstantial evidence in the literature to support this view (Vinson 1981).

In addition to attracting pollinators, plant characteristics evoke many responses from beneficial insects. For example, *Acacia* trees harbor ants that protect the plant from herbivores, and the plant in turn provides the ant colony with living space, shelter, and food (Janzen 1966). Manipulation of one trophic level may have undesired effects on another. Plant volatile chemistry has a complex effect on a plant's fitness in a given environment. Volatile chemicals may modify the behavior of associated insects by indicating that plants producing them are sources of food or shelter, environments to be avoided, or suitable starting points for searching for food, shelter, mates, or hosts. Some volatile plant products may also be toxic or have antifeedant activity and contribute directly to plant resistance to herbivory.

Plants benefit when parasitoids reduce herbivore stress; attraction of parasitoids and predators would be beneficial. However, toxic substances in plant tissues which repel, retard growth, reduce vigor, or kill susceptible herbivores may poison beneficial insects or cause physiological or metabolic changes in herbivore tissue which reduce its value as a food source for the third trophic level (Campbell and Duffey 1979, Barbosa et al. 1982, Barbosa and Saunders 1985). The conflict between attraction and defense (Fig. 5.2) must eventually affect the evolution of plant allelochemistry. The options for the plant are either to become highly attractive

PLANT CONFLICT

Poison to 2nd trophic level	OR	Attractive to 3rd trophic level
HERBIVORE		CARNIVORE
Avoid		Avoid
Excrete		Excrete
Tolerate		Tolerate
Sequester		Sequester
Metabolize		Metabolize

Figure 5.2 Conflict between attraction and defense.

to beneficial insects, poisonous to herbivores and thus possibly lose third-trophic-level protection, or, alternatively, achieve some compromise which exploits both protective mechanisms.

The evolution of tri-trophic-level interactions is a complicated problem to consider, particularly as it concerns the biology of the herbivore. Herbivores may cope with toxic plant substances by avoidance, sequestration, tolerance, or metabolic detoxification. These behaviors and physiological processes will affect the quality of the herbivore as a resource for associated predators or parasitoids (Vinson and Barbosa 1987; Fig. 5.2). Our studies of allelochemical interactions among cotton and its associated herbivores and their parasitoids were designed to determine how this three-trophic-level system operates in light of the biological problems outlined above.

2 ATTRACTION OF PARASITOIDS TO PLANTS: HISTORICAL BACKGROUND

Chemicals mediating parasitoid–host–plant interactions are classified as kairomones, allomones, and synomones (Nordlund and Lewis 1976) and appear to be key factors determining host location and governing the range of hosts attacked by parasitoids. One major task faced by a female parasitoid is location of a habitat containing host insects. Initially, a parasitoid may seek a certain environment regardless of the presence or absence of hosts (Doutt 1964, Vinson 1975). The host, however, occurs only in specific locations within that environment, and a female parasitoid must locate those areas of the habitat most likely to yield hosts. Factors that attract a parasitoid to a plant and retain it in the area have a positive selection value for the plant due to the parasitoid's beneficial effects in reducing herbivore survival and fitness. The foraging activity of parasitoids may be influenced by a number of factors in the plant or hosts.

Plants may produce chemicals that attract and retain parasitoids (Nettles 1979; Vinson 1975; 1984) or provide nutrition (Shahjahan 1974) and thereby attract parasitoids. A few attractants have been identified (Read et al. 1970, Camors and Payne 1972, Elzen 1983, Elzen et al. 1984a, Lecomte and Thibout 1984) in pioneering studies of parasitoid–plant interactions. Parasitoids are more often attracted to plants on which their hosts feed (Thorpe and Caudle 1938; Monteith 1955, 1964; Arthur 1962; Madden 1970) than to plants upon which their hosts do not feed. Damaged plants may provide stimuli for increased parasitoid searching (Nishida 1956, Bragg 1974, Vinson 1975). Other species of parasitoids may be attracted by fruiting or flowering plants (Voronin 1981) or to specific plant parts (Nishida 1956). Plants may also influence the general attractiveness of the herbivore (Mueller 1983) or the kairomonal activity of the herbivore's frass (Sauls et al. 1979, Nordlund and Sauls 1981, Elzen et al. 1984b) and hosts reared on different plants may vary in attractiveness (Nordlund and Sauls 1981). Retention of beneficial insects on a plant may require not only sources of nutrition, refuge sites, and an indication of the presence of host insects (Lewis et al. 1975), but also may involve additional chemical cues.

Our experiments on plant-produced attractants for beneficial insects and the effects of plant defense compounds on the quality of the herbivore as a reproductive resource for parasitoids focused on cotton (*Gossypium hirsutum* L.). Cotton was chosen because of its economic importance and the availability of many related genotypes (and species) as well as data on cotton volatiles (Minyard et al. 1965, 1966, 1967, 1968; Hedin et al. 1971a,b, 1972; Elzen 1983) and defensive chemistry (Bell et al. 1974, 1978; Stipanovic et al. 1977). *Heliothis virescens* (F.) was studied because of its economic importance in cotton and ability to consume other plants with different secondary plant chemistry (Neunzig 1968). *Campoletis sonorensis* (Cameron), an ichneumonid parasitoid of *H. virescens* and other noctuids, was used as the third-trophic-level organism.

3 CHEMICAL MEDIATION OF *CAMPOLETIS SONORENSIS* BEHAVIOR

3.1 Isolation and Identification of Biologically Active Cotton Volatiles

In view of the economic importance of cotton and the cost of its protection from herbivore damage, it is not surprising that much research has been directed toward understanding the natural defenses of the cotton plant and the factors responsible for recognition of the cotton plant by herbivores. Plant chemistry is involved both in recognition and protection; therefore, research in cotton chemistry has been a large part of this basic research. Minyard and coworkers first identified several monoterpenes (myrcene, α-pinene, β-pinene, limonene, and *trans*-ocimene) from steam distillate of cotton (Pair et al. 1982). Many other monoterpenes, sesquiterpenes, and miscellaneous nonterpenoid compounds have been reported from cotton preparations since that time (Minyard et al. 1966, 1967, 1968; Hedin et al. 1971a,b, 1972; Elzen 1983; Elzen et al. 1984a). Several representative volatile cotton terpenes, including all found to be active in our studies, are shown in Fig. 5.6.

Most previous studies were designed to determine the effect of cotton volatile chemistry on boll weevil behavior. In general, boll weevil feeding was stimulated by mixtures of cotton compounds (Keller et al. 1963; Hedin et al. 1968; Minyard et al. 1969; Gueldner et al. 1970; Thompson et al. 1970; McKibben et al. 1971, 1977; Hanny et al. 1973). In one case, a boll weevil repellent was produced during the purification process (Maxwell et al. 1963). New compounds continue to be identified (Hedin et al. 1972, Elzen 1983), and many have biological activities of various kinds. A recent report indicated that some cotton volatiles may be allergenic to susceptible humans (Lefkowitz and Lefkowitz 1984).

The methods used to identify plant compounds affecting insect behavior depend upon bioassay-directed purification and analysis (Tumlinson and Heath 1976). General procedures are (1) development of a bioassay by presenting the material of interest (in this case a plant) to the organism and defining the behavior which constitutes a positive response; (2) preparation and bioassay of active extracts of the material of interest; (3) purification of extracts by appropriate techniques, and bioassay of fractions and combinations of fractions, continuing the purification process until pure components which exhibit the original activity are isolated; (4) identification of the active chemical(s), using spectral, physical, or chemical means; and (5) preparation or purchase of authentic samples of the identified chemical(s) for bioassay to confirm activity.

3.2   *Campoletis* Bioassays and the Response to Plants and Plant Allelochemicals

Female *C. sonorensis* responded to the odor of cotton. Female wasps have been observed to approach cotton from a short distance, antennate the surface on contact, and then repeatedly thrust their ovipositor toward the plant surface. A series of experiments have provided important details of this behavior. The assay for contact chemicals consisted of counting antennation and probing frequencies toward each test sample, over a given period of time (Fig. 5.3). An assay for close-range plant volatile attraction employed a "petri dish Y tube" of new design (Elzen et al.

Figure 5.3 *Campoletis sonorensis* response to a cotton flower bud. From Elzen, G. W., H. J. Williams, and S. B. Vinson, 1984. Isolation and identification of cotton synomones mediating searching behavior by parasitoid *Campoletis sonorensis*. J. Chem. Ecol. 10:1254. Reprinted with permission of Plenum Publishing Corporation.

1983). Insects were placed in a petri dish with odor sources (i.e., a test extract and a solvent control) entering through two openings in the base. The number of insects moving into the tubes from the odor source over a given period of time was recorded. A wind tunnel, consisting of a 20-gallon aquarium connected to a variable-speed fan by a chamber containing activated charcoal (Elzen et al. 1986), was used to quantify longer-range attraction. A baffle at the upwind end of the chamber ensured linear airflow. Test samples were suspended at the upwind end of the wind tunnel, groups of parasitoids were released at the downwind end, and the number of insects landing on each sample in a given period was determined (Fig. 5.4).

Bioassays in the Y-tube apparatus showed that *C. sonorensis* females were attracted to cotton and actively entered the cotton-containing tube but were not trapped in the control tube. In addition, cotton was significantly more attractive over a short distance than were a number of other

Figure 5.4 Typical flight response of *C. sonorensis* to cotton in a wind tunnel.

plants, including bluebonnet, *Lupinus texensis,* and wild geranium, *Geranium carolinianum*. Parasitoids were also observed antennating and probing cotton to a signficantly greater degree than other plants, including tobacco, bluebonnet, and wild geranium. Parasitoids would antennate and probe cotton buds more extensively than cotton leaves and stems. This differential response may direct *C. sonorensis* to the areas of greatest host density (Elzen et al. 1983).

Extractions of green plant parts with diethyl ether consistently provided extracts to which parasitoids were "attracted" (Fig 5.5). Mass spectra (MS) indicated a range of molecular weights from 204 to 222 for the active materials. Many sesquiterpenes have molecular weights in this range and contribute to characteristic fragrances of plants (Formachek and Kubeczka 1982). Nuclear magnetic resonance (NMR) spectroscopy was used to determine structures of the compounds of interest. Several of the biologically active sesquiterpenes found in our cotton sample had previously been identified in another cultivar of cotton (Minyard et al. 1966, 1967; Hedin et al. 1971a,b, 1972). Two additional components (molecular weight 218 and 220) had not been reported in cotton and required

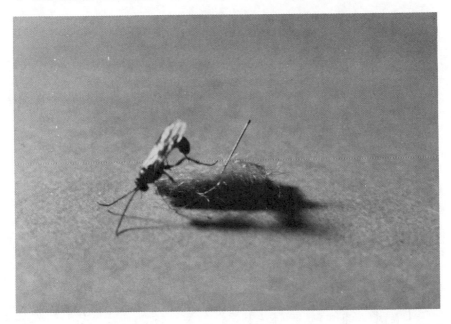

Figure 5.5 Response of *C. sonorensis* to an ethyl ether extract of cotton applied to felt.

further study for identification. The first component was present in the cotton extract in very low concentration, and the original NMR spectrum, taken on less than 10 μg of the active chemical, indicated that impurities were still present. Examination of the spectrum indicated that the material might be a disubstituted aromatic compound with one methyl group attached to the aromatic ring, and a longer fragment in the para position with two methyl groups attached to a double bond and one methyl at a tertiary alcohol center. From this information, a structure similar to that of bisabolol was proposed. The compound, called gossonorol (Fig. 5.6), was synthesized and found to be identical to the natural material.

The second component was subjected to NMR spectroscopy at 500 MHz. Careful analysis of the spectrum indicated that the compound was a polycyclic sesquiterpene containing seven-, five-, and three-membered rings. A review of the literature indicated that this compound, called spathulenol (Fig. 5.6), had been discovered earlier in camomile oil (Bowyer and Jefferies 1963, Juell et al. 1976, Motl et al. 1978). A third compound was identified as either copaene or ylangene, tricyclic terpenoids with closely related structures. Published data on the compounds did not

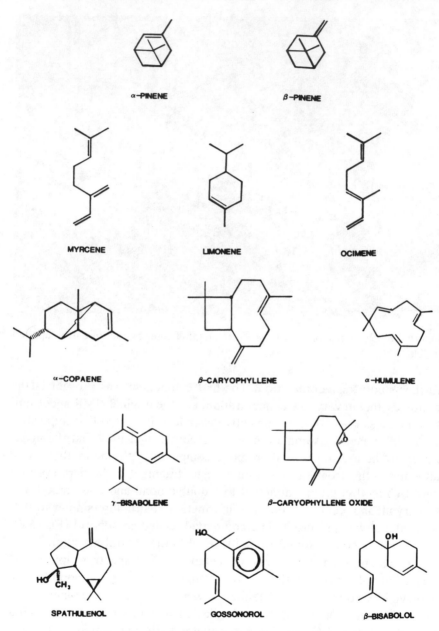

Figure 5.6 Representative cotton volatile terpenes.

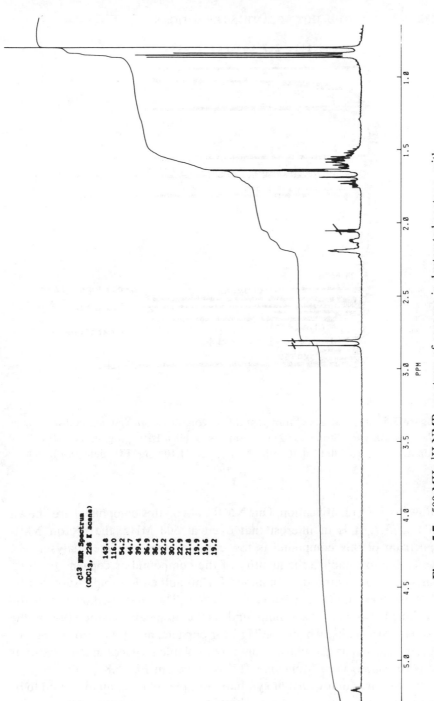

Figure 5.7 500-MHz ^{1}H NMR spectrum of copaene in deuterated acetone, with ^{13}C CDCl$_3$ NMR data in the inset.

c13 NMR Spectrum
(CDCl3, 228 K acene)

143.8
116.0
54.2
44.7
39.4
36.9
36.2
32.2
30.0
22.9
21.8
19.9
19.6
19.2

181

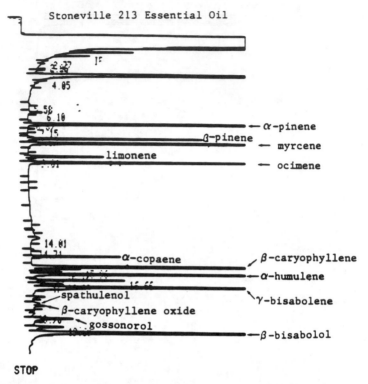

Figure 5.8 Typical gas chromatogram of volatiles from glanded cotton species *G. hirsutum* var. 'Stoneville 213' (25-m fused silica BP1, programmed 60°C for 4 minutes, then to 180°C at 10°/min, carrier He at 10 psig, FID detection).

allow positive identification. Our NMR data on this component are shown in Fig. 5.7. It is of interest that even at 500 MHz, the proton NMR spectrum of this compound is too complex for complete analysis. We were able to establish the identity of the compound as copaene by comparison with spectral data obtained for an authentic sample supplied by Ron Buttery, USDA, Albany, California. The best method for distinguishing between the two compounds is the ratio between the 119- and the 120-*m-z* MS peaks, which is 4.75 : 1 for copaene and 1.43 : 1 for ylangene. A typical capillary gas chromatogram of volatile terpenes in ether extracts of *G. hirsutum* var. 'Stoneville 213' is shown in Fig. 5.8.

When pure commercial or synthetic samples of compounds found to be attractive to *C. sonorensis* were bioassayed, none were as active as the original extract, but combinations of several of the compounds in near-natural ratios produced activity approaching that of the original. Individual compounds caused varying degrees of long-range attraction and antennation upon contact, indicating that the two activities may be mediated by different chemicals (Table 5.1).

While our results demonstrate the importance of cotton volatiles in providing host-habitat location cues for *C. sonorensis,* the role of herbivore-produced attractants is less clear. Female *C. sonorensis* were attracted to larvae reared on plants with greater frequency compared to larvae reared on an artificial medium (Elzen et al. 1984b). Synomone presence in hosts was examined by dissection of cotton-fed and diet-reared *H. virescens* larvae. Extracts of various body parts and frass were prepared by procedures developed for cotton volatiles and were analyzed by gas chromatography. Volatile cotton terpenes were present in whole body, gut, and frass extracts of cotton-fed *H. virescens,* but not in cuticle extracts (Elzen et al. 1984a). Whole-body extracts of artifical diet–fed *H. virescens* did not contain volatile terpenes. The influence of plant synomones on host selection by *C. sonorensis* was assessed with a four-choice olfactometer modified after Vet et al. (1983; see also Elzen et al. 1984b). Given a choice between cotton-fed and diet-fed *H. virescens, C. sonorensis* accepted (stung) a significantly greater proportion of the cot-

TABLE 5.1 Bioassay Results of Attraction of Female *C. sonorensis* to Volatile Terpenes Found in the Cotton Plant

Sample	Petri Dish: Mean Number of Antennal Palpations (±S.D.)	Mean Number Entering Y–Tube	
		Test Side (±S.D.)	Solvent Blank Side (±S.D.)
β-Caryophyllene	0	0	0
α-Humulene	12.8 (± 3.8)	0.7 (± 0.9)	0.1 (± 0.4)
γ-Bisabolene	11.8 (± 3.0)	7.9 (± 1.9)	0
β-Caryophylene oxide	5.1 (± 2.2)	3.7 (± 1.4)	0
Spathulenol	16.8 (± 2.2)	0	0.1 (± 0.4)
Gossonorol	13.1 (± 2.0)	5.1 (± 1.8)	0
β-Bisabolol	11.7 (±1.9)	0.6 (± 1.1)	0.4 (± 0.5)
All terpenes minus β-caryophyllene	20.4 (± 2.8)	9.0 (± 1.8)	0

ton-fed larvae. Parasitoids also responded with significantly more antennal searches of frass from cotton-fed hosts than frass from diet-fed hosts (Elzen et al. 1984b). When cotton, cotton plus hosts, hosts alone, and controls were presented in a four-choice test for long-range attraction, *C. sonorensis* females chose cotton plus hosts with greatest frequency, and cotton alone was three times as attractive as hosts alone, indicating the importance of the plant in attraction of this parasitoid.

4   GENETIC VARIABILITY IN COTTON VOLATILE CHEMISTRY

Considering the important role of cotton volatiles in the host location and selection process (Elzen et al. 1984a), it might be expected that quantitative and qualitative differences in cotton volatile profiles would affect the behavior of parasitoids and other associated insects. Of particular interest in our studies were cottons lacking subepidermal pigment glands (glandless cotton) which have been developed from glanded varieties and are therefore genetically similar to those varieties. Gas chromatographic quantification of volatile terpenes from five pairs of closely related glanded and glandless cotton cultivars and lines indicated that the glandless cottons lacked volatile terpenes or contained only minute quantities of terpenes, present in related glanded cottons. Volatile terpene location was determined by puncturing pigment glands of variety ACALA SJ-1 with glass capillaries and withdrawing the golden oil. Gas chromatographic (GC) analysis of this oil showed that it contained over 31,000 ppm volatile terpenes, while a 100-fold volume of fluid extracted from surrounding tissues contained none (Elzen et al. 1985). The added resistance to herbivore damage shown by glanded cotton varieties could be due, in part, to toxic or antifeedent qualities of these volatile chemicals. Resistance may also be aided by the attraction of beneficial insects.

The attraction and/or arrestment of beneficial predators and parasitoids caused by specific volatile compounds might suggest that related species or strains of a given plant would differ in attraction if the volatile chemical profile were altered. Some cultivated varieties of glanded cotton (*G. hirsutum*) have similar profiles of volatile terpenes. However, further studies of volatile chemicals in over 30 cotton species, cultivars, and lines have revealed distinct volatile profiles (Bell et al. 1987). For example, Egyptian cotton, *G. barbadense,* lacks myrcene, γ-bisabolene, and β-bisabolol, which are major components of upland cotton. Texas *G. hirsutum* race stocks 254 and 810 also lack γ-bisabolene and β-bisabolol, and race stocks 1055 and 1123 have greatly increased β-caryophyllene oxide contents. The African wild cotton, *G. anomalum*, and the Asiatic wild cottons, *G. arboreum* and *G. capitis-viridis*, lacked or had very reduced levels of γ-bisabolene and β-bisabolol. The wild cottons, *G. gossypioides*, *G. laxum*, and *G. raimondii*, lacked β-bisabolol, β-caryophyllene oxide, and α-humulene. No volatile terpenes were found in the wild cottons, *G. trilobum*, *G. thurberii*, and *G. turnerii*, which also lack hemigossypolone. Hemigossypolone and its derivatives do not occur in flower petals or stamenal tissues (Bell et al. 1987). In a study to determine chemical distribution in the plant, we discovered volatile terpenes from flower, ovary, bract, style and stigma, and calyx tissue, but not from petals, anthers, or pollen (Bell et al. 1987). Hybridization experiments involving *G. arboreum*, *G. anomalum*, and *G. hirsutum* showed that altered volatile terpene profiles could be produced in a predictable manner (Bell et al. 1987), which should be useful in improving resistance of cotton to pests. Further mapping of volatile terpene variation and determination of its significance in host plant resistance is now under way.

Campoletis sonorensis was released in the wind tunnel to characterize flight behavior of females on exposure to different cotton cultivars and lines. Females flew to either Stoneville 213 or CAMD-E cotton but never to moist filter paper used as controls. Parasitoids also flew to a partially purified CAMD-E ethyl ether extract applied to filter paper but never flew to an ethyl ether control. Given a choice of glanded and glandless cotton (three sets of glandless–glanded pairs), parasitoids flew to the glanded cultivars in significantly greater proportion (Elzen et al. 1986). These results indicate that *C. sonorensis* responds to volatile odor profiles produced by plants and may respond in significantly greater numbers to plants that produce a given set of volatile terpenes.

5   EFFECTS OF ALLELOCHEMICAL TOXINS ON PARASITOID AND HOST DEVELOPMENT

The secondary chemistry of plants used as a resource by a herbivore not only affects the physiology and behavior of the herbivore but also affects the quality of the herbivore as a resource for the beneficial insect. A given plant diet may cause the available quantity of nitrogen or other essential nutrients in herbivore tissue to be less than optimal for growth and development of beneficial parasitoids. For example, more male *Habrolepis rouxi* were produced when California red scale hosts were reared from gourds than from lemons (Flanders 1942). *Brachymeria intermedia* weighed more when reared from hosts fed red oak leaves than from those hosts fed other tree foliage (Greenblatt and Barbosa 1981). Other effects may be due to plant allelochemicals present in herbivore's diet that influence the quantity, quality, and utilization of the available nutrients that the herbivore represents (Althataway et al. 1976, Smith 1978). In some cases, plant allelochemicals may be toxic to the parasitoid (cf. Morgan 1910, Gilmore 1938). Thurston and Fox (1972) provided evidence that nicotine in tobacco hornworm diet affected percent emergence of *Cotesia congregatus* (=*Apanteles congregatus*). Compounds such as nicotine (Barbosa and Saunders 1985) or tomatine (Campbell and Duffey 1979, 1981) in a host's diet may affect parasitoids by direct exposure. In other cases, changes in host suitability due to the host's diet can influence the developmental rate, size, percent emergence, parasitization success, sex ratio, fecundity, and life span of parasitoids (Barbosa et al. 1982, Barbosa and Saunders 1985).

Allelochemical effects on parasitoid success and fitness are not well documented, but the allelochemical may (1) prevent normal nutrient utilization by a direct toxic effect on an organ, tissue, or cell of the parasitoid; (2) result in the formation of an unutilizable complex with a nutrient important to the parasitoid (Campbell and Duffey 1981); (3) inhibit enzyme systems important in nutrient digestion and utilization (Zucker 1983); (4) alter the detoxification systems of the host or parasitoid, leaving the parasitoid more vulnerable to toxins (Terriere 1984); (5) alter the energy and biochemical requirements of the host or parasitoid; or (6) alter the physiological and behavioral responses of the host in a manner that interferes with the quantity and quality of nutrients in the host (see Vinson and Barbosa 1987).

As stated earlier, *Gossypium* and related genera of the tribe Gossypieae, including *Cienfuegosia, Kokia,* and *Thespesia,* contain subepidermal pigment glands. The sesquiterpene dimer gossypol, a phenolic aldehyde, was identified as a major component of the glandular contents in studies beginning in the 1940s (Adams and Geissman 1960, Lukefahr and Fryxell 1967). The monomeric precursor of gossypol, called hemigossypol, was later isolated from root tissue and identified (Bell et al. 1975). The quinone hemigossopolone, derived from the oxidation of hemigossypol, and the Diels–Alder adducts between myrcene or ocimene and hemi-gossypolone, called heliocides, were later found in flower bud tissue (Gray et al. 1976, Stipanovic et al. 1977, Stipanovic et al. 1978a,b). *Gossypium raimondii* produces a related but unique compound called raimondal (Stipanovic et al. 1980). Figure 5.9 shows representative examples of the monomeric and dimeric terpene aldehydes found in cotton.

Gossypol has been the most intensively studied of the terpenoid aldehydes, due partly to the unusual biological effects produced on ingestion, but also due to its early discovery and ease of isolation. Gossypol has been shown to possess bactericidal, fungicidal, and male antifertility activity, perhaps at least partly caused by reduction of cilial motility (Berardi and Goldblatt 1980, Zatuchini and Osborn 1981, Manmade et al. 1983, Reyes et al. 1984). When a cotton line was discovered in 1954 which lacked pigment glands in foliar plant parts (McMichael 1954), entomologists determined that *Heliothis* spp. developed more rapidly on glandless plants than on closely related glanded varieties (Jenkins et al. 1966, Hedin et al. 1983). Gossypol, a major gland component, retarded insect growth when added to artificial diet (Stanford and Viehoover 1918, Gray et al.

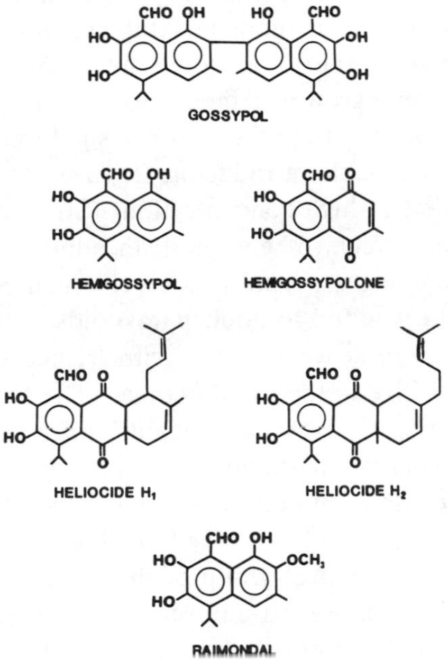

Figure 5.9 Representative terpene aldehydes found in cotton.

1976, Chan et al. 1978). Other compounds are found either in the cells surrounding the cotton glands or in the oil that makes up the gland contents (Chan and Waiss 1981, Elzen et al. 1985). and the added resistance to insect damage of glanded plants over related glandless varieties may not be totally due to the presence of gossypol and other phenolic terpene aldehydes.

Cotton secondary metabolic products might influence the biology of natural enemies; specifically, cotton gland chemicals might influence parasitoid biology. It seems probable that most responses of the parasitoid to plant allelochemical defenses would involve chemicals ingested by the host insect; therefore, the effects of gossypol on the herbivore were first determined. Several feeding studies had been performed to determine effects of diets containing gossypol on *H. virescens* larval development (Stanford and Viehoover 1918, Gray et al. 1976, Chan et al. 1978). Earlier work assumed that a nagative signoidal dose effect occurred (i.e., that there was no effect at small dosages and a logarithmic decrease in growth rate at larger dosages). However, in our studies we found that small quantities of gossypol stimulated herbivore growth, while larger dosages produced a negative sigmoidal curve, an overall response called a hormetic effect (Stipanovic et al. 1986). An "exponential-hormetic model" was developed to describe this reponse. At the highest gossypol levels, no significant increase in mortality was noted in test insects over the control group; the larvae simply grew at a much slower rate. We recently found that caryophyllene oxide, a fragrant cotton sesquiterpene attractive to *C. sonorensis,* was also capable of reducing *H. virescens* growth rate when added to artifical diets in high concentrations (Stipanovic et al. 1986).

To determine the effects of gossypol on adult parasitoids, a honey water solution containing gossypol at a level which completely arrested *H. virescens* growth was fed to adult parasitoids. After 1 week on this diet, female *C. sonorensis* were mated with treated or untreated males and allowed to parasitize in the usual manner. The sex ratio and percent emergence of offspring were not significantly different from the control groups in this preliminary experiment, and longevity was also unchanged.

To determine the effects of cotton allelochemicals on parasitoid development, we fed *H. virescens* larvae for 1 week on diet containing gossypol. These larvae and a control group of the same average weight were then parasitized by *C. sonorensis* and returned to their respective diets. Time of emergence and percent emergence did not differ in this preliminary experiment, but parasitoids reared in gossypol-treated hosts were significantly larger than controls at low gossypol dosages, while somewhat smaller than control insects when gossypol was in host diet at higher dosages. Our findings indicate that the hormetic effect of gossypol ingestion which was seen in *H. virescens* development carries over to parasitoids reared from them, apparently due to the feeding habits of the host larvae and effects on the final weight of parasitized host larvae. Exceptionally small parasitoids collected in cotton fields contain fewer eggs than average, which implies that exceptionally high gossypol levels would be detrimental to parasitoid development. In a related test, starvation of the host larvae before or after parasitism did not produce smaller parasitoids, indicating that the size reduction noted in the field-collected parasitoids was probably due to the presence of gossypol and other allelochemicals in the cotton plant diet of the host insects and not to the detrimental effects of host starvation.

To determine whether effects of gossypol in host diet were due to gossypol directly or to secondary effects due to changes in host physiology, *H. virescens* larvae were fed artificial diet containing gossypol, then extracted and found to lack gossypol internally, although metabolites were apparently present. Since reduction in size occurred for parasitoids reared in gossypol diet–fed hosts which on analysis had been shown to be essentially gossypol free, it seems likely that either a gossypol metabolite or some change in host physiology caused by gossypol ingestion is responsible for size changes. Later studies in which larvae fed cotton containing high levels of gossypol were extracted indicated the presence of some gossypol in larval body tissue (Montandon et al. 1986); perhaps other plant chemicals interact to change gossypol metabolism in the insect.

6 SUMMARY

Host plant resistance, the inherited ability of a given plant to resist the damaging effects of herbivores or pathogens to a greater extent than related plants when growing under the same conditions, is a desirable feature in plants grown for human use. In the case of damage caused by herbivorous insects, resistance would be increased by changes in any chemical, physical, or temporal factors in the plant's biology which decrease the ability of a given herbivore to utilize the plant as a food source. Plant characteristics which contribute to resistance include surface texture, growing and fruiting period, leaf and fruit nutrient content, and concentration, location, and identity of chemical toxins in plant tissue. Volatile plant chemicals can add to herbivore resistance by acting as toxins, repellents, or antifeedents. Physical and chemical features of the plant which attract and retain predators and parasitoids also contribute to host plant resistance.

It would seem that enhancing a plant characteristic which causes herbivore stress would increase host plant resistance and would therefore be beneficial. However, when the interactions of other organisms with the plant and the herbivore are considered, the beneficial effects of host plant resistance factors become less distinct. Factors which adversely affect the behavior or biology of a given herbivore are not necessarily neutral in their interaction either with other potential herbivores or with natural enemies which also protect the plant from herbivore damage. A given change in plant biology will contribute to host plant resistance only if the sum of its effects on all associated organisms allows production of the plant product of interest at lower cost. Several major insect pests with associated predators and parasitoids affect the production of most commercial plant species, and the beneficial or harmful effects on these natural enemies to be expected from a given change in plant biology may be difficult to predict. A complete study of the effects of chemical, physical, and temporal features of several plant species on important associated organisms, both beneficial and harmful, should make such predictions feasible.

The finding that volatile chemicals in cotton affect the behavior of parasitoids associated with the plant, along with the observation that the same chemicals affect the health and behavior of herbivores which damage the plant, indicate that the field of plant–insect chemical ecology is a useful area for further research. It is conceivable that a given chemical may act both as a plant orientation compound or feeding stimulant and poison for a given herbivore. Members of other trophic levels may be affected by the chemical through their association with the plant, by the chemical and its metabolites found in the body of their prey, or by physiological reactions of the prey caused by ingestion of the chemical. The metabolic complexity of most plants, including cotton, makes the number of interactions large.

The possibility of using naturally occurring toxic or attractive compounds from plants to reduce herbivore damage and increase the effectiveness of biological control agents is attractive and several approaches have been proposed. For example, Lewis et al. (1975, 1976, 1982) demonstrated the potential of spraying crops with kairomones in order to attract and retain parasitoids for increased biological control. However, the use of kairomones to influence the behavior of natural populations of beneficial insects may present some ecological problems (Vinson 1977). Potential benefits also exist for synomone application to crops as proposed by Altieri et al. (1981). Ecological damage might be minimized if chemical applications were coupled with mass releases of *in vitro*–reared beneficials (Vinson 1986). Spraying of crops with toxic or repellent plant products or their analogs to increase their concentration and effectiveness is possible (Tingle and Mitchell 1984), although such procedures might also present problems (Stipanovic 1982).

Another approach for using secondary plant chemistry to reduce herbivore damage and increase beneficial insect efficiency involves altering production of such chemicals by the plant. With classical breeding techniques, changes in the secondary metabolite profile of plants can be made which alter the biology of associated organisms, although such changes are generally limited to altering levels of chemicals already present in the species or in adding new chemicals from closely related taxa. Modern genetic engineering techniques may make profound alterations in plant chemistry possible, and it is vital that our understanding of the effects of such changes on all associated organisms keep pace with our ability to produce them. This will certainly require further ecological studies, but given sufficient thought, it is probable that the benefits of this approach will greatly outweigh the risks.

ACKNOWLEDGMENTS

This paper is approved as TA 21146 by the Director of the Texas Agricultural Experiment Station. Funding for the research was provided by the USDA Competitive Grants Program, grants 81-CRCR-1-0647 and 85–CRCR-1-1601.

REFERENCES

Adams, R., and T. A. Geissman. 1960. Gossypol, a pigment of cottonseed. Chem. Rev. 60:555–574.

Althataway, M. M., S. M. Hammad, and E. M. Hegazi. 1976. Studies on the dependence of *Microplitis rufiventris* Kok. parasitizing *Spodoptera littoralis*

(Boisduval) on its own food as well as on food of its host. Z. angew. Ent. 83:3–13.

Altieri, M. A., W. J. Lewis, D. A. Nordlund, R. C. Gueldner, and J. W. Todd. 1981. Chemical interactions between plants and *Trichogramma* wasps in Georgia soybean fields. Prot. Ecol. 3:259–263.

Arthur, A. P. 1962. Influence of host tree on abundance of *Itoplectis conquisitor* (Say) (*Hymenoptera: Ichneumonidae*), a polyphagous parasite of the European pine shoot moth, *Rhyacionia buoliana* (Schiff) (Lepidoptera: Olethreutidae). Can. Ent. 94:337–347.

Barbosa, P., and J. A. Saunders. 1985. Plant allelochemicals: linkages between herbivores and their natural enemies. In G. A. Cooper-Driver, T. Swain, and E. E. Conn (eds.), Chemically Mediated Interactions between Plants and Other Organisms. pp. 107–137. Plenum, New York.

Barbosa, P., J. A. Saunders, and M. Waldvogel. 1982. Plant-mediated variation in herbivore suitability and parasitoid fitness. In J. H. Visser and A. K. Minks (eds.), Proceedings of the 5th International Symposium on Insect–Plant Relationships. pp. 63–71. Pudoc, Wageningen, The Netherlands.

Bell, A. A., R. D. Stipanovic, C. R. Howell, and M. E. Mace. 1974. Terpenoid aldehydes of *Gossypium:* isolation, quantitation, and occurrence. pp. 40–41. Proceedings of the Beltwide Cotton Production Research Conference, Dallas, Tex.

Bell, A. A., R. D. Stipanovic, C. R. Howell, and P. A. Fryxell. 1975. Hemigossypol, 6-methoxyhemigossypol, and 6-deoxyhemigossypol. Phytochemistry 14:225–231.

Bell, A. A., R. D. Stipanovic, D. H. O'Brien, and P. A. Fryxell, 1978. Sesquiterpenoid aldehyde quinones and derivatives in pigment glands of *Gossypium*. Phytochemistry 17:1297–1305.

Bell, A. A., R. D. Stipanovic, G. W. Elzen, and H. J. Williams, Jr. 1987. Structural and genetic variation of natural pesticides in pigment glands of cotton (*Gossypium*). In G. R. Waller (ed.), Allelochemicals: Role in Agriculture, Forestry, and Ecology. American Chemical Society Symposium Series 330. pp. 477–490. American Chemical Society, Washington, D. C.

Berardi, L. D., and L. A. Goldblatt. 1980. Gossypol. In I. E. Liener (ed.), Toxic Constituents of Plant Foodstuff. pp. 183–237. Academic Press, New York.

Boethel, D. G., and R. D. Eikenbary (eds.). 1986. Interactions of Plant Resistance and Parasitoids and Predators of Insects. Wiley, New York.

Bowyer, R. D., and P. R. Jefferies. 1963. Structure of spathulenol. Chem. Ind. 1963:1245–1246.

Bragg, D. 1974. Ecological and behavioral studies of *Phaeogenes cynarae:* ecology, host specificity, search and oviposition, and avoidance of super-parasitism. Ann. Ent. Soc. Amer. 67:931–936.

Camors, F. B., Jr., and T. L. Payne. 1972. Response of *Heydenia unica* (Hymenoptera: Pteromalidae) to *Dendroctonus frontalis* (Coleoptera: Scolytidae) pheromones and a host tree terpene. Ann. Ent. Soc. Amer. 65:31–33.

Campbell, B. C., and S. S. Duffey. 1979. Tomatine and parasitic wasps: potential incompatibility of plant antibiosis with biological control. Science 205:700–702.

Campbell, B. C., and S. S. Duffey. 1981. Alleviation of α-tomatine-induced toxicity to the parasitoid, *Hyposoter exiguae,* by phytosterols in the diet of host, *Heliothis zea.* J. Chem. Ecol. 7:927–946.

Chan, B. G., and A. C. Waiss, Jr. 1981. Evidence for acetogenic and shikimic pathways in cotton glands. pp. 49–52. Proceedings of the Beltwide Cotton Production Research Conference, New Orleans, LA.

Chan, B. G., and A. C. Waiss, Jr., R. G. Binder, and C. A. Elliger. 1978. Inhibition of lepidopterous larval growth by cotton constituents. Ent. Exp. Appl. 24:294–300.

DeBach, P. (ed.). 1964. Biological Control of Insect Pests and Weeds. Reinhold, New York.

Denno, R. F., and M. S. McClure (eds.). 1983. Variable Plants and Herbivores in Natural and Managed Systems. Academic Press, New York.

Doutt, R. L. 1964. Biological characteristics of entomophagous adults. In P. DeBach (ed.), Biological Control of Insect Pests and Weeds. pp. 145–167. Reinhold, New York.

Elzen, G. W. 1983. Isolation, identification, and bioassay of plant chemicals mediating searching behavior by an insect parasitoid. Ph.D. dissertation. Texas A & M University, College Station, Tex. Diss. Abstr. Int. (B) 45:1984.

Elzen, G. W., H. J. Williams, S. B., Vinson. 1983. Response by the parasitoid *Campoletis sonorensis* (Hymenoptera: Ichneumonidae) to chemicals (synomones) in plants: implications for host habitat location. Environ. Ent. 12:1872–1876.

Elzen, G. W., H. J. Williams, S. B. Vinson. 1984a. Isolation and identification of cotton synomones mediating searching behavior by parasitoid *Campoletis sonorensis.* J. Chem. Ecol. 10:1251–1254.

Elzen, G. W., H. J. Williams, and S. B. Vinson. 1984b. Role of diet in host selection of *Heliothis virescens* by parasitoid *Campoletis sonorensis* (Hymenoptera: Ichneumonidae). J. Chem. Ecol. 10:1535–1541.

Elzen, G. W., H. J. Williams, A. A. Bell, R. D. Stipanovic, and S. B. Vinson. 1985. Quantification of volatile terpenes of glanded and glandless *Gossypium hirsutum* L. cultivars and lines by gas chromatography. J. Agric. Food Chem. 33:1079–1082.

Elzen, G. W., H. J. Williams, and S. B. Vinson. 1986. Wind tunnel flight responses by the hymenopterous parasitoid *Campoletis sonorensis* to cotton cultivars and lines. Ent. Exp. Appl. 42:285–289.

Flanders, S. E. 1942. Abortive development in parasitic Hymenoptera, induced by the food plant of the insect host. J. Econ. Ent. 35:834–835.

Formachek, V., and K. H. Kubeczka. 1982. Essential Oils Analysis by Capillary Gas Chromatography and Carbon-13 NMR Spectroscopy. Wiley Heyden, Chichester, West Sussex, England.

Gilbert, L. E., and P. H. Raven (eds.). 1975. Coevolution of Animals and Plants. University of Texas Press, Austin, Tex.

Gilmore, J. U. 1938. Notes on *Apanteles congregatus* (Say) as a parasite in tobacco hornworms. J. Econ. Ent. 31:712–715.

Gray, J. R., T. J. Mabry, A. A. Bell, R. D. Stipanovic, and M. J. Lukefahr. 1976. *Para*-hemigossypolone: a sesquiterpene aldehyde quinone from *Gossypium hirsutum*. J. Chem. Soc. 2:109–110.

Greenblatt, J. A., and P. Barbosa. 1981. Effects of host's diet on two pupal parasitoids of the gypsy moth: *Brachymeria intermedia* (Nees) and *Coccygomimus turionellae* (L.). J. Appl. Ecol. 18:1–10.

Gueldner, R. C., A. C. Thompson, D. D. Hardee, and P. A. Hedin. 1970. Constituents of the cotton bud. XIX. Attractancy to the boll weevil of the terpenoids and related plant constituents. J. Econ. Ent. 63:1819–1821.

Hanny, B. W., A. C. Thompson. R. C. Gueldner, and P. A. Hedin. 1973. Constituents of cotton seedlings: an investigation of the preference of male boll weevils for the epicotyl tips. J. Agric. Food Chem. 21:1004–1006.

Hedin, P. A., L. R. Miles, A. C. Thompson, and J. P Minyard. 1968. Constituents of a cotton bud. Formulation of a boll weevil feeding stimulant mixture. J. Agric. Food Chem. 16:505–515.

Hedin, P. A., A. C. Thompson, R. C. Gueldner, and J. P. Minyard. 1971a. Isolation of bisabolol from the cotton bud. Phytochemistry 10:1693–1694.

Hedin, P. A., A. C. Thompson, R. C. Gueldner, and J. P. Minyard. 1971b. Constituents of the cotton bud. Phytochemistry 10:3316–3331.

Hedin, P. A., A. C. Thompson, R. C. Gueldner, and J. M. Ruth. 1972. Isolation of bisabolene oxide from the cotton bud. Phytochemistry 11:2118–2119.

Hedin, P. A., J. W. Jenkins, D. H. Collum, W. H. White, W. L. Parrott, and M. W. MacGowan. 1983. Cyanidin-3-β-glucoside, a newly recognized basis for resistance in cotton to the tobacco budworm *Heliothis virescens* (Fab.) (Lepidoptera: Noctuidae). Experientia 39:799–801.

Hoy, M. A., and D. C. Herzog. 1985. Biological Control in Agricultural IPM Systems. Academic Press, New York.

Janzen, D. H. 1966. Coevolution between ants and acacias in Central America. Evolution 20:249–275.

Jenkins, J. N., F. G. Maxwell, and H. N. Lafener. 1966. The comparative preference of insects for glanded and glandless cotton, J. Econ. Ent. 59:352–356.

Juell, S. M. K., R. Hansen, and H. Jork. 1976. Neue Substanzen aus ätherischen Ölen verschiedener *Artemesia* species. 1. Mitt. Spathulenol, ein azulenogener Sesquiterpenalkohol. Arch. Pharmacol. 309:458–466.

Keller, J. C., F. G. Maxwell, J. N. Jenkins, and T. B. Davich. 1963. A boll weevil attractant from cotton. J. Econ. Ent. 56:110–111.

Lawton, J. H., and S. McNeill. 1979. Between the devil and the deep blue sea: on the problem of being a herbivore. In B. D. Turner and L. R. Taylor (eds.), Population Dynamics (20th Symposium of the British Ecological Society). pp. 233–244. Blackwell Scientific, Oxford.

Lecomte, C., and E. Thibout. 1984. Étude olfactometrique de l'action de diverses substances allelochimiques végétales dans la recherche de l'hôte par *Diadromus pulchellus* (Hymenoptera: Ichneumonidae). Ent. Exp. Appl. 35:295–303.

Lefkowitz, S. S., and D. L. Lefkowitz. 1984. A severe response to substances released from cotton. J. Amer. Med. Assoc. 251:1835–1836.

Lewis, W. J., R. L. Jones, D. A. Nordlund, and A. N. Sparks. 1975. Kairomones and their use for the management of entomophagous insects. I. Evaluation for increasing rate of parasitization by *Trichogramma* spp. in the field. J. Chem. Ecol. 1:343–347.

Lewis, W. J., R. L. Jones, H. R. Gross, Jr., and D. A. Nordlund. 1976. The role of kairomones and other behavioral chemicals in host finding by parasitic insects. Behav. Biol. 16:267–289.

Lewis, W. J., D. A. Nordlund, R. C. Gueldner, P. E. A. Teal, and J. H. Tumlinson. 1982. Kairomones and their use for management of entomophagous insects. XIII. Kairomonal activity for *Trichogramma* spp. of abdominal tips, excretions, and a synthetic sex pheromone blend of *Heliothis zea* (Boddie) moths. J. Chem. Ecol. 8:1323–1331.

Lukefahr, M. J., and P. A. Fryxell. 1967. Content of gossypol in plants belonging to genera related to cotton. Econ. Bot. 21:128–131.

Madden, J. L. 1970. Physiological aspects of host tree favourability for the woodwasp, *Sirex noctilo* F. Proc. Ecol. Soc. Aust. 3:147–149.

Manmade, A., P. Herlihy, J. Quick, R. P. Duffley, M. Burgos, and A. P. Hoffer. 1983. Gossypol. Synthesis and *in vitro* spermicidal activity of isomeric hemigossypol derivatives. Experientia 39.1276–1277.

Maxwell, F. G., J. N. Jenkins, and J. C. Keller. 1963. A boll weevil repellent from the volatile substance of cotton. J. Econ. Ent. 56:894–895.

McKibben, G. H., P. A. Hedin, R. E. McLaughlan, and T. B. Davich. 1971. Development of the bait principle for control of boll weevils: addition of terpenoids and related plant constituents. J. Econ. Ent. 64:1493–1495.

McKibben, G. H., E. R. Mitchell, W. P. Scott, and P. A. Hedin. 1977. Boll weevils are attracted to volatile oils from cotton plants. Environ. Ent. 6:804–806.

McMichael, S. C. 1954. Glandless boll character in upland cotton and its use in the study of natural crossing. Agric. J. 46:527–588.

Minyard, J. P., J. H. Tumlinson, P. A. Hedin, and A. C. Thompson. 1965. Constituents of the cotton bud. Terpene hydrocarbons. J. Agric. Food Chem. 13:599–602.

Minyard, J. P., J. H. Tumlinson, A. C. Thompson, and P. A. Hedin. 1966. Constituents of the cotton bud. Sesquiterpene hydrocarbons. J. Agric. Food Chem. 14:332–336.

Minyard, J. P., J. H. Tumlinson, A. C. Thompson, and P. A. Hedin. 1967. Constituents of the cotton bud. The carbonyl compounds. J. Agric. Food Chem. 15:517–525.

Minyard, J. P., A. C. Thompson, and P. A. Hedin. 1968. Constituents of the cotton bud. Bisabolol, a new sesquiterpene alcohol. J. Org. Chem. 33:909–911.

Minyard, J. P., D. D. Hardee, R. C. Gueldner, A. C. Thompson, G. Wiygul, and P. A. Hedin. 1969. Constituents of the cotton bug-compounds attractive to the boll weevil. J. Agric. Food Chem. 17:1093–1097.

Montandon, R., H. J. Williams, W. L. Sterling, R. D. Stipanovic, and S. B. Vinson. 1986. Comparison of the development of *Alabama argillacea* (Huebner) and *Heliothis virescens* (F.) (Lepidoptera: Noctuidae) fed glanded and glandless cotton. Environ. Ent. 15:128–131.

Monteith, L. G. 1955. Host preference of *Drino bohemica* Mesn. (Diptera: Tachinidae), with particular reference to olfactory response. Can. Ent. 87:509–530.

Monteith, L. G. 1964. Influence of the health of the food plant of the host on host-finding by tachinid parasites. Can. Ent. 96:1477–1482.

Morgan, A. C. 1910. Observation reported at the 236th regular meeting of the Entomological Society of Washington. Proc. Ent. Soc. Wash. 12:72.

Motl, O., M. Repzak, and P. Sedmera. 1978. Weitere Bestandteile des Kamomillenols. 2. Mitt. Arch. Pharmacol. (Weinheim) 311:75–76.

Mueller, T. F. 1983. The effect of plants on the host relations of a specialist parasitoid of *Heliothis* larvae. Ent. Exp. Appl. 34:78–84.

Nettles, W. C., Jr. 1979. *Eucelatoria* sp. females: Factors influencing response to cotton and okra plants. Environ. Ent. 8:619–623.

Neunzig, H. H. 1968. The Biology of the Tobacco Budworm and the Corn Earworm in North Carolina. North Carolina Agricultural Experiment Station Technical Bulletin 196. North Carolina Agricultural Experiment Station, Raleigh, N.C.

Nishida, T. 1956. An experimental study of the ovipositional behavior of *Opius fletcheri* Silvestri (Hymenoptera: Braconidae), a parasite of the melon fly. Proc. Hawaii. Ent. Soc. 16:126–134.

Nordlund, D. A., and W. J. Lewis. 1976. Terminology of chemical releasing stimuli in intraspecific and interspecific interactions. J. Chem. Ecol. 2:211–220.

Nordlund, D. A., and C. E. Sauls. 1981. Kairomones and their use for the management of entomophagous insects. XI. Effect of host plants on kairomonal activity of frass of *Heliothis zea* larvae for the parasitoid *Microplitis croceipes*. J. Chem. Ecol. 7:1057–1061.

Pair, S. D., M. L. Laster, and D. F. Martin. 1982. Parasitoids of *Heliothis* spp. (Lepidoptera: Noctuidae) larvae in Mississippi associated with sesame interplantings in cotton. 1971–1974: implications of host habitat interactions. Environ. Ent. 1:509–512.

Price, P. W. 1981. Semiochemicals in evolutionary time. In D. A. Nordlund, R. L. Jones, and W. J. Lewis (eds.), Semiochemicals: Their Roles in Pest Control. pp. 251–279. Wiley, New York.

Price, P. W., C. E. Bouton, P. Gross, B. A. McPherson, J. N. Thompson, and A. E. Weis. 1980. Interactions among three trophic levels: influence of plants on interactions between insect herbivores and natural enemies. Annu. Rev. Ecol. Syst. 11:41–65.

Read, D. P., P. D. Feeny, and R. B. Root. 1970. Habitat selection by the aphid parasite *Diaeretiella rapae* (Hymenoptera: Braconidae) and hyperparasite *Charips brassicae* (Hymenoptera: Cynipidae). Can. Ent. 102:1567–1578.

Reyes, J., J. Allen, N. Tanphaichitr, A. R. Bellve, and D. J. Bennon. 1984. Molecular mechanisms of gossypol action on lipid membranes. J. Biol. Chem. 259:9607–9615.

Rhoades, D. F., and R. G. Cates. 1976. Toward a general theory of plant antiherbivore chemistry. In J. W. Wallace and R. L. Mansell (eds.), Biochemical Interaction between Plants and Insects, Vol. 10, Recent Advances in Phytochemistry. pp. 168–213. Plenum, New York.

Rosenthal, G. A., and D. H. Janzen (eds.). 1979. Herbivores: Their Interaction with Secondary Plant Metabolites. Academic Press, New York.

Sauls, C. E., D. A. Nordlund, and W. J. Lewis. 1979. Kairomones and their use for the management of entomophagous insects. VIII. Effects of diet on the kairomonal activity of frass from *Heliothis zea* (Boddie) larvae for *Microplitis croceipes* (Cresson) J. Chem. Ecol. 5:363–369.

Schultz, J. C. 1983. Impact of variable plant defense chemistry on susceptibility of insects to natural enemies. In P. A. Hedin (ed.), Plant Resistance to Insects. American Chemical Society Symposium Series 208, pp. 37–53. American Chemical Society, Washington, D. C.

Shahjahan, M. 1974. *Erigeron* flowers as food and attractive odor source for *Peristenus pseudopallipes,* a braconid parasitoid of the tarnished plant bug. Environ. Ent. 3:69–72.

Smith, D. A. S. 1978. Cardiac glycosides in *Danaus chrysippus* (L.) provide some protection against an insect parasitoid. Experientia 34:844–846.

Stanford, E. E., and A. Viehoover. 1918. Chemistry and histology of the glands of the cotton plant, with notes on the occurrence of similar glands in related plants. J. Agric. Res. 13:419–436.

Stipanovic, R. D. 1982. Function and chemistry of plant trichomes and glands in insect resistance. In P. A. Hedin (ed.), Plant Resistance to Insects. American Chemical Society Symposium Series 208. pp. 69–100. American Chemical Society, Washington, D. C.

Stipanovic, R. D., A. A. Bell, and M. J. Lukefahr. 1977a. Natural insecticides from cotton (*Gossypium*). In P. A. Hedin (ed.), Host Plant Resistance to Insects. American Chemical Society Symposium Series 62. pp. 197–214. American Chemical Society, Washington, D. C.

Stipanovic, R. D., A. A. Bell, D. H. O'Brien, and M. J. Lukefahr. 1977b. Heliocide H2: an insecticidal sesquiterpenoid from cotton (*Gossypium*). Tetrahedron Lett. 6:567–570.

Stipanovic, R. D., A. A. Bell, D. H. O'Brien, M. J. Lukefahr. 1978a. Heliocide H1: a new insecticidal C25 terpenoid from cotton (*Gossypium hirsutum*). J. Agric. Food Chem. 26:115–118.

Stipanovic, R. D., A. A. Bell, D. H. O'Brien, and M. J. Lukefahr. 1978b. Heliocide H3: an insecticidal terpenoid from *Gossypium hirsutum.* Phytochemistry 17:151–152.

Stipanovic, R. D., A. A. Bell, and D. H. O'Brien. 1980. Raimondal, a new sesquiterpenoid from pigment glands of *Gossypium raimondii.* Phytochemistry 19:1735–1738.

Stipanovic, R. D., H. J. Williams, and L. A. Smith. 1986. Cotton terpenoid inhibition of *Heliothis virescens* development. In M. A. Green and P. A. Hedin (eds.), Natural Resistance of Plants to Pests—Role of Allelochemicals. American Chemical Society Symposium Series 296. pp. 79–94. American Chemical Society, Washington, D. C.

Sweetman, H. L. 1958. Principles of Biological Control. Wm. C. Brown, Dubuque, Iowa.

Terriere, L. C. 1984. Induction of detoxification enzymes in insects. Annu. Rev. Ent. 29:71–88.

Thompson, A. C., B. H. Wright, D. D. Hardee, R. C. Gueldner, and P. A. Hedin. 1970. Constituents of the cotton bud. XVI. The attractancy response of the boll weevil to the essential oils of a group of host and nonhost plants. J. Econ. Ent. 63:751–753.

Thorpe, W. A., and H. B. Caudle. 1938. A study of the olfactory responses of insect parasites to the food plant of their host. Parasitology 30:523–528.

Thurston, R., and P. M. Fox. 1972. Inhibition by nicotine of emergence of *Apanteles congregatus* from its host, the tobacco hornworm. Ann. Ent. Soc. Amer. 65:547–550.

Tingle, F. C., and E. R. Mitchell. 1984. Aqueous extracts from indigenous plants as oviposition deterents for *Heliothis virescens* (F.). J. Chem. Ecol. 10:101–113.

Tumlinson, J. H., and R. R. Heath. 1976. Structure elucidation of insect pheromones by microanalytical methods. J. Chem. Ecol. 2:87–99.

Vet, L. E. M., J. C. van Lenteren, M. Heymans, and E. Meelis. 1983. An air flow olfactometer for measuring olfactory responses of hymenopterous parasitoids and other small insects. Physiol. Ent. 8:97–106.

Vinson, S. B. 1975. Biochemical coevolution between parasites and their hosts. In P. W. Price (ed.), Evolutionary Strategies of Parasitic Insects and Mites. pp. 14–48. Plenum, New York.

Vinson, S. B. 1977. Behavioral chemicals in augmentation of natural enemies. In R. L. Ridgway and S. B. Vinson (eds.), Biological Control by Augmentation of Natural Enemies. pp. 237–279. Plenum, New York.

Vinson, S. B. 1981. Habitat location. In D. A. Nordlund, R. L. Jones, and W. J. Lewis (eds.), Semiochemicals: Their Role in Pest Control. pp. 51–77. Wiley, New York.

Vinson, S. B. 1984. Parasitoid-host relationship. In W. J. Bell and R. T. Cardé (eds.), Chemical Ecology of Insects. pp. 205–236. Chapman & Hall, London.

Vinson, S. B. 1986. The role of behavioral chemicals for biological control. In J. M. Franz (ed.), Biological Plant and Health Protection. Forschritt Zool Bd. 32. pp. 75–87. Gustav Fischer, New York.

Vinson, S. B., and P. Barbosa. 1987. Interrelationships of nutritional ecology of parasitoids. In F. Slansky, Jr. and J. G. Rodriguez (eds.), Nutritional Ecology of Insects, Mites, and Spiders and Related Invertebrates. pp. 673 695. Wiley, New York.

Voronin, K. E. 1981. Ecological aspects of the behavior of *Telenomia* (Hymenoptera: Scelionidae). In V. P. Pristavko (ed.), Insect Behavior as a Basis for Developing Control Measures against Pests of Field Crops and Forests (English translation). pp. 36–41. Amerind, New Delhi.

Zatuchini, G. E., and C. K. Osborn. 1981. Gossypol: a possible male antifertility agent. Report of a workshop. Res. Front. Fert. Regul. 1:1–15.

Zucker, W. V. 1983. Tannins: does structure determine function? An ecological perspective. Amer. Nat. 121:335–365.

CHAPTER 6

Natural Enemies and Herbivore–Plant Interactions: Influence of Plant Allelochemicals and Host Specificity

P. Barbosa
University of Maryland
College Park, Maryland

CONTENTS

1 Introduction
2 Influence of allelochemicals on specialist and generalist herbivores
3 Parasitoid host specificity and the influence of plant allelochemicals
4 Interactions among herbivores, allelochemicals, and insect pathogens
5 Counterintuitive evolution or unpredictable and variable selective forces?
Acknowledgments
References

1 INTRODUCTION

Theories on the relationships between chemical plant defense and herbivory have guided recent research on ecological and evolutionary aspects of insect–plant interactions. However, few testable general hypotheses on the role of the third trophic level in these interrelationships have been proposed, despite growing awareness of its importance. Current theories on the evolutionary and ecological importance of plant chemical defense to insect herbivory (excluding the role of nutrients) to a great extent revolve around the hypotheses developed in the mid-to-late 1970s by three research laboratories. Although some of the assumptions and details of the conceptual models developed by Feeny (1970, 1975, 1976), by Rhoades and Cates (1976), Cates and Rhoades (1977), Cates (1980), and by Futuyma (1976) and Futuyma and Gould (1979) differed, the predictions formulated by these researchers were similar. The chemical defense theory they proposed ascribed certain defense characteristics to two categories of plants, each with particular life history traits, designated as *apparent* or *unapparent plants*. Major details of the theory can be summarized as follows.

Unapparent plants, or those not likely to be detected, are generally short-lived and occur in a diverse community. Unapparent plants are protected by chemicals that act qualitatively (i.e., as toxins) occurring in low concentration, which are of little "cost" to the plant and which act in an "all or none" fashion rather than in a dose-dependent manner. Large, dominant, long-lived (apparent) plants, on the other hand, possess chemicals that act quantitatively; that is, they possess digestibility-reducing compounds, which are usually present in high concentrations, are "costly" to the plant, act in a dose-dependent manner, and affect such basic functions as digestion, growth, and reproduction.

The chemical defense theory also predicts the nature of the selective forces exerted by each type of defense on specialist and generalist herbivores. Qualitative defenses are presumed to exert selective pressures to which specialized herbivores evolve counteradaptations: generalized herbivores are less likely to evolve counteradaptations to these defenses. Nonadapted polyphagous herbivores are expected to be deterred by toxins, while adapted herbivores are capable of detoxifying qualitative defenses (Cates 1980). Apparent plants will inevitably be found and fed upon by specialists as well as generalists. Thus, defenses based on chemicals that act in a quantitative fashion are evolved so that they are difficult to overcome and remain effective against generalists and specialist insect herbivores.

In recent years it has become clear that the defense theory requires some refinement. There are several predictions of the theory that have not been fully supported by available data (see discussion below). These include, but are not limited to, the predictions that insects cannot evolve resistance to (i.e., overcome) quantitative defenses and that many compounds exhibit characteristics of both categories. In addition, many of the inconsistencies and other difficulties that arise in current theory may exist because the role of important selective forces (for example, the natural enemies of herbivores) virtually have been ignored.

While quantitative defenses can be effective against herbivores and difficult to overcome, there are, nevertheless, major exceptions. Some insects are resistant to hydrolyzable tannins (Fox and Macauley 1977), are stimulated to feed by so-called quantitative defense compounds (Bernays 1981), or are unaffected by them. Herbivores can adapt to quantitative defenses without exhibiting reduced feeding efficiency, growth, or reproduction (Fox and Macauley 1977). Bernays et al. (1980) also demonstrated that grasshoppers that normally feed on tanniferous plants are adapted to hydrolyzable tannins. Although several herbivore species have been reported to counter the effects of tannins, the exact mechanisms remain unknown. Possible mechanisms include the absorption of tannin by the midgut peritrophic membrane (Bernays and Chamberlain 1980), alkalinity of the midgut (Berenbaum 1980, 1983; Martin and Martin 1983; Schultz and Lechowicz 1986), and the presence of surfactants in the gut (Martin and Martin 1984).

Many compounds exhibit characteristics belonging to both qualitative and quantitative defense categories. Lincoln and Langenheim (1976) and Langenheim et al. (1978) demonstrated that in certain plant species terpenes occurred in low concentrations with high intraspecific variation, as expected for qualitative defenses. However, in other species (*Eucalyptus* trees, for example) terpenes can account for as much as 20% of leaf dry weight (Morrow and Fox 1980), as one would expect of quantitative defenses. On the other hand, some herbivores respond to toxins in a dose-dependent fashion. For example, *Spodoptera exiguae* shows a dose-dependent response to the leaf sesquiterpene, caryophyllene (Stubblebine and Langenheim 1977, Langenheim et al. 1980). Still other compounds, typically considered qualitative, such as cardiac glycosides or glucosinolates, have both toxic and digestiblity-reducing effects (Chew and Rodman 1979). Thus, for herbivorous insects the distinction between qualitative and quantitative defenses may be difficult to maintain.

In summary, these and other difficulties in the appropriateness and/or consistency of current terminology and concepts are sufficient to suggest that a modification of current theory is needed. Although other theories recently have been proposed, which focus on either resource allocation by plants (Mooney 1983) or resource availability in the habitat (Janzen 1979, McKey 1979, Coley et al. 1985), the plant apparency/predictability theory may still be viable, in a modified form. I suggest that many of the problems with the predictions of current theory arose because of the development, over the last decade, of hypotheses that excluded the role of the third trophic level.

Recently, Bergman and Tingey (1979), Campbell and Duffey (1979), and Price et al. (1980) have stressed that progress cannot be made without careful consideration of the third trophic level. Nevertheless, very few generalizations have been developed on the role of organisms of the third trophic level in herbivore–plant interactions. It is as yet unclear how and what interactions among plants, herbivores, and their natural enemies might modify current theory or produce new hypotheses. In part, the scarcity of unifying concepts is a result of the lack of robust empirical and experimental data (see Barbosa and Saunders 1985).

A great deal more data will be needed before new theories can be generated. However, available data may provide some interesting generalizations, which can serve as working hypotheses. This chapter is a review of available data on the influences of allelochemicals within herbivore hosts on parasitoids and other natural enemies. One dominant trend that becomes apparent in such a review is a clear dichotomy in the susceptibility of herbivores and their natural enemies (e.g., parasitoids), based on the host specificity of the species (i.e., whether it is monophagous or polyphagous). A second important trend illustrated is the dichotomy in the effects that two allelochemicals (typically classified as qualitative defenses) have on herbivores and their natural enemies. These are important issues because current theory predicts that the nature of the defense evolved by plants is driven by their apparency and by the benefits of the dichotomy in the effects of qualitative and quantitative defensive compounds. In addition, as noted previously, the reliability of the qualitative/quantitative dichotomy has been a subject of debate. A final, related question is whether the observed interactions among plant allelochemicals, herbivores, and their natural enemies are consistent with current plant defense theory; i.e., do they support the underlying assumption that the evolution of plant defense is in response to insect herbivory? Or are they counterintuitive and inconsistent with the above assumptions?

In general, the caveat expressed by Bergman and Tingey, Price et al., and others was based on extensive data on the influence of plants and plant extracts on the physiology, behavior, and ecology of parasitoids (see other chapters in this book). A great many of these data demonstrate the importance of chemical cues from plants in the host-habitat and host-finding behavior of parasitoids. Plant cues can enhance parasitoid search and oviposition behavior and thus result in increases in host mortality. Similarly, other more subtle interactions (such as reduced feeding in parasitized herbivores) can also benefit the plant. In contrast, recent research which demonstrates that plant allelochemicals ingested by herbivores can have a significant negative impact on the survival, development, morphology, and size of parasitoids, appears counterintuitive. A scenario more parsimonious with current theory would be one in which the effects of allelochemicals are detrimental to herbivores but parasitoids are unaffected or benefited. A review of available data on the influence of allelochemicals within herbivore hosts, on insect parasitoids, and on other natural enemies provides insight into factors which may explain the discrepancies observed. For example, the effects of allelochemicals can depend on the host specificity of either herbivore or parasitoid (which, for polyphagous parasitoid species, implies exposure to more than one host plant allelochemical or to variation in the type and concentration of allelochemicals in herbivore hosts) and on the degree of variation in both constituitive and induced levels of defensive chemicals. Both broad host species range and unpredictable variation in defensive chemical concentration may play a part in minimizing the ability of parasitoids (and possibly other natural enemies) to adapt to plant allelochemicals, producing what appears to be a counterintuitive interaction.

It should be noted that a discussion of the influence of allelochemicals on insect predators and its relation to current theory is excluded from this chapter. There are few, if any, appropriate data to develop even general working hypotheses on the subject [see Barbosa and Saunders (1985) and other chapters in this volume].

2   INFLUENCE OF ALLELOCHEMICALS ON SPECIALIST AND GENERALIST HERBIVORES

The influence that plant allelochemicals have on insect parasitoids, in large part, depends on the effects of those chemicals on their herbivore hosts. If, for example, the effects of a given allelochemical are more severe on specialist than on generalist herbivores, one would expect their parasitoids to be differentially affected by the allelochemical. The effects of rutin and nicotine on generalist and specialist herbivores provide good examples of such a response to two distinct qualitative defenses. Rutin is a quercitin glycoside, a widespread phenolic found in virtually every plant family. Concentrations of this flavonoid in plants range from 0.007 to 1.0%. It is biologically active in several herbivore species (Shaver and Lukefahr 1969, Neville and Luckey 1971, Kogan 1977, Isman and Duffey 1982), although it exhibits relatively low toxicity compared to other qualitatively acting compounds. In cultivated tobacco it is found in average concentrations ranging from 0.008 to 0.61% (Krewson and Naghski 1953). Nicotine, on the other hand, is the major alkaloid in tobacco and occurs in concentrations ranging from near 0% to 1.9% wet weight in domestic cultivars and from about 0.3% to 2% wet weight in wild species (Vandenberg and Matinger 1970, Sisson and Saunders 1983). When *Manduca sexta* (the tobacco hornworm) and *Trichoplusia ni* (the cabbage looper) are reared on allelochemical-free synthetic diets or on diets with rutin or nicotine (at 0.4, 0.6, 0.8, 1.0, 1.2, and 1.4% wet weight for the hornworm, and 0.016, 0.031, 0.061, 0.125, 0.150, 0.50, and 1.0% for *T. ni*), several interesting differences can be observed. The generalist, *T. ni,* is much more sensitive to both allelochemicals than is the specialist, *M. sexta.* concentrations that produce minimal effects on the tobacco hornworm produce extensive mortality of cabbage looper larvae. In *T. ni,* very high mortality is observed at low concentrations of nicotine (Fig. 6.1). There is no dose-dependent mortality due to nicotine: it is virtually an all-or-none response. Since only individuals feeding on diet with a concentration of 0.008% or less nicotine reached the pupal stage, no trends depicting the relationship between increasing dietary nicotine and larval development time or pupal weight exist. On the other hand, concentrations as high as 0.5% rutin, in the diet of *T. ni,* have very little effect on mortality, and even at 1.0% survivorship is still 30% (Fig. 6.1). There is a significant dose-dependent relationship between either larval development time or pupal weight and increasing rutin concentration (Figs. 6.2 and 6.3). Finally, as with the specialist herbivore, analyses of covariance demonstrate that the effects of increasing concentrations of rutin or nicotine on the generalist, *T. ni,* are distinct from each other, at least for larval development time and pupal weight (Barbosa et al., unpublished data). The differences in the way nicotine and rutin affect herbivores are consistently demonstrated in an array of developmental parameters (Fig. 6.1 to 6.7).

For the specialist tobacco hornworm, the effects of two defensive compounds, typically considered qualitative defenses, can be quite distinct. As the concentration of dietary nicotine increases, the weight of *Manduca sexta* fourth instars and pupae decreases and larval development time is

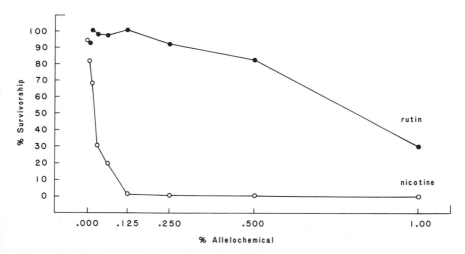

Figure 6.1 Influence of increasing concentrations of rutin and nicotine on *T. ni* survivorship.

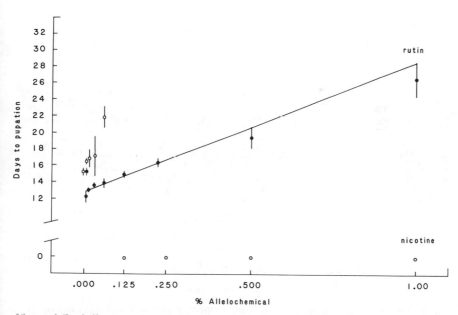

Figure 6.2 Influence of increasing concentrations of rutin and nicotine on *T. ni* larval development time.

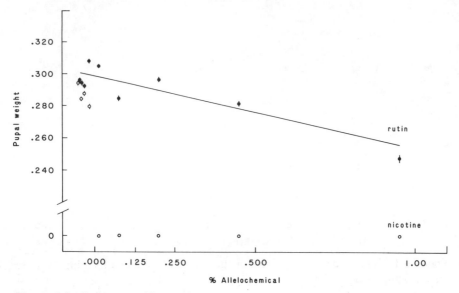

Figure 6.3 Influence of increasing concentrations of rutin and nicotine on *T. ni* pupal weight.

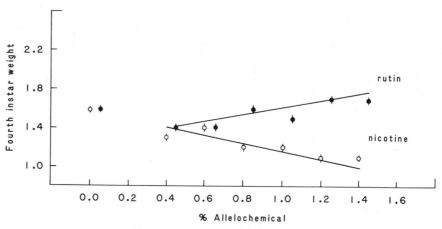

Figure 6.4 Influence of increasing concentrations of rutin and nicotine of *M. sexta* fourth instar weight.

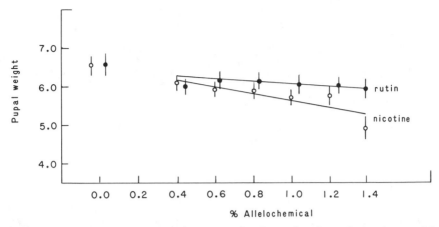

Figure 6.5 Influence of increasing concentrations of rutin and nicotine on *M. sexta* pupal weight.

prolonged (Figs. 6.4 to 6.6). On the other hand, with increasing concentration of rutin there are significant increases in fourth instar weight and little or no differences in pupal weight or larval development time (Figs. 6.4 to 6.6) (Krischik and Barbosa 1988). Survival of larvae, exposed to dietary rutin, differs little from control larvae, while larvae exposed to dietary nicotine exhibit variable survivorship, which on average is slightly lower than that of the control (Fig. 6.7). Perhaps most important is the fact that

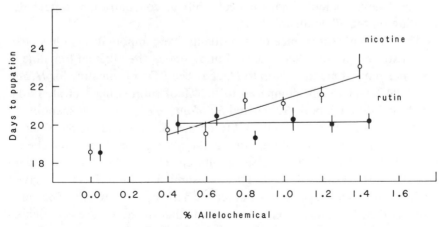

Figure 6.6 Influence of increasing concentrations of rutin and nicotine on *M. sexta* larval development time.

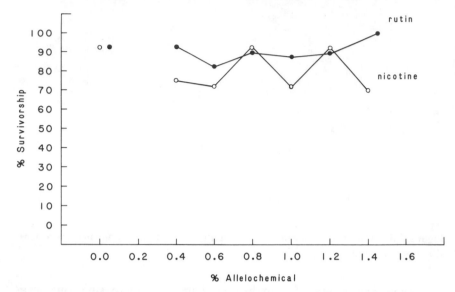

Figure 6.7   Influence of increasing concentrations of rutin and nicotine on *M. sexta* survivorship.

the effects of rutin and nicotine on fourth instar weight, pupal weight, and larval development time (as determined by analysis of covariance) are significantly different for each compound. That is, for this specialist herbivore the magnitude of the change in fourth instar and pupal weight and in larval development time, for each shift in concentration, differs depending on the allelochemical.

The differential influence of two qualitative compounds also has been demonstrated in other herbivores. For example, the alkaloid tomatine is six times more toxic than rutin to *H. zea*. The LD_{50} of tomatine for *H. zea* was 0.04% wet weight compared to 0.24% of rutin (Elliger et al. 1981). Similarly, although both *H. zea* and *Spodoptera exigua* are susceptible to the detrimental effects of rutin and tomatine, *S. exigua* is the more sensitive to both chemicals and tomatine is the more toxic of the allelochemicals (Duffey et al. 1986). These results closely parallel those found in studies of *M. sexta* and *T. ni*. Thus, in summary, the effects of a given allelochemical can depend on whether the herbivore is a relative specialist or generalist. In addition, regardless of whether the herbivore is a specialist or a generalist, the effects of two toxins typically designated as qualitative defenses can be dramatically distinct.

3 PARASITOID HOST SPECIFICITY AND THE INFLUENCE OF PLANT ALLELOCHEMICALS

A second major issue of interest is the question: Does the host specificity of parasitoids play a role in the effects of plant allelochemicals on the parasitoids' biology, physiology, and survival? Although there are not enough data to answer these questions, they do suggest hypotheses that merit consideration. Currently, data are available on six parasitoid species, which can be listed in increasing order of polyphagy as follows: *Cotesia congregata, Campoletis sonorensis, Cotesia marginiventris, Euplectrus plathypenae, Hyposoter annulipes,* and *H. exiguae.*

   C. congregata, the relatively monophagous parasitoid of the specialist *Manduca sexta,* shows the least detrimental effects of exposure to nicotine. The extent of its susceptibility is reflected in changes in development and survival (Table 6.1). Of the *C. congregata* larvae which emerge from their hosts, a greater proportion fail to form cocoons if they have emerged from nicotine-fed hornworms. However, there are no significant differences in pupal mortality or the average total number of adults resulting per host. Larval and pupal development as well as the size of adult *C. congregata* are also unaffected by nicotine (Barbosa et al. 1986). In field experiments using low- and high-nicotine cultivars, results paralleled those obtained in the laboratory. While the total number of *C. congregata* larvae per *M. sexta* feeding on each of the two cultivars is not different, significantly more male and female *C. congregata* adults are produced from hosts feeding on the low-nicotine variety. The difference is observed because a significantly greater proportion of larvae fail to emerge from hosts on high-nicotine tobacco and a high proportion of those that do emerge die prior to spinning cocoons. There are no differences in longevity or dry weight of *C. congregata* adults (Thorpe and Barbosa 1986). In another species, *Campoletis sonorensis,* a specialist on noctuids, gossypol incorporated in the diet of its *Heliothis* hosts has no detrimental effects. In fact, at very low concentrations larger adults were produced than produced from hosts on gossypol-free diets (S. B. Vinson, personal communication; see also Chapter 5).

   Other experiments, in which *S. frugiperda* was used as a host for *C. marginiventris, E. pathypenae,* and *H. annulipes,* and *Heliothis zea* served as host for *H. exiguae,* have evaluated the influence of plant allelochemicals on polyphagous parasitoid species. Of the three parasitoids utilizing the generalist *S. frugiperda,* the most polyphagous parasitoid

TABLE 6.1 Comparison of the Effects of Nicotine on Four Species of Insect Parasitoids[a]

	Parasitoid Species (Host)			
Life History Parameter	Cotesia congregata (Manduca sexta)	Hyposoter annulipes (Spodoptera frugiperda)	C. marginiventris (S. frugiperda)	Euplectrus platyhypenae (S. frugiperda)
Survival (% parasitism, % emergence, % cocoons formed)	−	−	0	−
Larval development	0	+	0	0
Pupal development	0	0	+	0
Total development	0	+	+	0
Adult size (dry weight)	0	−	*	*

Source: Based on data from Barbosa et al. (1986), El-Heneidy et al. (in press), and unpublished data.

[a] 0 = No effect; − = had an effect that reduced the parameter; + = had an effect that prolonged the parameter; * = data unavailable at this time. Nicotine was incorporated into the diet at a sublethal concentration of 0.1% (wet weight) for *M. sexta* and 0.025% for *S. frugiperda*.

appeared to suffer the most severe effects (Table 6.1). The fitness of *H. annulipes* is reduced to a greater degree than that of *C. congregata*. Of the fall armyworms reared on nicotine-containing diet and exposed to parasitism, only about 46% produced parasitoid larvae, compared to almost 76% of the hosts reared on nicotine-free diet. Larval parasitoid survival is reduced, development time is prolonged, and adult size is significantly reduced due to exposure to dietary nicotine (Thorpe and Barbosa 1986). Thus, fewer parasitoids are produced and those produced are smaller. The results are in marked contrast to those for *C. congregata,* where weight differences between adults reared on hosts with dietary nicotine and those reared on hosts without dietary nicotine were not significant. The decline in adult parasitoid weight associated with the presence of nicotine in hosts may have important negative consequences for the fecundity and overall fitness of *H. annulipes*.

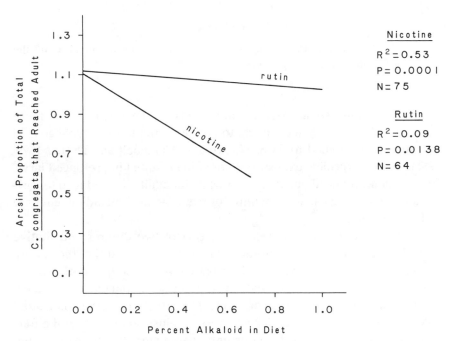

Figure 6.8 Influence of increasing concentrations of rutin and nicotine on the proportion of total *Cotesia congregata* adults.

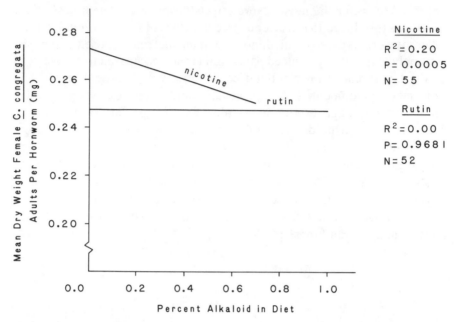

Figure 6.9 Influence of increasing concentrations of rutin and nicotine on the weight of female *Cotesia congregata* adults.

The most extensive and detrimental effects yet demonstrated are those caused by the plant allelochemical tomatine on the most polyphagous of the parasitoids studied to date, *H. exiguae* (Campbell and Duffey 1979, 1981). For this species, exposure to tomatine resulted in prolonged larval development, reduced pupal eclosion, smaller adult size, reduced longevity, and morphological abnormalities of major organs, including antennae and ovipositors.

Comparative data on the relative effects of two different qualitative allelochemicals are limited, but the data available suggest that the pattern is similar to that in herbivores. The negative influence of nicotine on *C. congregata* is exhibited in a greater variety of developmental parameters and is more severe than that of rutin. The effects on larval parasitoid survival and adult size (Figs. 6.8 and 6.9) are illustrative of several differences observed in various developmental parameters (Barbosa et al., unpublished data).

4 INTERACTIONS AMONG HERBIVORES, ALLELOCHEMICALS, AND INSECT PATHOGENS

A final issue is the role of plant allelochemicals in the interactions between herbivores and their pathogens. Can plant allelochemicals alter the interactions between an insect pathogen and its herbivore host? Results of experiments addressing this question suggest that while nicotine and rutin do indeed alter vegetative growth of the pathogen *Bacillus thuringiensis,* the effects of each allelochemical is distinct. Although the effect of both chemicals on colony growth are significant, the effects of nicotine are more severe than those produced by rutin. Indeed, above a nicotine concentration of 0.5%, no *B. thuringiensis* colonies are formed (Krischik et al. 1988a). Another important aspect of the interaction among herbivores, their pathogens, and allelochemicals is whether allelochemicals (like nicotine and rutin) alter the pathogenicity of organisms such as *B. thuringiensis,* on specialist and generalist herbivores (such as *M. sexta* and *Trichoplusia ni*).

Although survivorship of *M. sexta* larvae reared on *B. thuringiensis*–free nicotine diet can be between 80 and 100%, when diets contain both nicotine and *B. thuringiensis,* results differ. That is, while lower nicotine concentrations do not provide protection from *B. thuringiensis* (0.001%) (and 100% of the larvae die), concentrations above 0.4% nicotine dramatically enhance survival (Fig. 6.10) (Krischik et al. 1988a). In summary, the specialist herbivore, *M. sexta,* can not only tolerate toxins such as nicotine (Parr and Thurston 1972, Krischik and Barbosa 1988) but can do so at higher concentrations than tolerated by nonadapted insects (Yang and Guthrie 1969) and gain protection from entomogenous bacteria. As in previous illustrations, the influence of rutin on the specialist, *M. sexta,* is distinct from that of nicotine. While survivorship of larvae ranged from about 60% to above 90% on a rutin diet without *B. thuringiensis,* no larvae survived on a rutin diet with *B. thuringiensis* (Fig. 6.11) (Krischik et al. 1988b).

In contrast, for the generalist herbivore, *T. ni,* the interaction between *B. thuringiensis* and nicotine was totally different. When neonate *T. ni* larvae were reared on diets with increasing concentrations of *B. thuringiensis,* the presence of (0.01%) dietary nicotine resulted in enhanced mortality. For example, at 300 IU/G diet, the presence of nicotine in the

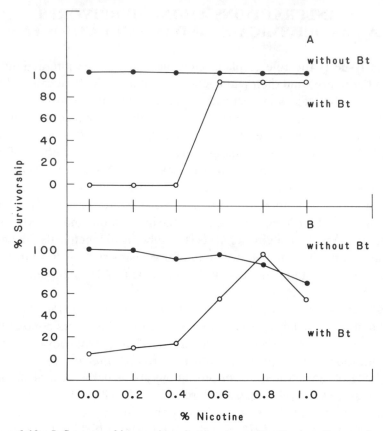

Figure 6.10   Influence of increasing nicotine concentrations on the survivorship of *M. sexta* larvae reared on diet with and without *B. thuringiensis*.

diet doubled the percent mortality of larvae (Fig. 6.12) (Reichelderfer et al., unpubl.). Rutin, on the other hand, appeared to provide some protection (i.e., enhanced survivorship) for *T. ni* larvae at low concentrations, although survivorship dropped dramatically above a rutin concentration of 0.010% (Figs. 6.13 and 6.14; see also Krischik et al. 1988b). The issue of the importance of the relative host specificity of pathogens and the effects of allelochemicals cannot be discussed at this time, because the data simply are not available. Indeed, the issue of specificity among insect pathogens and the ecological consequences of specificity has yet to be addressed, by anyone, in a comprehensive fashion. In summary, each of the two qualitative defenses, nicotine and rutin, mediates pathogenicity

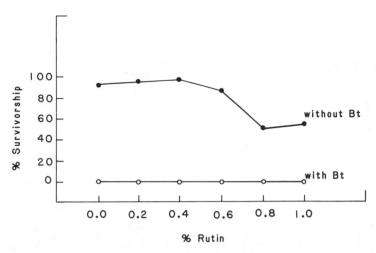

Figure 6.11 Influence of increasing rutin concentrations on the survivorship of *M. sexta* larvae reared on diet with and without *B. thuringiensis*.

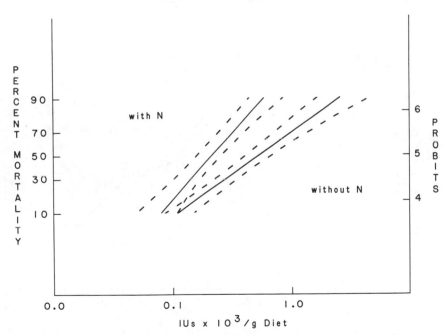

Figure 6.12 Influence of nicotine on mortality of *I. ni* reared on diets with increasing concentrations of *B. thuringiensis*.

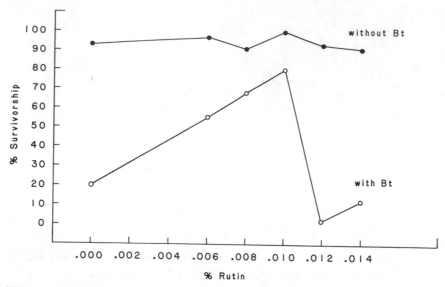

Figure 6.13 Influence of increasing rutin concentrations on the survivorship of *T. ni* larvae reared on diet with and without *B. thuringiensis*.

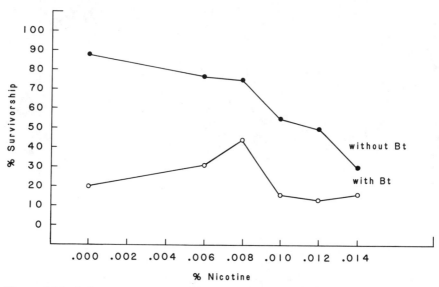

Figure 6.14 Influence of increasing nicotine concentrations on the survivorship of *T. ni* larvae reared on diet with and without *B. thuringiensis*.

218

differently. In addition, the influence of nicotine on the pathogenicity of *B. thuringiensis* differs in specialist and generalist herbivores.

5 COUNTERINTUITIVE EVOLUTION OR UNPREDICTABLE AND VARIABLE SELECTIVE FORCES?

Although some of the patterns described above are consistent with current theory, others, such as the susceptibility of herbivore natural enemies to the antibiotic effects of allelochemicals, appear counterintuitive and require alternative theories or modifications of current theory. Susceptibility to antibiotic effects of plant chemicals appears to be more pronounced in polyphagous species. This differential susceptibility of specialists and generalists may reflect different responses to the type, concentration and number of allelochemicals, as well as the degree of fluctuation in allelochemical concentrations to which each is exposed. Although the generalist is able to survive and, to some degree, survive exposure to a variety of allelochemicals, it does not do so as effectively as the specialist exposed to one specific allelochemical.

Inter- and intraspecific differences as well as daily and seasonal variation in host plant quality can have a significant impact on herbivore performance, survival, and the quality of the herbivore as hosts for natural enemies (Barbosa and Saunders 1985). Variation in allelochemicals and thus the quality of food plants for herbivores should be, for the most part, reflected in corresponding changes in the quality of herbivore hosts. Still another source of variation and an important ecological factor affecting natural enemies is induced changes in allelochemical concentrations in plants and often, in turn, in host herbivores' quality. They are particularly important forces because changes in defensive chemistry may be caused by a variety of agents and the nature of these changes may be unpredictable; thus making adaptation to specific levels of allelochemicals unlikely, particularly for polyphagous natural enemies.

Indeed, induction may be a very common source of allelochemical variation. Although well known by plant pathologists (Cruickshank and Mandryk 1960; Hare 1966; Kuc 1966, 1982; Sequeira 1983), the phenomenon of induced resistance in plants has recently become a topic of great interest to entomologists. The factors responsible for the induction of defensive allelochemicals have been thoroughly reviewed (Hare 1966,

Kuc 1966, Keen and Brueger 1977, Cohen and Kuc 1981, Davies and Schuster 1981, Hart et al. 1983, Ryan 1983, Sequeira 1983) and appear to be important sources of variation in plants. Recently, several studies have demonstrated that insect herbivory, as well as invasion of pathogens, induce resistance in plants (Green and Ryan 1972, 1973; Haukioja and Niemela 1977; Rhoades 1979; Wallner and Walton 1979; Carroll and Hoffman 1980; Schultz and Baldwin 1982; Kogan and Paxton 1983; Tallamy 1985). Regardless of the agent, induction due to pathogens, herbivores, or mechanical damage causes changes which are either localized (in the area adjacent to the damage) or systemic (occurring throughout the plant). Although exceptions exist, in general, induced responses to insect attack are usually systemic in nature (Green and Ryan 1972, 1973; Rhoades 1979, 1983a,b; Edwards and Wratten 1980, 1983; Tallamy 1985), while infection by plant pathogens results in either systemic (Kuc and Richmond 1977, Jenns and Kuc 1979, Guedes et al. 1980, Sequeira 1983) or localized induced changes (see Sequeira 1983).

The site of induction often varies. While many of the available examples are of induction resulting from damage to leaves, induction of resistance may occur as a result of damage to other plant parts, such as roots, by agents like phytopathogens (Roddick 1974, Cohen and Kuc 1981) or nematodes (Hanounik and Osborne 1975), or to flower heads (of, e.g., tobacco) (Waller and Nowacki 1978). Other factors may influence the nature and extent of induction. In the Solanaceae, for example, the magnitude of a wounding response is related to the location of the wound on the plant (i.e., whether on young upper leaves or lower older leaves) and to the severity of the wound (Ryan 1983).

There is an extremely wide variation in the rate at which plants respond to insect or pathogen attack (Rhoades 1979). The sources of variation include differences in plant species attacked, inducing agent, and the specific chemical that is mobilized or altered. The mobilization of defensive chemicals may occur in minutes or in hours. So-called qualitative chemical defenses may be induced more rapidly than so-called quantitative defenses (Loper 1968, Green and Ryan 1972, Benz 1977, Rhoades 1979, Carroll and Hoffman 1980, Tallamy 1985), but the latter defenses appear to be induced for months or even years (Thielges 1968; Baltensweiler et al. 1977; Benz 1977; Higgins et al. 1977; Niemela et al. 1979; Rhoades 1979, 1983a,b; Wallner and Walton 1979; Davies and Schuster 1981; Schultz and Baldwin 1982). Finally, although little is known about rates of relaxation (the rate at which elevated allelochemical concentrations decrease to preinduction levels), they appear to vary with the severity and duration of the response and with the nature of the chemical induced (Shumway et al. 1970, 1976; Caruso and Kuc 1977, 1979; Keen and Brueger 1977; Walker-Simmons and Ryan 1977; Walker-Simmons et al. 1984).

The issue of host specificity also may be of importance in the induction phenomenon. It has been noted that induction of short and long duration should have very different effects on specialist and generalist herbivores (Rhoades 1979, 1983a,b). Qualitative defenses which (according to current theory; see Feeny 1975, Rhoades and Cates 1976) are inexpensive to produce and easy to transport to the site of injury should thus be employed against generalists rather than specialists. Support for this contention is found in data on the specialist Colorado potato beetle, which is unaffected by rapidly induced potato defenses (Benz 1977). Alternatively, quantitative defenses which are postulated to be effective against both generalists and specialists should be induced for long periods, as they appear to be (see references above). Finally, depending on whether the herbivore is an adapted or a nonadapted species, induced levels of allelochemicals may have a variety of biological and behavioral effects. Any given chemical may act as a phagostimulant, a phagodeterrent, an arrestant, an attractant, or a repellant; or it may reduce consumption rate, reduce digestive efficiency, or may be toxic, depending on the concentration of the chemical, the herbivore species, the age of the herbivore, the mode of exposure, the ability of the herbivore to metabolize, sequester or egest the chemical, and so on.

Thus, in conclusion, the type and concentration of allelochemical that an herbivore and its parasitoid may have to contend with may depend on a series of factors and the sequence in which they occur. Variables that determined whether an induced response is systemic or localized can also determine whether or not a resident herbivore (and species of the next trophic level) will be affected by enhanced allelochemical concentrations. Similarly, the rapidity of the induction response and its duration may or may not affect the herbivore (and thus its parasitoid), depending on the length of feeding bouts and the tenure of the herbivore on the plant. Since the magnitude of some induction responses depend on the tissue injured, the specificity of herbivore feeding preferences can be very critical in determining whether the herbivore (and its natural enemies) are exposed to and affected by induced plant defenses. Finally, the extent of overall variation may be so great that the probability that natural enemies (e.g., parasitoids) could adapt to allelochemicals in the food plants of their herbivore hosts is very low, particularly if they are polyphagous parasitoids. Parasitoids or other natural enemies which are generalists are routinely exposed not only to highly variable allelochemicals but to numerous classes of compounds, and thus the latter are less likely to serve as directional selective forces.

It is clear from the examination of available data that while aspects of the interactions among herbivores, plant allelochemicals, and natural enemies are consistent with current theory, other aspects diverge significantly from theoretical expectations. As noted above, herbivores and natural enemies, such as parasitoids, that are relatively monophagous appear not only to evolve counteradaptations to qualitative defenses but may benefit from the presence of these compounds. Conversely, polyphagous species are less able to evolve counteradaptations. These patterns would appear to be consistent with those predicted by current theory. However, the data presented also make it clear that the action of all qualitative defenses is not invariant. The effects of two qualitative chemicals on herbivores, parasitoids, or pathogens can be quite distinct. In addition, the effect of any given qualitative defense may depend on the host specificity of herbivore or parasitoid exposed to the compound. Obviously, the patterns discussed and the speculations developed are preliminary conclusions. However, it is hoped that this analysis may stimulate further research that will help determine the importance of three-trophic-level interactions in the development of theory on insect-plant interactions.

ACKNOWLEDGMENTS

Scientific contribution No. 7811, article No. A-4791 of the Maryland Agricultural Experiment Station, Department of Entomology. I acknowledge and express appreciation for the support of the University of Maryland Computer Center and funding of research which stimulated many of the ideas expressed here (BSR-8400614). I am particularly grateful for the untiring support and the beautifully creative and stimulating minds of my laboratory team, which includes J. Bentz, P. Gross, J. Kemper, J. Kester, V. Krischik, and G. Morrison, Finally, many thanks for the support and guidance of my co-editor, Deborah Letourneau.

REFERENCES

Baltensweiler, W., G. Benz, P. Bovey, and V. Delucchi. 1977. Dynamics of larch bud moth populations. Annu. Rev. Ent. 22:79–100.

Barbosa, P., and J. A. Saunders. 1985. Plant allelochemicals: linkages between herbivores and their natural enemies. In G. A. Cooper-Driver, T. Swain, and E. E. Conn (eds.), Chemically Mediated Interactions between Plants and Other Organisms. pp. 107–137. Plenum, New York.

Barbosa, P., J. A. Saunders, J. Kemper, R. Trumbule, J. Olechno, and P. Martinat. 1986. Plant allelochemicals and insect parasitoids: effects of nicotine on *Cotesia congregata* and *Hyposoter annulipes*. J. Chem. Ecol. 12:1319–1328.

Benz, G. 1977. Insect-induced resistance as a means of self defense in plants. Eucarpia/IOBC Work. Group. (eds.), Resistance, Insects, Mites. pp. 155–159. Bull. SROP. 1977/1978.

Berenbaum, M. 1980. Adaptive significance of mid-gut pH in larval Lepidoptera. Amer. Nat. 115:138–146.

Berenbaum, M. 1983. Effects of tannins on growth and digestion in two species of papilionids. Ent. Exp. Appl. 34:245–250.

Bergman, J. M., and W. M. Tingey. 1979. Aspects of interaction between plant genotypes and biological control. Bull. Ent. Soc. Amer. 25:275–279.

Bernays, E. A. 1981. Plant tissues and insect herbivores: an appraisal. Ecol. Ent. 6:353–360.

Bernays, E. A., and D. J. Chamberlain. 1980. A study of tolerance of ingested tannin in *Schistocerca gregaria*. J. Insect Physiol. 26:415–420.

Bernays, E. A., D. J. Chamberlain, and P. McCarthy. 1980. The differential effects of ingested tannic acid on different species of Acridoidea. Ent. Exp. Appl. 28:158–166.

Campbell, B. C., and S. F. Duffey. 1979. Tomatine and parasitic wasps: potential incompatibility of plant antibiosis with biological control. Science 20:700–702.

Campbell, B. C., and S. F. Duffey. 1981. Alleviation of α-tomatine induced toxicity to the parasitoid, *Hyposoter exiguae* by phytosterols in the diet of the host, *Heliothis zea*. J. Chem. Ecol. 7:927–946.

Carroll, C. R., and C. A. Hoffman. 1980. Chemical feeding deterrent mobilized in response to insect herbivory and counteradaptation by *Epilachna tredecimnota*. Science 209:414–416.

Caruso, F., and J. Kuc. 1977. Protection of watermelon and muskmelon against *Colletotrichum lagenurium* by *Colletotrichum larenarium*. Phytopathology 67:1285–1289.

Caruso, F., and J. Kuc. 1979. Induced resistance of cucumber to anthracnose and angular leaf spot by *Pseudomonas lachrymans* and *Colletotrichum lagenarium*. Physiol. Plant Pathol. 14:191–201.

Cates, R. G. 1980. Feeding patterns of monophagous, oligophagous and polyphagous insect herbivores: the effect of resource abundance and plant chemistry. Oecologia (Berlin) 46:22–31.

Cates, R. G., and D. Rhoades. 1977. Patterns in the production of antiherbivore chemical defenses in plant communities. Biochem. Syst. Ecol. 5:185–193.

Chew, F. S., and J. E. Rodman. 1979. Plant resources for chemical defense. In G. A. Rosenthal and D. H. Janzen (eds.), Herbivores: Their Interaction with Secondary Plant Metabolites. pp. 271–307. Academic Press, New York.

Cohen, Y., and J. Kuc. 1981. Evaluation of systemic resistance to blue mold induced in tobacco leaves by prior stem inoculation with *Peronospora hyoscyami tabacina*. Phytopathology 71:783–787.

Coley, P. D., J. P. Bryant, and F. S. Chapin. 1985. Resource availability and plant antiherbivore defense. Science 230:895–899.

Cruickshank, I., and M. Mandryk. 1960. The effect of stem infestation of tobacco with *Peronospora tabacina* on foliage infection to blue mold. J. Aust. Inst. Agric. Sci. 26:369–372.

Davies, E., and A. Schuster. 1981. Intercellular communication in plants: evidence for a rapidly generated, bidirectionally transmitted wound signal. Proc. Natl. Acad. Sci. USA 78:2422–2426.

Duffey, S. S., K. A. Bloem, and B. C. Campbell. 1986. Consequences of sequestration of plant natural products in plant-insect-parasitoid interactions. In D. J. Boethel and R. D. Eikenbarry (eds.), Interactions of Plant Resistance and Parasitoids and Predators of Insects. pp. 31–60. Wiley, New York.

Edwards, P. J., and S. D. Wratten. 1980. Ecology of Insect-Plant Interactions. Edward Arnold, London.

Edwards, P. J., and S. D. Wratten. 1983. Wound induced defences in plants and their consequences for patterns of insect grazing. Oecologia (Berlin) 59:88–93.

El-Heneidy, A. H., P. Barbosa, and P. Gross. 1988. Influence of host's diet on larval parasitoids of the fall armyworm, *Spodoptera frugiperda. 3. Hyposoter annulipes. Ent. Exp. Appl. (in press).*

Elliger, C. A., Y. Wong, B. G. Chan, and A. C. Waiss, Jr. 1981. Growth inhibitors in tomato (*Lycopersicon*) to tomato fruitworm (*Heliothis zea*). J. Chem. Ecol. 7:753–758.

Feeny, P. P. 1970. Seasonal changes in oak leaf tannins and nutrients as a cause of spring feeding by winter moth caterpillars. Ecology 51:565–581.

Feeny, P. P. 1975. Biochemical coevolution between plants and their insect herbivores. In L. E. Gilbert and P. H. Raven (eds.), Coevolution of Animals and Plants. pp. 3–19. University of Texas Press, Austin, Tex.

Feeny, P. P. 1976. Plant apparency and chemical defense. In J. W. Wallace and R. L. Mansell (eds.), Biochemical Interactions between Plants and Insects. pp. 1–40. Plenum, New York.

Fox, L. R., and B. J. Macauley. 1977. Insects grazing on *Eucalyptus* in response to variations in leaf tannins and nitrogen. Oecologia (Berlin) 29:145–162.

Futuyma, D. J. 1976. Food plant specialization and environmental predictability in Lepidoptera. Amer. Nat. 110:285–292.

Futuyma, D., and F. Gould. 1979. Associations of plants and insects in a deciduous forest. Ecol. Monogr. 49:33–50.

Green, T. R., and C. A. Ryan. 1972. Wound-induced proteinase inhibitor in plant leaves: a possible defense mechanism against insects. Science 175:776–777.

Green, T. R., and C. A. Ryan. 1973. Wound-induced proteinase inhibitor in tomato leaves. Plant Physiol. 51:19–21.

Guedes, M. E. M., S. Richmond, and J. Kuc. 1980. Induced systemic resistance to anthracnose in cucumber as influenced by the location of the inducer inoculation with *Colletotrichum lagenarium* and onset of flowering and fruiting. Physiol. Plant Pathol. 17:229–233.

Hanounik, S. B. and W. W. Osborne. 1975. Influence of *Meloidogyne incognita* on the content of amino acids and nicotine in tobacco grown under gnotobiotic conditions. J. Nematol. 7:332–335.

Hare, R. C. 1966. Physiology of resistance to fungal diseases in plants. Bot. Rev. 32:95–137.

Hart, S. V., M. Kogan, and J. D. Paxton. 1983. Effect of soybean phytoalexins on the herbivorous insects Mexican bean beetle and soybean looper. J. Chem. Ecol. 9:657–673.

Haukioja, E., and P. Niemela, 1977. Retarded growth of a geometrid larva after mechanical damage to leaves of its host tree. Ann. Zool. Fenn. 14:48–52.

Higgins, K. M., J. E. Browns, and B. A. Haws. 1977. The black grass bug (*Labops hesperius* Uhler): its effect on several native and introduced grasses. J. Range Manag. 30:380–384.

Isman, M. R., and S. S. Duffey. 1982. Toxicity of tomato phenolic compounds to the fruitworm, *Heliothis zea*. Ent. Exp. Appl. 31:370–376.

Janzen, D. H. 1979. New horizons in the biology of plant defenses. In G. A. Rosenthal and D. H. Janzen (eds.), Herbivores: Their Interactions with Secondary Plant Metabolites. pp. 331–350. Academic Press, New York.

Jenns, A. E. and J. Kuc. 1979. Graft transmission of systemic resistance of cucumber to anthracnose induced by *Colletotrichum lagenarium* and tobacco necrosis virus. Phytopathology 69:753–756.

Keen, N. I., and B. Brueger. 1977. Phytoalexins and chemicals that elicit their production in plants. In P. A. Hedin (ed.), Host Plant Resistance to Pests.

American Chemical Society Symposium Series 62, pp. 1–26. American Chemical Society, Washington, D. C.

Kogan, M. 1977. The role of chemical factors in insect/plant relationships. In Proceedings of the 15th International Congress on Entomology. pp. 211–227. Entomological Society of America, College Park, Md.

Kogan, M., and J. Paxton. 1983. Natural inducers of plant resistance to insects. In P. A. Hedin (ed.), Plant Resistance to Insects. American Chemical Society Symposium Series 208. pp. 153–171. American Chemical Society, Washington, D. C.

Krewson, C. F., and J. Naghaski. 1953. Occurrence of rutin in plants. Amer. J. Pharm. 117:190–200.

Krischik, V. A., and P. Barbosa. 1988. The effects of a toxic and a non-toxic allelochemical on generalist and specialist herbivores. (in preparation).

Krischik, V. A., P. Barbosa, and C. Reichelderfer. 1988a. Three trophic level interactions: allelochemicals, *Manduca sexta,* and *Bacillus thuringiensis* var. *kurstaki*. Environ. Ent. 17:476–482.

Krischik, V. A., P. Barbosa, and C. Reichelderfer. 1988b. Three trophic level interactions: differences in the effects of two plant allelochemicals on the virulence of *B. t.* (in preparation).

Kuc, J. 1966. Resistance of plants to infectious agents. Annu. Rev. Microbiol. 20:337–364.

Kuc, J. 1982. Plant immunization-mechanisms and practical implications. In R. K. S. Wood and E. Tjamos (eds.), Active Defensive Mechanisms in Plants. pp. 157–178. Plenum, New York.

Kuc, J., and S. Richmond. 1977. Aspects of the protection of cucumber against *Colletotrichum lagenarium* by *Colletotrichum lagenarium*. Phytopathology 67:533–536.

Langenheim, J. H., W. H. Stubblebine, D. E. Lincoln, and C. E. Foster. 1978. Implications of variation in resin composition among organs, tissues and populations in the tropical legume *Hymenaea*. Biochem. Syst. Ecol. 6:299–313.

Langenheim, J. H., C. E. Foster, and R. B. McGinley. 1980. Inhibitory effects of different quantitative compositions of *Hymenaea* leaf resins on a generalist herbivore, *Spodoptera exigua*. Biochem. Syst. Ecol. 8:385–396.

Lincoln, D. E., and J. H. Langenheim. 1976. Geographic patterns of monoterpenoid composition in *Satureja douglasii*. Biochem. Syst. Ecol. 4:237–248.

Loper, G. M. 1968. Effects of aphid infestation on the coumestrol content of alfalfa varieties differing in aphid resistance. Crop Sci. 8:104–106.

Martin, J. S., and M. M. Martin. 1983. Precipitation of ribuluse-1,5-bisphosphate carboxylase oxygenase by tannic acid, quebrachos, and oak foliage extracts. J. Chem. Ecol. 9:285–294.

Martin, M. M., and J. S. Martin. 1984. Surfactants: their role in preventing precipitation of proteins by tannins in insect guts. Oecologia (Berlin) 61:342–345.

McKey, D. 1979. The distribution of secondary compounds within plants. In G. A. Rosenthal and D. H. Janzen (eds.), Herbivores: Their Interactions with Secondary Plant Metabolites. pp. 59–133. Academic Press, New York.

Mooney, H. A., S. L. Gulman, and N. D. Johnson. 1983. Physiological constraints on plant chemical defense. In P. Hedin (ed.), Plant Resistance to Insects. American Chemical Society Symposium Series 208. pp. 21–26. American Chemical Society, Washington, D.C.

Morrow, P. A., and L. R. Fox. 1980. Effects of variation in *Eucalyptus* essential oil yield on insect growth and grazing damage. Oecologia (Berlin) 45:209–219.

Neville, P. F., and T. D. Luckey. 1971. Bioflavonoids as a new growth factor for the cricket, *Acheta domesticus*. J. Nutr. 101:1217–1224.

Niemela, P., A. M. Aro, and E. Haukioja. 1979. Birch leaves as a resource for herbivores. Damage-induced increase in leaf phenols with trypsin-inhibiting effects. Rev. Kevo. Subarct. Res. Stn. 15:37–40.

Parr, J. C., and R. Thurston. 1972. Toxicity of nicotine in synthetic diets to larvae of the tobacco hornworm. Ann. Ent. Soc. Amer. 65:1185–1188.

Price, P. W., C. E. Bouton, P. Gross, B. A. McPheron, J. N. Thompson, and A. E. Weis. 1980. Interactions among three trophic levels: influence of plants on interactions between insect herbivores and natural enemies. Annu. Rev. Ecol. Syst. 11:41–65.

Rhoades, D. F. 1979. Evolution of plant chemical defenses against herbivores. In G. A. Rosenthal and D. H. Janzen (eds.), Herbivores: Their Interaction with Secondary Plant Metabolites. pp. 3–54. Academic Press, New York.

Rhoades, D. F. 1983a. Responses of alder and willow to attack by tent caterpillars and webworms: evidence for pheromonal sensitivity of willows. In P. Hedin (ed.), Plant Resistance to Insects. American Chemical Society Symposium Series 208. pp. 55–68. American Chemical Society, Washington, D.C.

Rhoades, D. F. 1983b. Herbivore population dynamics and plant chemistry. In R. F. Denno and M. S. McClure (eds.), Variable Plants and Herbivores in Natural and Managed Systems. pp. 155–220. Academic Press, New York.

Rhoades, D. F., and R. H. Cates. 1976. Toward a general theory of plant antiherbivore chemistry. In J. W. Wallace and R. L. Mansell (eds.), Biochemical Interactions Between Plants and Insects. pp. 168–213. Plenum, New York.

Roddick, J. G. 1974. The steroidal glycoalkaloid α-tomatine. Phytochemistry 13:9–25.

Ryan, C. A. 1983. Insect-induced chemical signals regulating natural plant protection responses. In R. F. Denno and M. S. McClure (eds.), Variable Plants and

Herbivores in Natural and Managed Systems. pp. 43–60. Academic Press, New York.

Schultz, J. C., and I. T. Baldwin. 1982. Oak leaf quality declines in response to defoliation by gypsy moth larvae. Science 217:149–150.

Schultz, J. C., and M. J. Lechowicz. 1986. Host plant, larval age, and feeding behavior influence midgut pH in the gypsy moth (*Lymantria dispar*). Oecologia (Berlin) 71:133–137.

Sequeira, L. 1983. Mechanisms of induced resistance in plants. Annu. Rev. Microbiol. 37:51–79.

Shaver, T. N., and N. J. Lukefahr. 1969. Effect of flavonoid pigments and gossypol on growth and development of the bollworm, tobacco hornworm and pink bollworm. J. Econ. Ent. 62:643–646.

Shumway, K., J. M. Rancour, and C. A. Ryan. 1970. Vacuolar protein bodies in tomato leaf cells and their relationship to storage of chymotrypsin inhibitor I protein. Planta 93:1–14.

Shumway, K., V. V. Yank, and C. A. Ryan. 1976. Evidence for the presence of proteinase inhibitor I in vacuolar bodies of plant cells. Planta 129:161–165.

Sisson, V. A., and J. A. Saunders. 1983. Catalog of the tobacco introductions in the U.S. Department of Agriculture's tobacco germplasm collection (*Nicotiana tabacum*). USDA, ARS. ARM-5-27. pp. 1–27.

Stubblebine, W. H., and J. H. Langenheim. 1977. Effects of *Hymenaea courbaril* leaf resin in the generalist herbivore *Spodoptera exigua* (beet armyworm). J. Chem. Ecol. 3:633–647.

Tallamy, D. W. 1985. Squash beetle feeding behavior: An adaptation against induced cucurbit defenses. Ecology 66:1574–1579.

Thielges, B. A. 1968. Altered polyphenol metabolism in the foliage of *Pinus sylvestris* associated with European pine sawfly attack. Can. J. Bot. 46:724–726.

Thorpe, K. W., and P. Barbosa. 1986. Effects of consumption of high and low nicotine tobacco by *Manduca sexta* on survival of gregarious endoparasitoid *Cotesia congregata*. J. Chem. Ecol. 12:1329–1337.

Vandenberg, P., and D. F. Matzinger. 1970. Genetic diversity and heterosis in *Nicotiana*. III. Crosses among tobacco introductions and flue cured varieties. Crop Sci. 10:437–440.

Walker-Simmons, M., and C. A. Ryan. 1977. Immunological identification of proteinase inhibitors I and II in isolated tomato leaf vacuoles. Plant Physiol. 60:61–63.

Walker-Simmons, M., H. Hollander-Czytko, J. K. Andersen, and C. A. Ryan. 1984. Wound-signals in plants: a systemic plant wound signal alters plasma membrane integrity. Proc. Natl. Acad. Sci. USA 81:3737–3741.

Waller, G. R., and E. K. Nowacki. 1978. Alkaloid Biology and Metabolism in Plants. Plenum, New York.

Wallner, W. E., and G. S. Walton. 1979. Host defoliation: a possible determinant of gypsy moth population quality. Ann. Ent. Soc. Amer. 72:62–67.

Yang, R. S. H., and F. E. Guthrie. 1969. Physiological responses of insects to nicotine. Ann. Ent. Soc. Amer. 62:141–146.

PART IV

KEY ROLES OF PLANT ALLELOCHEMICALS IN SURVIVAL STRATEGIES OF HERBIVORES

Chapters 7 (Pasteels, Rowell-Rahier, and Raupp) and 8 (Bowers) concern the ecology of unpalatable prey. It is well known that plant defense compounds can be exploited by certain phytophagous insects. Detrimental effects of plant allelochemicals do not become evident until herbivores are attacked or ingested by a member of the third trophic level. As sources of chemical substances that are sequestered and processed by plant-feeding insects for defense against their natural enemies, plant allelochemicals can be critical determinants of herbivore survival. Thus, the interactions of predator and prey are linked closely to parameters of host plant quality. Discussions of effects of plant allelochemicals on parasitoids, developed in the preceding section, strongly emphasized physiological processes. In Part IV, these physiological and metabolic processes, such as modes of sequestering, excreting, or secreting plant substances by herbivores, are linked not only to physiological effects on natural enemies, but to predator behavior, coevolved mimicry systems, and patterns of population distribution.

Defensive substances in insects can act as deterrents and toxicants to a wide variety of target organisms: vertebrates, invertebrates, and microorganisms (Blum 1981). Herbivores become unpalatable via plant allelochemicals in a number of ways, including the retention of noxious compounds in the foregut or hindgut for subsequent discharge, selective sequestration of chemicals that can be expelled from specialized glands, and the incorporation of toxic or emetic compounds into the hemolymph or tissues. The sequestration of plant chemicals, used as a means of defense against predators, is often accompanied by elaborate structural or physiological adaptations. Aposematic coloration, reflex bleeding, and eversible glands that extrude noxious substances are among features that enhance the efficacy of allelochemically based protective mechanisms (Dettner 1987). These adaptations, and the behavioral responses by their predators, make many of the predator–prey interactions seem more ecologically complex than those of host interference with the success of internal parasitoids via plant toxins, even though the deleterious effects of plant allelochemicals on parasitoids (e.g., Campbell and Duffey 1979, Duffey et al. 1986; Chapters 5 and 6) are often more severe.

The combination of herbivore and plant properties that determines the efficacy of plant allelochemicals as a basis for herbivore defense include (1) the frequency with which the compounds occur in host plants, (2) the existence of a mechanism to prevent deleterious effects of the chemical on the herbivore, (3) the relative ability of the herbivore to sequester and process different plant compounds, (4) the presence of a system for delivery to or detection by predators, and (5) the relative costs to the herbivore as compared to biosynthesis of defensive compounds *de novo*. Recent studies of the availability of secondary compounds (both concentration and diversity) in different plant species, but also within plants through time, between plant tissues, between individuals, and between populations, have added a great deal of complexity to what seemed to be elaborate but unambiguous defensive systems. The ecological consequences are particularly interesting in mimicry systems, in which the basis of the interrelationships is an unpalatable model. How does such variation affect the stability of mimicry systems? Early discoveries of extreme cases of host plant variability with respect to allelochemicals, in which the same herbivore population contains models and mimics (Brower et al. 1968) or where different species switch roles, from model to mimic, and vice versa, depending upon their diet of local host plants (Marsh et al. 1977), are now being accepted as common (Pasteur 1982; see also Chapter 8).

We might expect a set of unique behaviors in herbivores that depend upon chemicals in their host plants as a means of defense. In Chapter 8, Bowers discusses the relationship between ovipositional patterns, larval survival rates, and adult defense for a number of lepidopteran species that sequester plant allelochemicals as larvae and also as adults. She also reviews the available information on costs of defensive chemical storage for Lepidoptera that sequester chemicals from plants. Pasteels et al. (Chapter 7) suggest that the diverse array of defense mechanisms, at least in chrysomelid beetles, is evolutionarily labile and strongly dependent upon the types and availability of plant chemicals within host plants, the relative energetics of allomone biosynthesis versus sequestration of plant secondary compounds, and selective pressures from natural enemies.

Part IV, then, provides critical analyses of the available data on chemical effects between and among trophic levels for a variety of plant–herbivore–enemy systems. The first two chapters cross subdisciplines of physiology, behavior, ecology, and evolutionary biology to analyze the movement of plant allelochemicals through the food chain in the context of defense and exploitation ecology. Promising avenues of research are targeted in fascinating and controversial areas such as mimicry, competition between phytophagous insects, herbivore distribution patterns, and variability in natural systems. Brattsten (Chapter 9) departs from the previous theme of plant allelochemical effects on natural enemies as mortality factors imposing selective pressure upon herbivores. Instead, she addresses a direct and abiotic selection pressure imposed upon herbivores. This final chapter is an analysis of the effects of plant allelochemicals on the survival of herbivores exposed to synthetic insecticides. In contrast to the selective pressures by predators and parasitoids, human-generated selective pressures in the form of insecticides are more comparable to the chemical stresses imposed upon herbivores by secondary chemicals in their host plants. Previous chapters addressed plant allelochemical effects on susceptibility or resistance of herbivores to their natural enemies; Brattsten's timely and intriguing discussion focuses on behavioral, physiological, and biochemical pathways that link effects of plant allelochemicals to herbivore susceptibility or resistance to synthetic insecticides.

D. K. LETOURNEAU

REFERENCES

Blum, M. S. 1981. Chemical Defenses of Arthropods. Academic Press, New York.

Brower, L. P., W. N. Ryerson, L. L. Coppinger, and S. C. Glazier. 1968. Ecological chemistry and the palatability spectrum. Science 161:1349–1351.

Campbell, B. C., and S. F. Duffey. 1979. Tomatine and parasitic wasps: potential incompatibility of plant-antibiosis with biological control. Science 205:700–702.

Dettner, K. 1987. Chemosystematics and evolution of beetle chemical defenses. Annu. Rev. Ent. 32:17–48.

Duffey, S. S., K. A. Bloem, and B. C. Campbell. 1986. Consequences of sequestration of plant natural products in plant-insect-parasitoid interactions. In B. J. Boethel and R. D. Eikenbary (eds.), Interactions of Plant Resistance and Parasitoids and Predators of Insects. pp. 31–60. Wiley, New York.

Marsh, N. A., C. A. Clarke, M. Rothschild, and D. N. Kellett. 1977. *Hypolimnas bolina* (L.) a mimic of danaid butterflies, and its model *Euploea core* (Cram) store cardioactive substances. Nature 268:726–728.

Pasteur, G. 1982. A classificatory review of mimicry systems. Annu. Rev. Ecol. Syst. 13:169–199.

CHAPTER 7

Plant-Derived Defense in Chrysomelid Beetles

J. M. Pasteels
Université Libre de Bruxelles
Bruxelles, Belgium

M. Rowell-Rahier
Zoologisches Institut der Universität Basel
Basel, Switzerland

M. J. Raupp
University of Maryland
College Park, Maryland

CONTENTS

1   Introduction: diversity and distribution of chemical defenses in leaf beetles
2   Influence of the host plant on the defense secretion of Chrysomelidae
3   Impact of host plant variation on the spatial and temporal distributions of leaf beetles
4   Efficacy of the different groups of chemical compounds and their distribution among developmental stages of herbivores
5   Protection against predators and parasitoids
6   Protection against competing herbivores
7   Summary and Conclusions
    References

1   INTRODUCTION: DIVERSITY AND DISTRIBUTION OF CHEMICAL DEFENSES IN LEAF BEETLES

Phytophagous insects are particularly vulnerable to predators and parasitoids. Unless they are concealed within the plant tissue or otherwise protected, their location on the leaf surface and relative immobility makes them visible and highly vulnerable to enemies (Price 1984). Additionally, their relatively low energy conversion rates (Southwood 1973) require that they spend long periods of time feeding. Escape is thus difficult and these organisms are potentially easily located and captured. Therefore, it is not surprising that a large array of defensive mechanisms must have evolved in phytophagous insects. This is certainly true for the family of leaf beetles, Chrysomelidae, in which a large diversity of behavioral, morphological, mechanical, and chemical protective devices has been described (reviewed in DeRoe and Pasteels 1982).

In this chapter we focus our attention on chemical defense and more particularly, on the influence of the host plant on defense. After a brief review of the distribution and diversity of chemical defenses in the family Chrysomelidae, we discuss some of the factors that might have exerted selective pressure on the chemical nature of the defense. These include host plant secondary chemistry and the enemies of leaf beetles. Section 3 is devoted to a discussion of how the host plant influences the spatial and temporal distribution of leaf beetles on individual plants and within the community; clearly, these patterns have emerged not only as a consequence of the unequal distribution of plant allelochemicals but also the modification of the probabilities of encountering potential enemies.

Chemical defense is unevenly distributed in the family Chrysomelidae. It has been unambiguously demonstrated in only four subfamilies (Criocerinae, Chrysomelinae, Galerucinae, and Alticinae) out of 19 (Table 7.1). However, only the west European fauna has been investigated consistently.

Defensive compounds are stored in various organs and their modes of release differ not only from species to species, but also between the various developmental stages. Further, the chemical nature of the compound is often quite different in closely related taxa or in different life stages of the same species.

Homologous glands are found in the different groups on the elytra and pronotum of members of the four subfamilies (DeRoe and Pasteels 1982), but their role in defense has been demonstrated only in the Criocerinae

TABLE 7.1 Chemical Defense in Leaf Beetles

Taxa	Storage Sites	Modes of Release	Chemical Classes	References
		Adults		
Criocerinae	Pronotal and elytral glands	Secretion covers the integument	Unknown	DeRoe and Pasteels (1982)
Chrysomelinae				
Chrysomelini, Chrysolinina (20 studied species)	Pronotal and elytral glands	Secretion covers the integument	Cardenolides or polyoxygenated steroids	Pasteels et al. (1984), Daloze et al. (1985)
Chrysomela brunsvicensis	Unknown	Unknown	Hypericin	Rees (1969)
Chrysomelina and Phyllocectina (13 studied species)	Pronotal and elytral glands	Secretion covers the integument	Isoxazolinone derivatives and hydrocarbons	Pasteels et al. (1984)
Doryphorina				
Leptinotarsa decemlineata[a]	Pronotal and elytral glands	Secretion covers the integument	γ-Glutamyl dipeptide	Daloze et al. (1986)
Galerucinae				
Diabrotica (3 species)	Hemolymph	Wounds	Cucurbitacins	Ferguson and Metcalf (1985)
Acalymma vittatum	Hemolymph	Wounds	Cucurbitacins	Ferguson and Metcalf (1985)
Alticinae				
Dibolia borealis	Unknown	Unknown	Iridoid glycosides	Bowers (in press)

TABLE 7.1 (*Continued*)

Taxa	Storage Sites	Modes of Release	Chemical Classes	References
		Larvae[b]		
Chrysomelinae				
Chrysomelini, Chrysolinina (3 species studied)	Unknown	Unknown	Cardenolides	Daloze and Pasteels (1979)
Chrysomelina and Phyllodectina (22 species studied)	9 pairs of thoracic and abdominal glands	Defensive droplet at the tip everted reservoirs	Methylcyclopentanoid monoterpenes	Pasteels et al. (1984)
			or	
			Salicylaldehyde and benzaldehyde (in some)	Pasteels et al. (1984)
			or	
			Juglone	Matsuda and Sugawara (1980)
			or	Blum et al. (1972)
			Phenylethyl esters	Takizawa (1976)
Phytodectina (some)	1 pair of postabdominal glands	Defensive droplet at the tip of the everted reservoirs	Unknown	
Paropsina (3 species studied)	1 pair of postabdominal glands	Defensive secretion at the tip of everted reservoirs	HCN and benzaldehyde	Moore (1967)

Galerucinae				
Diabrotica (3 species)	Hemolymph	Reflex bleeding	Cucurbitacins	Ferguson and Metcalf (1985)
Acalymma vittatum	Hemolymph	Wounds?	Cucurbitacins	Ferguson and Metcalf (1985)
Pyrrhalta luteola	Hemolymph	Wounds	Anthraquinones and anthrones	Howard et al. (1982a)
Chrysomelinae				
Chrysomelini, Chrysolina (2 species studied)	Egg plasma?	Wounds?	Cardenolides	Daloze and Pasteels (1979)
Chrysomelina and Phyllodectina (10 species studied)	Egg plasma	Wounds	Isoxazolinone derivatives, salicin (in some)	Pasteels et al. (1986)
Gastrophysa cyanea	Egg plasma	Wounds	Oleic acid	Howard et al. (1982b)
Galerucinae				
Diabrotica (2 species studied) Acalymma vittatum	Egg plasma?	Wounds?	Cucurbitacins	Ferguson and Metcalf (1985)

[a] A toxic protein, leptinotarsin, is present in the hemolymph of all development stages of the Colorado potato beetle and in the larvae of *Leptinotarsa juncta* (Hsiao and Fraenkel 1969, Parker 1972). According to Hsiao and Fraenkel 1969, the protein is only toxic by injection and has no oral toxicity, which indicates that it has no natural function as a toxin.

[b] The larval content of several Alticinae (*Diamphidia* and *Polyclada* species) are used as arrow poison by Bushmen in southern Africa. The toxic principle is a low-molecular-weight compound which is closely attached or bound to a protein protecting it from inactivation (Mebs et al. 1982). To our knowledge a natural defensive function has not yet been demonstrated for the toxin.

239

(Pasteels, unpublished observations) and the Chrysomelinae. In the latter subfamily, the glands of the different subtribes of the Chrysomelini store and secrete biosynthetically unrelated compounds (Table 7.1). These glands are absent in some Galerucinae and Alticinae, illustrating how one sophisticated mode of defense can be replaced by another in the course of evolution. Reflex bleeding, for example, occurs most frequently in the larvae and adults of the Galerucinae, but it also occurs in other subfamilies [e.g., in the well-known bloody-nose beetles (*Timarcha* spp., Chrysomelinae)]. In *Timarcha,* the elytral glands degenerate when secondary sclerotization obliterates the hemolymph supply to the elytra and transforms the fused elytra into a completely inert carapace (Pasteels, unpublished observations).

Eversible abdominal glands are observed occasionally in chrysomeline larvae. Among these, however, the glands of the tribe Paropsina and Gonioctenina are not homologous with those of the Chrysomelina and Phyllodectina (Hennig 1938, More 1967, Takizawa 1976). Here again unrelated compounds are secreted by related taxa (Pasteels et al. 1984; see also Table 7.1).

Defecation and regurgitation of the crop content are defensive reflexes in many chrysomelids as well as in other insects, such as grasshoppers and caterpillars (Eisner 1970). These reflexes evolved spectacularly in larvae of Criocerinae. In some genera (e.g., *Lilioceris* and *Lema*) the abdomen is permanently covered by viscous excrement. In other genera (e.g., *Crioceris*) disturbed larvae regurgitate a copious quantity of fluid, which eventually covers its entire body (Pasteels, unpublished observations). The protective effect could be solely mechanical, but as we will see later, protection could also be due to plant toxins in the fluid. The diversity of egg allomones matches that observed in other developmental stages (Table 7.1; see also Pasteels et al. 1986).

The overall diversity of chemical defensive mechanisms in the leaf beetles suggest that such defenses appeared independently many times in the course of evolution, and that they are evolutionarily labile and strongly dependent on specific selective pressures. One of the important factors limiting chemical defense could be its cost, which is unfortunately difficult to assess (Rowell-Rahier and Pasteels 1986). If the cost is high or the benefit low, the balance between cost and benefit could easily fluctuate between species and habitats and explain the complex distribution of chemical defenses observed today.

2   INFLUENCE OF THE HOST PLANT ON THE DEFENSE OF CHRYSOMELIDAE

If the metabolic cost of allomone biosynthesis is high, an obvious way to minimize cost of chemical defense is to use presynthesized secondary compounds if they are available in food plants. The sequestration of plant chemicals is well documented in insects (Duffey 1980). There are numerous examples of herbivorous insects which utilize plant toxins from their host plants for their own defense (Rothschild 1973). This kind of host plant influence is more probable in specialized herbivores. Most Chrysomelidae are highly specialized feeders, some of them on toxic plants. It is thus not very surprising that in several cases host plants strongly influence chemical defense in this group.

The secretion of salicylaldehyde by larvae of several chrysomeline species offers an excellent example of the use of plant secondary chemicals in insect defense and is reviewed below. Further insights are provided by a review of other known or suspected examples of the use of plant metabolites by the Chrysomelinae as precursors of defensive secretions.

Salicin, along with other phenolglucosides, is a characteristic compound found in the leaves and bark of many willows and poplars (Salicaceae). Wain (1943) and Pavan (1953) suggested that it was used as a substrate for the production of salicylaldehyde by larvae of *Phratora vitellinae* and *Chrysomela populi*. Pasteels et al. (1983b) verified this hypothesis experimentally for larvae of two species that feed on leaves of salicaceous trees (*P. vitellinae* and *C. tremulae*).

P. vitellinae larvae are specialized foliage feeders on willow species rich in salicin. Under experimental conditions, however, they can be induced to feed on leaves of *Salix caprea*. *Salix caprea* is a willow species whose leaves are mechanically protected by trichomes on their lower surfaces and do not contain salicin. The leaves of *S. caprea* are acceptable to *P. vitellinae* larvae only after the trichomes have been removed. Larvae reared on these leaves, which lack the putative precursor of salicylaldehyde, produce no defensive fluid when disturbed. If, however, salicin is deposited on the leaf surface prior to feeding, the defensive secretion of salicylaldehyde is restored (Pasteels et al. 1983b, Rowell-Rahier and Pasteels 1982). These experiments demonstrate conclusively that the plant phenolglucoside, salicin, is the substrate for the production of salicylaldehyde by *P. vitellinae* larvae. Moreover, incorporation experiments with increasing quantities of salicin showed that the concentration of salicylaldehyde in the secretion is positively correlated with the amount of salicin ingested by the larvae (Pasteels et al. 1983b).

The larvae of *Chrysomela* species also secrete salicylaldehyde (sometimes mixed with benzaldehyde). Pasteels et al. (1983b) fed larvae of *C. tremulae* labeled salicin and showed that the plant phenolglucoside is a precursor of the defensive secretion. The transformation of salicin, the glucoside of salicylalcohol, into salicylaldehyde requires the hydrolysis of the original glucoside molecule by β-glucosidase followed by the oxidation of the resulting aglycone. Incubation of different parts of dissected *C. tremulae* larvae with labeled salicin showed that the β-glucosidase activity occurs principally in larval secretory glands. Additionally, a relatively low β-glucosidase activity was detected in the gut. A glucose molecule must result from the hydrolysis of each salicin molecule. However, little glucose is excreted and quantitative analysis of the secretion (produced by larvae of *P. vitellinae* and *C. tremulae*) revealed that the concentrations of the two compounds were far from equimolar. Apparently, the glucose moiety split from the original glucoside molecule was recovered by the larvae, probably by diffusion through the gland membrane into the hemolymph.

The use of salicin as a precursor of salicylaldehyde in defensive secretions may thus have many advantages for leaf beetle larvae. First, it provides them with a cheaper defense than the autochthonously synthesized monoterpene secreted by many related species. Second, it enables them to mobilize an otherwise unexploited nutritional resource. The evolution of a high β-glucosidase activity in the gut, allowing the hydrolysis of the plant phenolglucosides and the utilization of their sugar constituent, may have been prevented by the potential toxicity of the resulting aglycone. Exocrine glands, such as those of the Chrysomelinae larvae are, on the other hand, highly suitable as a site for a high β-glucosidase activity.

In known cases where insects metabolize or sequester plant secondary compounds for defense, they do so, at least in part, to detoxify the compounds. In Chrysomelinae larvae, the use of a plant secondary compound appears to be advantageous not only as a potential detoxification mechanism but in reducing the metabolic cost of defense.

Rowell-Rahier and Pasteels (1986) evaluated and compared the metabolic cost of different chemical defensive secretions in larvae of a group of closely related chrysomelines, feeding on *Salix,* including species secreting monoterpenes and species secreting salicylaldehyde. The results confirmed that under laboratory conditions the biosynthesis of autogenous chemical defenses such as monoterpenes entails small but finite costs. In contrast, the use of a plant precursor costs nothing or can even be beneficial when this precursor is a glucoside, as in the case of the willow-feeding Chrysomelinae, which then recover the glucose. The benefits or costs of the defensive secretion were reflected in a gain or a loss in weight of beetles, which were forced experimentally to renew their supply of secretion frequently. However, we do not know how fitness is related to the weight of the adult beetles. It is also unclear what the situation is in the field, but it is very likely that the larvae are often disturbed by parasites or predators or even by the movement of leaves by wind (Raupp and Denno 1983). If this is true, the laboratory situation where the secretion is collected daily might indeed reflect the field situation.

The use of salicin by chrysomeline larvae has yet another selective advantage for these organisms. Pasteels et al. (1986) showed that salicin was sequestered in the eggs of the species secreting salicylaldehyde as larvae, but not in the eggs of other species feeding on salicaceous plants and producing monoterpenes in the larval secretion. The quantity of salicin sequestered was, at least in some *Chrysomela,* equivalent to the LD_{50} dose for the ant, *Myrmica rubra.* This sequestered salicin is utilized by larvae prior to hatching to fill the glandular reservoirs with salicylaldehyde. This must confer a considerable advantage over, for example, the species secreting monoterpene; the glands of these larvae are empty at hatching and need several hours to fill and become functional. In the few first hours following eclosion the immobile and clustered neonates are exposed and highly vulnerable to predation and cannibalism.

The advantages of the use of salicin by chrysomeline larvae are clear. However, salicin is not the only phenolglucoside present in the leaves of the Salicaceae. Other acetylated salicin derivatives (e.g., salicortin, fragilin, and tremulacin) are present in the leaves of some of the willows and poplars on which the beetles feed. Incorporation experiments, similar to those made with *P. vitellinae* larvae for salicin, show that at least one of these glucosides, salicortin, is utilized by larvae as a substrate for the production of salicylaldehyde. The advantages of the utilization of salicortin, however, are more dubious than in the case of salicin. Salicortin is probably metabolized to salicin and another compound, in the gut of larvae. The exact chemical nature of this compound is not yet known and might depend on conditions (pH, enzymes present, etc.) in the gut. The resulting salicin is then available for use by larvae. However, the remaining metabolite seems to be toxic, since the mortality of larvae ingesting salicortin in the incorporation experiment was greater than that of control larvae.

The quantitative relationship between the content of salicin and salicortin in the leaves of various willow species is currently under study. Its effect on the distribution of beetles is considered later in this chapter. Known or suspected cases of sequestration and utilization of plant secondary chemicals by leaf beetles are summarized in Table 7.2.

The secretion of salicylaldehyde by the dorsal glands of several chrysomeline larvae feeding on *Salix* and *Populus* plants is notable since most of the other species with similar glands produce cyclopentanoid monoterpenes. These compounds are believed to represent the original defenses of ancestral Chrysomelinae (Pasteels et al. 1982, 1984). Host plant influences in the willow feeders led to the replacement of one group of defensive compounds by another. That this could be advantageous for the insect is supported by the fact that this shift has occurred independently, at least twice and perhaps three times. It has occurred in the ancestor of the *Chrysomela* species and in *Phratora vitellinae*. The presence of salicylaldehyde in *Plagiodera versicolora* is somewhat unclear. It was first claimed by Hollande (1909) and later in one population of *Plagiodera* from Germany (Pasteels et al. 1984). However, only monoterpenes were detected in samples originating from the United States, Japan, England, Belgium, southern France, and Switzerland (Rowell-Rahier and Pasteels 1986). The odor of these compounds can be very misleading: plagiolactone, found in all samples of *Plagiodera,* smells much like salicylaldehyde. If confirmed, the utilization of host salicin by *Plagiodera* must have evolved in a very local population.

The secretion of juglone by the larvae of *Gastrolina depressa* must have followed an evolutionary trajectory parallel to that of salicylaldehyde discussed above. Juglone could easily be derived by a β-glucosidase hydrolysis from a 1,4,5-trihydroxynaphthalene present in the leaves of the walnut tree (*Juglans*) on which the beetles feed.

A well-documented example of sequestration of plant allelochemicals has been recently described by Ferguson and Metcalf (1985) for several chrysomelid cucumber beetles. The polyphagous beeltes *Diabrotica balteata, D. undecimpunctata howardi,* and *D. virgifera* were rejected by Chinese mantids (*Tenodera aridifollis sinenesis*) only when fed with bitter fruit (*Cucurbita andreana* × *C. maxima*) rich in cucurbitacins, whereas

TABLE 7.2 Sequestration and Utilization of Host Plant Chemicals in Leaf Beetles

Taxa (Number of Species Studied)	Developmental Stages	Defensive Allomones from Host Plant Origin	Food Plants	References
Chrysomela spp. (7 species studied)	Larvae	Salicylaldehyde	*Salix* and *Populus*	Pasteels et al. (1984)
Phratora vitellinae	Larvae	Salicylaldehyde	*Salix* and *Populus*	Pasteels et al. (1984)
Plagiodera versicolora	Larvae	Salicylaldehyde[a]	*Salix*	Pasteels et al. (1984), Hollande (1909)
Chrysomela spp. (5 species studied)	Eggs	Salicin	*Salix* and *Populus*	Pasteels et al. (1986)
Phratora vitellinae	Eggs	Salicin	*Salix* and *Populus*	Pasteels et al. (1986)
Gastrolina depressa	Larvae	Juglone	*Juglans*	Matsuda and Sugawara (1980)
Chrysolina hyperici	Adults	Hypericin	*Hypericum*	Rees (1969)
Diabrotica spp. (3 species studied)	Adults and eggs	Cucurbitacins D metabolite	Cucurbitaceae	Ferguson and Metcalf (1985)
Acalymna vittatum	All stages	Cucurbitacins D metabolite	Cucurbitaceae	Ferguson and Metcalf (1985)
Dibolia borealis	Adults	Iridoid glycosides	*Chelone* (Scrophulariaceae)	Bowers (in press)

[a] The secretion of salicylaldehyde in some populations of *P. versicolora* remains uncertain (see Rowell-Rahier and Pasteels 1986).

Acalymma vittatum, whose larvae have an obligate relationship with cucurbit hosts were always rejected, even when the adults were fed with
pollen devoid of cucurbitacins. The protection is due to cucurbitacins
sequestered by the larvae and transferred to the adults. Beetles fed on
bitter squash fruit accumulate large amounts of cucurbitacin D conjugate
in the hemolymph as a long-term storage product and not as a transient
metabolite. Sequestration of cucurbitacins is thus opportunistic in the
polyphagous species, but a permanent defensive mechanism in the cucurbit specialist.

The sequestration of hypericin in some *Chrysolina* species is not as
well documented. Several species of *Chrysolina* are strict specialists on
Hypericum species. Rees (1969) briefly reported that the hypericin content of adult *Chrysolina brunsvicensis* is too high to be present only in the
gut. Preliminary observations, however, failed to demonstrate any storage of hypericin in the hemolymph or pronotal and elytral defensive secretions in two species feeding on *Hypericum* (*C. varians* and *C. hyperici*), but large concentrations of hypericin were found in their feces
(Pasteels and Daloze, unpublished data). It could be that hypericin in the
feces or in the crop content plays a role in the protection of the beetles
against enemies. The fate and role of hypericin is currently being investigated. In any case, hypericin does not replace the autogenous chemical
defense. All *Chrysolina* species feeding on *Hypericum* possess well-de-
veloped and functional defensive glands. The cardenolide sarmentogenin
was identified in one species, *C. dydimata* (Daloze and Pasteels 1979),
and highly toxic polyoxygenated steroids in another, *C. hyperici* (Daloze
et al. 1985).

Sequestered plant toxins do not necessarily replace preexisting defensive secretions, but they can reinforce them. Salicin, for example, is
incorporated in the eggs of some species of Chrysomelina even though
these eggs are already protected by isoxazolinone derivatives (Pasteels et
al. 1986). Sequestration of plant toxins seems so highly advantageous that
it is not its existence that is surprising but rather the fact that many
chrysomelid beetles feeding on toxic plants do not utilize the plant toxins
for their own defensive purpose. Indeed, even within species feeding on
salicaceous plants, not all sequester salicin (list in Pasteels et al. 1984).
Furthermore, salicin is apparently not utilized by the adult beetles for
their own defense even in species that utilize it in the larval and egg
stages. The doryphorine Colorado potato beetle (*Leptinotarsa decemlineata*) does not sequester solanaceous alkaloids, such as those contained

in its host plants (Rothschild 1973), but rather secretes a toxic dipeptide most likely autogenously synthesized (Daloze et al. 1986). Unlike the monarch butterfly, *Danus plexippus,* various lygaeid bugs, and the cerambycid, *Tetraopes* (Duffey and Scudder 1972, Scudder et al. 1986), the aposematic chrysomelid beetles, *Labidomera clivicollis* and *Chrysocus cobaltinus,* do not sequester significant amounts of cardenolides from their asclepiad host plant (Rothschild 1973, Isman et al. 1977). As a chemical class, cardenolides are highly effective defensive substances and some chrysomelines (*Chrysolina* and *Oreina*) go to the expense of biosynthesizing them *de novo* (Pasteels and Daloze 1977). Larvae of *Paropsis* and *Chrysophtharta* feed on *Eucalyptus* and biosynthesize HCN and benzaldehyde (Moore 1967), but unlike the larvae of the sawfly *Perge affinis* (Pergidae) do not sequester the host plant terpenes (Morrow et al. 1976). Adults of *Chrysochloa* feeding on *Senecio* also do not sequester alkaloids. Not all plant compounds are suitable for sequestration by and defense of insects. For example, oxalic acid could be an effective antiherbivore defensive substance for the *Rumex* plants that produce it. However, it is doubtful that oxalic acid could advantageously replace the monoterpenes produced by larvae of *Gastrophysa viridula* that consume *Rumex* leaves. These herbivores must rely on more volatile compounds to deter predators at a distance before being injured. Additionally, sequestration of plant toxins presents the insect with several problems: specific transport across nonpermeable barriers, and movement against concentration gradients, without being intoxicated (see review in Duffey 1980). Such processes might be as costly as the production of autogenous compounds or could require several independent but complementary mechanisms, unlikely to be evolved frequently.

3 IMPACT OF HOST PLANT VARIATION ON THE SPATIAL AND TEMPORAL DISTRIBUTIONS OF LEAF BEETLES

Variation in the biochemical characteristics of host plants and their importance as key factors in explaining the distribution of herbivores is a widely studied area (Visser and Minks 1982, Ahmad 1983, Denno and McClure 1983). In the beginning of this section we review the evidence showing how one group of plant secondary compounds, the phenolglucosides, could influence herbivore distribution. First, we discuss the effect of interspecific and interindividual variation in foliar phenolglucoside content on the distribution of beetles. Second, we consider the impact of spatial and temporal variation in secondary chemistry of the host plants on the beetles. Finally, we discuss how sequestered plant toxins, as opposed to precursors of defensive secretions, can influence leaf beetle distribution.

According to Thieme and Benecke (1971) and Palo (1984), phenols are the only group of secondary metabolites present in significant amounts in the leaves of the Salicaceae. However, they are not present in the leaves of all willow and poplar species. Interspecific variation in the total phenolglucoside content of willow leaves seems to have an impact on the distribution of the different type of herbivores on different *Salix* species.

Rowell-Rahier (1984a) showed that the degree of dietary specialization of the insects feeding on willow leaves is related to the presence or absence of phenolglucosides in these leaves. Data from published food plant lists for weevils (Hoffman 1958), sawflies (Lorenz and Krauss 1957), and moth larvae (South 1948) indicate that the insects feeding on willow species with leaves rich in phenolglucosides are more specialized than those feeding on species devoid of these compounds. Due to their toxic properties (Marks et al. 1961, Edwards 1978, Vickery and Vickery 1981), the phenolglucosides seem to have a protective function for plants. Only the more specialized herbivore species can overcome them or even exploit them. The hypothesis of a defensive function of the phenolglucosides is also supported by the fact that willow species with phenolglucosides in their leaves have relatively similar fauna, as shown by cluster analysis of the published data cited above. These herbivores must have evolved adaptation to the phenolglucosides. On the other hand, the willow species that lack this group of compounds in their leaves do not demand particular adaptation from their herbivores, and as predicted, they do not show any special similarity in their faunas. Thus, the presence or absence of phenolglucosides fundamentally affects the distribution of herbivorous insects.

The distribution of particular phenolglucosides also determines host plant specificity of specialized herbivores. In their recent study Tahvanainen et al. (1985) showed that the concentration and composition of the different phenolglucoside blends are species specific in *Salix*. For four common chrysomelid species *(Phratora vitellinae, Plagiodera versicolora, Lochmaea capreae,* and *Galerucella lineola)*, the patterns of food plant selection observed in multiple-choice preference experiments are closely related to the phenolglucoside spectra of the willow tested. Indeed, the second choice of the beetles was always the willow species which was chemically the most similar to the preferred host plant. The phenolglucoside blends can have both stimulatory and inhibitory influences on the leaf beetles. *P. versicolora* and *L. capreae*, two species that do not use phenolglucosides as defensive precursors, prefer willow species with medium to low phenolglucoside concentrations and do not feed on plants with high phenolglucoside values. *G. lineola* seems specialized on leaves rich in salidrosid but is inhibited by salicin or salicortin (Tahvanainen et al. 1985). *P. vitellinae*, on the other hand, is closely associated with *Salix* species rich in salicin (Rowell-Rahier 1984b,c) to which the larvae have adapted to the point of using it for their own defense.

These data suggest that the phenolglucoside mixtures of the different salicaceous plants might become an important selective pressure influencing the evolution of the leaf beetles. Tahvanainen et al. (1985) concluded their study by asking what effect leaf beetles had on the willows. A possible answer to this question is found in the work of Smiley et al. (1985). They studied the relationship between the Californian Sierra Nevada beetle (*C. aenicollis*) and its salicaceous host plants (*S. lasiolepis, S. orestera,* and *S. planifolia*). The larvae of these beetles belong to the group of species which use salicin as a precursor of their defensive secretions. The willow leaves they encounter in the field present highly variable concentrations of salicin, salicortin, and tremulacin. No other phenolglucosides could be detected in significant amounts. In leaves of *S. lasiolepsis,* the salicin concentration varied between neighboring plants by 100-fold, from 0.05% dry weight to 5%. Damage due to herbivory is significantly higher on the plants rich in salicin. Beetle larvae placed on high-salicin plants have a high survival rate, possibly because of their better defense and the positive influence of the extra glucose derived from the glucoside (see above). These results, according to Smiley et al. (1985), demonstrate how a defensive plant chemical can become a hazard for a plant. This situation may be further exacerbated by the fact that production of defensive chemical precursors in the plant may be stimulated by herbivory. Recently, it has been demonstrated that the production of phenolic glucosides including salicin may be stimulated by mechanical damage or by herbivore feeding on poplars (Kimmerer, personal communication). It seems likely that inducible responses in salicaceous plants may actually favor herbivory by producing defensive chemical precursors utilized by specialized chrysomelids. Smiley et al. (1985) suggested that high levels of herbivory, due chiefly to one extremely well adapted species, might have put enough pressure on the willows to favor the appearance of plants containing no salicin, which would be less attacked by the beetles.

Inter- and intraspecific variation in the salicin content of leaves have a clear impact on the distribution of the leaf beetles. This impact is positive for some adapted species (*P. vitellinae* and *C. aenicollis*), negative for others (*G. lineola*), and may depend on concentration: being positive at low concentrations and negative at high ones, for still other species such as *P. versicolora* and *L. capreae* (Tahvanainen et al. 1985). One should probably be cautious when evaluating the effect of these variations on leaf beetles and should not overemphasize their importance compared to that of other chemical or physical factors (Raupp and Denno 1983; Rowell-Rahier 1984b,c). Variation in herbivory observed by Smiley et al. (1985) was conducted in a high-altitude habitat (3000 m) where *C. aenicollis* were extremely abundant. Other insect herbivores were not present in any significant numbers; and the evolution of large variation in salicin content of the leaves was thus possible. In a central European lowland site (300 m), with an abundant fauna of herbivorous insects, the variation in salicin content between different *S. nigricans* can be as much as threefold (2.5% to 7.5% dry weight; Rowell-Rahier, unpublished data). This variation does not correlate, in any way, with the abundance of the beetle *P. vitellinae* in the field. For this species at least, salicin is not a phagostimulant (Rowell-Rahier and Pasteels 1982); the range of variation in salicin content of leaves is such that it probably does not affect the quality and quantity of larval defensive secretion.

As far as the phenolglucosides are concerned, variation occurs not only between plants but also spatially and temporally within the same plant. For example, willow trees produce new leaves at the apex of the yearly shoots for the greater part of the growing season. Thus, the insects feeding on willows are continuously faced with a linear succession of leaves, from young ones at the apex of the shoot to older ones at its base. The phenolglucoside concentration decreases with the age of leaves (Thieme 1965, Rowell-Rahier 1984b, Horn 1986). The effects of this within-plant variation on chrysomelid distribution has yet to be elucidated. The bulk of evidence reviewed thus far indicates that species relying on plant precursors must, of course, consume hosts containing the requisite chemicals to be protected. The distribution of these beetles should reflect patterns of secondary chemical precursors found in their host. Similar relationships have been reported for other leaf beetles sequestering plant compounds (e.g., the diabotricines feeding on cucurbitacins and *Chrysolina brunsviscensis* specialized on *Hypericum*).

The Diabroticini chrysomelids that sequester cucurbitacins strongly discriminate between plant species, cultivars, and tissues as a function of their cucurbitacin content. Metcalf et al. (1980) found cucurbitacin sensitive chemoreceptors on the maxillary palps of *D. undecimpunctata*. Adult beetles were able to detect low concentrations of cucurbitacins. Several studies revealed substantial within-plant variation in cucurbitacin levels. Fruits, roots, and cotyledons are especially rich in these compounds (Rhem et al. 1957, Howe et al. 1976, Metcalf et al. 1980, Ferguson et al. 1983). Furthermore, these and other studies have revealed strong correlations between beetle abundance and levels of cucurbitacin found in tissues, cultivars, or species (Chambliss and Jones 1966, Howe et al. 1976, Metcalf et al. 1982, Ferguson et al. 1983). Within-plant variation in patterns of allelochemicals and leaf beetle feeding may also result from prior exposure or wounding. Metcalf et al. (1980, 1982) found crumpled cucurbit leaves extremely effective in arresting movement and eliciting feeding diabroticines. Apparently, disruption of the integument exposes dissolved cucurbitacins to the atmosphere, where they are more readily detected by beetles. Aggregation of beetles at the site of the injury results. A wounding effect may be amplified by systemic, induced changes within the plant. Recently, Tallamy (1985) demonstrated that mechanical damage increases the cucurbitacin levels in damaged and adjacent undamaged leaves of *Cucurbita pepo* 'Black'. Leaf sections with elevated cucurbitacin levels were preferentially consumed by *Diabrotica undecimpunctata howardi*. Herbivory may expose cucurbitacins to the atmosphere and induce higher cucurbitacin concentrations in the leaves. In this way, herbivory, by the Diabroticini which utilize cucurbitacins as cues for cessation of movement and feeding, may be self-amplifying.

At the level of the individual plant there is also a convincing example that the distribution of leaf beetles is strongly correlated with patterns of plant allelochemicals. Rees (1969) reported that adults of *Chrysolina brunsviscensis* feed on *Hypericum hirsutum* and accumulate hypericin (see above), an allelochemical toxic to mammals (Thomson 1957). Adult beetles have tarsal chemoreceptors capable of detecting hypericin at low concentration. Hypericin was not uniformly distributed throughout *Hypericum hirsutum*. Neither was there a uniform distribution of beetles on the plant. Beetles were most abundant on flower heads and the upper third of the plant where hypericin concentration was the greatest (Rees 1969).

Finally, many factors other than secondary metabolites can affect the distribution of insects on their host plants. Among them is the searching behavior of natural enemies, which is not necessarily identical on all parts of the plant. It has been suggested that herbivorous insects tend to occur in enemy-free sites on the plant (Lawton 1986). In this context, Raupp and Denno (1983) have observed that adults of the generalist predatory ladybug, *Hippodamia convergens,* search willow leaves contaminated by feces of *Plagiodera versicolora* more frequently and for longer periods of time than leaves lacking fecal deposits. This could, in part, affect the distribution of the different stages of leaf beetles on willows. Adults usually feed at the tip of twigs on young leaves, but eggs are deposited on older leaves, more deeply located in the foliage, away from adult fecal deposits.

4   EFFICACY OF THE DIFFERENT GROUPS OF CHEMICAL COMPOUNDS AND THEIR DISTRIBUTION AMONG DEVELOPMENTAL STAGES OF HERBIVORES

The distribution of the different chemical groups of defensive compounds among the developmental stages of the beetles could provide some indication of the efficacy of chemical defense(s) against specific natural enemies. Volatile irritants such as aldehydes, monoterpenes, phenylethylesters, or HCN are only found in larval secretions. The adult specifics, on the other hand, are characterized by nonvolatile systemic poisons (see Table 7.1). This dichotomy in the distribution of defensive compounds is especially striking since it occurs between developmental stages of the same species (e.g., Chrysomelina). Pasteels et al. (1984) proposed that such differences might be due to the fact that the major enemies of larvae and adults are different, even though both stages are found on the same host plant.

Adult beetles are usually observed on the upper surface of young leaves and are often brightly colored. They are highly visible and birds are most probably among their main predators. The incorporation of specific bitter tasting toxin associated with aposematic coloration has often been shown to be an efficient defensive strategy against visually oriented predators (e.g., birds), which are capable of rapid learning. The heavy exoskeleton of adults provides good mechanical protection against predaceous insects and the defensive secretion acts as a deterrent, after such a predator has attempted to bite the beetle. Contact with predaceous insects thus seems to be much less hazardous for the adult than for the soft-bodied larva. A single wound can be fatal for the latter. Therefore, volatile irritants acting before the predators encounter their prey may be more effective defenses for chrysomelid larvae. This hypothesis is supported by the fact that some leaf beetle larvae transform the bitter-tasting salicin found in the host plants (see above) into the much more volatile salicylaldehyde. Salicylaldehyde is a much better repellent than salicin against ants (Pasteels et al. 1983a). Nonspecific volatile irritants are not very effective against birds unless they are produced in very large quantity or are applied to the sensitive mucosa of birds (Pasteels et al. 1983b, Boevé and Pasteels 1985). Most of the larvae are cryptic and generally much less visible than the adults. The larvae feed on the undersurface of the leaves, which can be very close to the ground in the case of species feeding on herbaceous plants. Only larvae of the more robust species are aposematic, and these often feed in groups (Wallace and Blum 1969). Aggregated larvae of *C. aenicollis* are more resistant to predation than are solitary larvae (Smiley 1986), possibly by pooling the effect of individual secretions. The difference in predatory guilds and concomitant defensive requirements could explain why the adults feeding on salicaceous plants do not secrete salicylaldehyde and their larvae do.

Eggs are unable actively to discharge volatiles. Therefore, they must rely on feeding deterrents to discourage predators. In the eggs, salicin is used as a deterrent and not transformed into salicylaldehyde. Eggs are laid in clusters and the sacrifice of one egg might be enough to deter the predator and so protect the other eggs. The viscous maternal secretion that covers eggs might also act as a mechanical barrier as well as a protection against dehydration. The bright coloration of some of the eggs might be an aposematic signal. Berenbaum and Miliczky (1984) demonstrated that some insect predators can be conditioned to avoid bright colors.

Chrysomelids employ two methods to extend in time the effectiveness of host-derived defenses beyond the stage that actually acquires the defensive compound from the plant. Sequestration is one such method employed by Diabroticini and Chrysomelini alike. Ferguson and Metcalf (1985) found that striped cucumber beetles, *Acalymma vittatum*, which were never exposed to cucurbitacins in the adult stage, were rejected by the chinese preying mantis. Apparently, larvae ("root worms") sequester cucurbitacins, which provide protection to adults. These authors also indicated that relatively high levels of cucurbitacins are deposited in cucumber beetle eggs and that this may be an important defense against egg predators, such as ants. In addition to sequestering host-derived defenses, leaf beetles possess morphological and behavioral traits enabling them to optimally exploit plant-associated secretions optimally. Hollande (1909) and Hinton (1951) noted that in some chrysomelid species the pupae remain attached to the cuticle of the last instar. These species pupate on the leaves. When disturbed, pupae compress fluid-filled reservoirs in the old larval integument, thereby exposing droplets of secretion. The droplets are retracted after the disturbance abates. Wallace and Blum (1969) reported similar behavior of *Chrysomela scripta* and found that pupae thus protected were avoided by the Argentine ant, *Iridomyrmex humilis*. Furthermore, they observed teneral *C. scripta* adults resting on cast larval and pupal integuments. Adults removed from larval integuments smelled strongly of salicylaldehyde. These unique behaviors and morphologies allow three developmental stages, larva, pupa, and adult, to benefit from a defensive chemical produced by a single stage (Wallace and Blum 1969).

So far, our discussion of the role of defensive secretion against leaf beetle enemies has emphasized predators. Unfortunately, nothing is known of the effectiveness of defensive secretions against parasitoids. Many defensive secretion could also have antibiotic properties. Larval secretion stored in the exuvia, for example, could play a protective role against microorganisms for the species pupating in the soil. Similarly, elytral and pronotal secretions could protect the adult against pathogens during hibernation. All these speculations need to be investigated experimentally.

5 PROTECTION AGAINST PREDATORS AND PARASITOIDS

As noted above, the most obvious function of liquids discharged in response to disturbance is protection against various kinds of enemies. Many enemies of leaf beetles are known, but the efficacy of chemical defense against them has not been assessed thoroughly. A comprehensive list of enemies of the Chrysomelidae is given by Jolivet (1950). According to that list, all life stages are attacked by predators or parasites. These include many predatory insects, spiders, mollusks, vertebrates, parasitic nematodes, bacteria, and fungi. The larvae of the chrysomelid beetles that employ plant-derived defenses also have a rich complex of natural enemies. In Europe, Jolivet (1950) reported 42 insect species in 17 families and five orders preying on them. Studies by Smereka (1965), Head et al. (1977), and Burkot and Benjamin (1979) revealed a similar diversity preying on salicylaldehyde-producing North American chrysomelid larvae. Furthermore, Webster (1913) reported 24 species of avian predators of cucurbitacin sequestering Diabroticine beetles.

On the basis of such compilations it is difficult to show that beetles which are chemically protected are better off than those which are not. There are three main reasons for this. First, these lists include many anecdotal observations with very few systematic studies. Second, many species without chemical defense have other means of protection, such as crypsis or rapid escape mechanisms (DeRoe and Pasteels 1982). Third and most important, there is not quantitative and comparative data of beetles with and without defenses that were exposed to predators and parasites in these reports.

There is no doubt, however, that some chemically defended chrysomelid beetles are heavily preyed upon without any apparent harmful effect on the predators. For example, ladybird larvae (Coccinellidae) have been observed feeding on *Plagiodera versicolora* larvae (Whitehead and Duffield 1982). Predation by spiders (*Xysticus* sp.) and neuropteran larvae (*Anisochrysa prasina*) on *Phratora vitellinae* and *P. tibialis* larvae was observed in the field (Rowell-Rahier, unpublished observation). Kanervo (1946) cited no fewer than 64 predators and two parasites of the different stages of *Chrysomela* (= *Linaeidea*) *aenea* feeding on *Alnus* and producing cyclopentanoid monoterpenes in the larval stages and isoxazolinone in the adults and probably the eggs. However, it is no more justified to conclude that the secretions do not serve as defenses than to conclude that plant secondary compounds are not defenses since they do not completely prevent herbivory. The success of some leaf beetles enemies only shows us that insect chemical defense is no more globally effective than any other mode of defense.

There is little, but intriguing, evidence suggesting that some predators could be specialized on chemically defended prey, in the same way that some herbivores are specialized to a toxic food plant. Fabre (1891) described in detail how wasps (*Odynerus nidulator*) made trips back and forth to capture larvae of *C. populi,* which were the only prey found in their nest. Perhaps these wasps are capable of developing a search image of their prey, possibly including chemical information. Smereka (1965) and recently Smiley (1986) reported the same kind of behavior for the eumenid wasp, *Symmorphus cristatus,* which seems to be a specialized feeder on *Chrysomela* larvae. According to Smiley (personal communication) the wasp carries the larvae by grasping them by the ventral side and then rubs the dorsal side (where the defensive glands are located) on the leaf surface, emptying the gland reservoirs. Circumvention of the defenses of the larva probably results from this behavior.

The existing demonstrations of chemical defense of leaf beetles are based primarily on laboratory experiments using generalist predators. Various cyclopentanoid monoterpenes and salicylaldehyde were shown to be highly repellent to different ant species (Wallace and Blum 1969, Blum et al. 1978, Sugawara et al. 1979, Pasteels et al. 1983b). Ants have been reported as important predators of leaf beetle eggs, larvae, and pupae in both temperate and tropical systems (Smereka 1965, Risch 1981). Rejection of *Diabrotica* spp. by Chinese preying mantids is directly correlated with the amount of cucurbitacins sequestered from the host plants (Ferguson and Metcalf 1985). Anthraquinones and anthrones in the blood of *Pyrrhalta luteola* reduce feeding by fire ants (Howard et al. 1982a). The blood of the species exhibiting reflex bleeding is deterrent, irritant, or even toxic for some vertebrates (Cuenot 1896, Hollande 1909).

Results of several laboratory studies on the effectiveness of plant-derived defenses against predators are summarized in Table 7.3. These investigations provide convincing evidence supporting the role of host-derived secretions as antipredator defenses. The leaf beetles secreting salicylaldehyde have been the most thoroughly studied. The repellent activity of this substance was confirmed when foraging ants avoided paper disks impregnated with salicylaldehyde (Matsuda and Sugawara 1980) or sucrose solutions containing salicylaldehyde in low concentrations (Pasteels et al. 1983b). Undoubtedly, ants have been used in a majority of studies because they are easy to care for and observe under laboratory conditions. All ant genera assayed were repelled by or avoided larvae or their defensive secretion. The secretions of two beetles, *Chrysomela vigintipunctata costella* and *Gastrophysa depressa,* contain other chemicals in addition to salicylaldehyde. *Chrysomela vigintipunctata costella* secretes a combination of salicylaldehyde and benzaldehyde and *G. depressa* secretes juglone. In both instances *Lasius niger* workers were repelled by larvae or avoided leaf disks treated with benzaldehyde or juglone (Matsuda and Sugawara 1980).

Ants also avoid salicylaldehyde under field conditions. Ants in the genus *Prenolepis* regularly tend aggregations of the aphid *Tuberolachnus salignus,* on small willow branches. Workers actively remove honeydew and guard aphids in these aggregations. In 10 aggregations ranging in size from 17 to 64 aphids, the number of attendant ants was recorded for two consecutive 1-minute intervals. Between the second and the third minutes, approximately 1.5 μm of salicylaldehyde was applied to the aphid aggregation. During the next 3 minutes, no ants were observed tending the aphid colonies (Raupp, unpublished data; Figure 7.1). Clearly, salicylaldehyde had turned a highly favorable resource into one totally unsuitable for *Prenolepis* workers. This phenomenon might be more widespread, as suggested by the fact that within a patch of *Rumex* plants there was a clear dichotomy between plants infested by aphids and ants and those eaten by *Gastrophysa viridula* larvae secreting volatile monoterpenes (Rowell-Rahier, unpublished observations).

Are some groups of chemical compounds "better" defensive compounds than others, and does the use of host plant toxins improve the quality (the level) of chemical defense? Very few comparative studies are available which could answer this question, and there is probably no simple answer. Efficiency of chemical defense clearly depends on the type of enemies confronted. This can only be studied in the natural environment of the insect, not in laboratory experiments. Moreover, even for

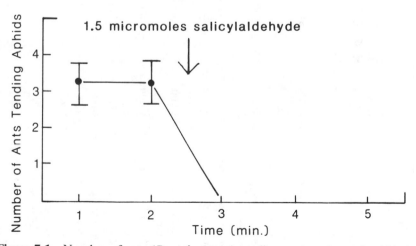

Figure 7.1 Number of ants (*Prenolepis* sp.) tending aggregations of aphids (*Tuberolachnus salignus*) before and after the introduction of salicylaldehyde to the aggregation. Dots represent means; vertical lines, standard errors.

TABLE 7.3 Responses of Predators to Plant Derived Chrysomelid Defenses in the Laboratory

Chrysomelid Species (Stage)	Host Plant(s)	Chemical Defense(s)	Predator (Family)	Predator Response(s)	Reference
Chrysomela scripta (larva, pupa)	*Salix* sp.	Salicylaldehyde[a]	*Iridomyrmex humilis*, *Crematogaster clara* (Formicidae)	Avoidance, repulsion	Wallace and Blum (1969)
Chrysomela vigintipunctata costella (larva)	*Salix babylonica*	Salicylaldehyde[a] + benzaldehyde	*Lasius niger* (Formicidae)	Repulsion	Matsuda and Sugawara (1980)
Chrysomela populi (larva)	*Populus sieboldi*	Salicylaldehyde[a]	*Lasius niger* (Formicidae)	Repulsion	Matsuda and Sugawara (1980)
Gastrolina depressa (larva)	*Juglans mandshurica* var. *sieboldiana*	Juglone[a]	*Lasius niger* (Formicidae)	Repulsion	Matsuda and Sugawara (1980)
Phratora vitellinae (larva)	*Salix* spp.	Salicylaldehyde[a]	*Tenthredo olivacea* (Tenthredinidae)	Consume	Pasteels and Gregoire (1984)
Phratora vitellinae (larva)	*Salix nigricans*	Salicylaldehyde[a]	*Myrmica rubra* (Formicidae)	Repulsion	Pasteels et al. (1983b)
Chrysomela tremulae (larva)	*Populus trichocarpa*	Salicylaldehyde[a]	*Myrmica rubra* (Formicidae)	Repulsion	Pasteels et al. (1983b)
Chrysomela populi (egg)	*Populus trichocarpa*	Salicin[b]	*Myrmica rubra* (Formicidae)	Deterred	Pasteels et al. (1986)
Chrysomela tremulae (egg) *Phratora vitellinae* (egg)					

Herbivore	Plant	Defense	Predator	Response	Reference
Chrysomela saliceti (egg) Chrysomela 20-punctata (egg)	Salix purpurea	Salicin[b]	Myrmica rubra (Formicidae)	Deterred	Pasteels et al. (1986)
Diabrotica balteata (adult)	Cucurbita andreana × C. maxima	Cucurbitacins[b]	Tenodera aridifolia sinensis (Mantidae)	Reject	Ferguson and Metcalf (1985)
Diabrotica undecimpunctata howardi (adult)	Cucurbita andreana × C. maxima	Cucurbitacins[b]	Tenodera aridifolia sinensis (Mantidae)	Reject	Ferguson and Metcalf (1985)
Diabrotica virgifera virgifera (adult)	Cucurbita andreana × C. maxima	Cucurbitacins[b]	Tenodera aridifolia sinensis (Mantidae)	Reject	Ferguson and Metcalf (1985)
Acalymma vittatum (adult)	Cucurbita andreana × C. maxima	Cucurbitacins[b]	Tenodera aridifolia sinensis (Mantidae)	Reject	Ferguson and Metcalf (1985)
Diabrotica undecimpunctata howardi (adult)	Cucumus sativus	Cucurbitacins[b]	Perimyscus maniculatus (Cricetidae)	Consume	Gould and Massey (1985)
			Colinus virginianus (Phasianidae)	Consume	Gould and Massey (1985)
			Bufo americanus (Bufonidae)	Consume	Gould and Massey (1985)
			Bufo fowleri (Bufonidae)	Consume	Gould and Massey (1985)

[a] Secreted defense.
[b] Sequestered defense.

259

a given predator or parasite, the defensive efficiency of a compound is not immutable but rather, it can change with the circumstances. Pasteels and Gregoire (1984) demonstrated that female sawflies (*Tenthredo olivacea*), when given a choice between larvae of *Phratora vitellinae* (secreting salicylaldehyde) and *Plagiodera versicolora* (secreting cyclopentanoid monoterpenes), preferred to feed on the species with which they were more familiar. Selective predation on larvae defended by one or the other type of chemical was in this case, the result of previous experience and conditioning of the predator. If this kind of response is widespread among predators, strong selective pressure for novelty in the compounds used for defense may help explain the diversity of defensive mixtures observed between sympatric leaf beetle species or even within a single species. Although there was no evidence that the sawflies used the defensive secretion as a cue to locate their prey, the possibility should not be excluded.

The efficacy of a chemical can also differ according to the parameter estimated. In Table 7.4, threshold values of feeding deterrency and toxicity to the ant, *Myrmica rubra,* are compared for several defensive compounds of egg and adult leaf beetles. Major differences between compounds exist, but it should be stressed that there is no correlation between deterrency and toxicity. Interestingly, salicin appeared to be both deterrent and more toxic than the autochthonously produced isoxazolinone derivatives. Its incorporation into the eggs of some of the beetle species feeding on *Salix* should reinforce the egg defense against predators such as ants.

The efficacy of sequestered chemicals as defenses differs with the type of predator studied. Ferguson and Metcalf (1985) assayed four species of Diabroticini that sequester cucurbitacins. The Chinese mantid, *Tenodera aridifolia sinensis,* rejected varying proportions of beetles previously fed bitter squash fruit and consumed all beetles which had had no prior exposure to cucurbitacins. Bioassays of the effectiveness of sequestered cucurbitacins against vertebrate predators contrast with those described for mantids. Gould and Massey (1985) studied predation by four species of vertebrates, the toads, *Bufo americanus* and *B. fowleri,* the mouse, *Peromyscus maniculatus,* and the quail, *Colinus virgianus.* These predators were offered different groups of *D. undecimpunctata howardi* adults. One group of beetles had consumed cucumbers without cucurbitacins and the other had eaten cucumbers with cucurbitacins. They found no evidence that sequestered cucurbitacins protected the beetles from vertebrate attack.

TABLE 7.4 Deterrency Activity and Toxicity of Various Egg and Adult Defensive Allomones toward the Ant *Myrmica rubra*

Compounds	Origin	Concentration,[a] Amount (μg/individual)	Deterrency Threshold[b]	Oral Toxicity: Concentration,[c] LD 50 (μg/ant)	Reference
Salicin	Eggs of *Chrysomela populi*	4.10^{-2} M, 6 μg	10^{-3} M	10^{-2} M, 5 μg	Pasteels et al. (1986)
Isoxazolin-5-one derivatives[d]	Eggs of *Chrysomela populi* adult secretion *Chrysomela populi*	3.10^{-2} M, 6 μg; 10^{-1} M, 35 μg	10^{-3} M	Not toxic at 10^{-2} M	Pasteels et al. (1986)
Polyoxygenated steroid glucosides[e]	Adult secretion of *Chrysolina hyperici*	10^{-2} M, 21 μg	10^{-4} M	10^{-3} M, 0.1 μg	Pasteels, unpublished data
Cardenolides[f]	Adult secretion of *Chrysolina herbacea*	10^{-1} M, 138 μg	10^{-3} M	10^{-2} M, 9 μg	Pasteels, unpublished data
γ-Glutamyl dipeptide[g]	Adult secretion of *Leptinotarsa decemlineata*	1.710^{-1} M, 13 μg	Not deterrent at 10^{-2} M	10^{-2} M, 7 μg	Daloze et al. (1986)

[a] Estimated by calculation.

[b] In sucrose 10^{-1} M.

[c] Lowest toxic concentration in sucrose 10^{-1} M.

[d] 2-(3-D-Glucopyranosyl)-3-isoxazolin-5-one and 2-[6'-(3"-nitropropanoyl)-β-D-glucopyranosyl]-3-isoxzolin-5-one. Deterrency threshold and oral toxicity are given for the second compound, which is the major constituent.

[e] 3β-O-β-D-Glucopyranosyl-5-stigmastane-20, 25, 28-triol-6, 16-dione-28-acetate and its corresponding 25-acetate. Deterrency threshold and oral toxicity are given for the 25-acetate derivative (33% w/w).

[f] Several xylosides and aglycones. Deterrency threshold and oral toxicity are given for bipendogenin xyloside (25% w/w).

[g] γ-L-Glutamyl-L-2-amino-3(Z), 5-hexadienoic acid.

Relatively few studies have measured the effect of host-derived defenses on beetle predators under field conditions. A notable exception is the work of Smiley et al. (1985), who demonstrated differential survival of *Chrysomela aenicollis* larvae on willow clones containing different levels of salicin. Larvae reared on salicin-rich clones were able to produce salicylaldehyde, whereas those on salicin poor clones were not. Survivorship of the salicylaldehyde-producing cohort was significantly greater.

6   PROTECTION AGAINST COMPETING HERBIVORES

A second category of antagonist affecting leaf beetles includes the other herbivores utilizing the same food resource. Natural enemies probably exert a much greater selection pressure on defenses of phytophagous insects than do competitors sharing the food resource (Price 1984, Strong et al. 1984). There are no systematic studies unambiguously demonstrating the existence of inter- or intraspecific competition in leaf beetles. Nevertheless, it is not rare to see plants completely devastated by aggregations of not very mobile and rapidly reproducing leaf beetles. If the plant resources are widely scattered in the environment, it is possible that local competition can occur and that mechanisms aimed at reducing it have evolved. Bernays (1982) suggested that small herbivores such as insects might selectively feed on plant parts unlikely to be consumed by larger herbivores, thereby escaping disturbance or death. An alternative explanation is possible if small herbivores have the ability to defend the resource they occupy. Raupp et al. (1986) found that larvae of the leaf beetle *Plagiodera versicolora* are able to repel larvae of the butterfly *Nymphalis antiopa*. The effect was clearly mediated by the larval defensive secretion. In addition, *Nymphalis* larvae consumed a smaller quantity of willow leaves that bore *Plagiodera* larvae than those without beetle larvae. The cyclopentanoid monoterpenes secreted by *Plagiodera* are not believed to be derived from precursors found in the host. However, the effectiveness of these secretions in repelling other herbivores led to investigations of the role of the plant-derived secretion salicylaldehyde in mediating the behavior of other herbivores utilizing willows. Salicylaldehyde interrupts directed movement in at least two other willow-feeding herbivores. A micropipette contaminated with salicylaldehyde arrested linear forward movement in 70% of the gypsy moth larvae (*Lymantria dispar*) tested. Eighty-one percent of the poplar tentmaker (*Ichthyura inclusa*) larvae tested exhibited a similar response. For both lepidopterans, more than 70% attempted to avoid the pipette by turning their head and thorax. Less than 10% of the same larvae responded in these ways to untreated pipettes. Not only does salicylaldehyde alter movements in Lepidoptera, it also appears to be a feeding deterrent for at least one species. Leaf disks treated with 1 μm of salicylaldehyde were less likely to be eaten by larvae of the mourningcloak, *N. antiopa*, compared to untreated disks in the same arena. A similar, though not significant trend was observed for larvae of *L. dispar* and *I. inclusa* (Raupp, unpublished data; Fig. 7.2).

Chemically mediated interactions may affect the distribution of the beetles under field conditions. Rowell-Rahier (unpublished data) observed that *P. vitellinae* avoid *S. nigricans* trees that were previously colonized by *C. 20-punctata*. The latter species appears early in spring and well-defended late third instar larvae and pupae are present on the leaves when *P. vitellinae* starts to oviposit. Such an epideictic function of the larval secretion could be widespread in the leaf beetles.

Intraspecific competition could also favor the production of volatiles by the larval stages. Many leaf beetles deposit their eggs in batches. Because the number of eggs per batch can be large, a very high density of larvae can result on a single leaf or plant. Renner (1970) and Raupp et al. (1986) observed that the adults of *Gastrophysa viridula* and *Plagiodera*

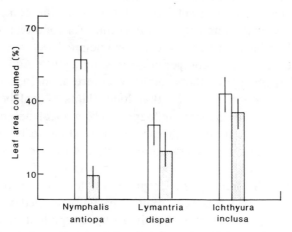

Figure 7.2 Leaf area consumed by three species of Lepidoptera offered leaf disks with (shaded bars) or without (open bars) salicylaldehyde. Bars represent means; vertical lines, standard errors.

versicolora are repelled by the secretion of their larvae, suggesting that fewer new eggs would be laid in the close vicinity of ones already hatched. Clearly, the possibility that host-derived secretions mediate interactions among herbivores deserves greater attention.

7 SUMMARY AND CONCLUSIONS

Extremely diverse selective pressures, acting simultaneously, have shaped and continue to shape the chemical defense of crysomelid beetles. Natural enemies are major factors that can in themselves be very complex. They include a large array of vertebrate and invertebrate predators, parasitoids, and pathogens. Although we understand very little of how beetles have responded to the pressures exerted by these enemies, undoubtedly their diversity has contributed to the evolution of the diversity of defensive strategies, not only between species and individuals, but also within a species or an individual where apparently redundant mechanisms of defense are observed. Selection pressure for diversity is also expected to be high if an enemy can adapt either genetically or behaviorally (i.e., a conditioning process) to particular modes of defense and thus become specialized on a chemically defended prey. The existence of such mechanisms deserve to be investigated in more detail.

In addition to defense per se, deterrent compounds may act in some cases as a mechanism to limit intra- and interspecific competition. It is difficult to assess the intensity of competition for food or how important its impact has been on the patterns of chemical defense. The use of defensive secretion as a competition-reducing agent could simply be an exaptation resulting from the selective pressure of the natural enemies.

There is no doubt, however, that food plants of leaf beetles have a tremendous direct or indirect influence on chemical defense. There is growing evidence that some chrysomelids, feeding on toxic plants, are able to utilize plant secondary compounds for their own defense. The form and consequences of this utilization are variable. The transformation of salicin into salicylaldehyde exemplifies an elaborate form of utilization of a plant compound which is detoxified, partially utilized as nutrient, and used as a precursor of a highly effective volatile repellent. Such utilization has led to a shift in the defensive chemistry of the beetles, resulting in an increase in the overall diversity of defensive compounds found in the taxon. However, the fact that several species produce salicylaldehyde has led to a decrease in the interspecific diversity of defensive compounds in the beetles feeding on salicaceous plants. A simpler form of utilization of plant allelochemicals is their sequestration or concentration in the gut. In this way autogenous defense can be reinforced. For example, salicin is incorporated in eggs already protected by isoxazolinones, and a high concentration of hypericin is found in the gut of species already defended by polyoxygenated sterols. In some species the sequestration of plant compounds can also be the only known chemical defensive mechanism, opportunistic in polyphagous herbivores and obligatory in specialists. Diabotricine and alticine beetles are good examples of herbivores that utilize plant allelochemicals in simple ways.

It should be stressed that despite the ever-increasing evidence for utilization of plant toxins by leaf beetles for defense, there are as many negative reports even in aposematic and otherwise chemically protected insects. Obviously, some compounds are easier to sequester than others, but as yet, no general pattern has emerged.

We predict that the more elaborate the form of plant allelochemical utilization, the less likely it will be, since it probably requires many specific and independent adaptations. In this context, it is striking to note that within the genus *Phratora,* not all the species feeding on phenolglucoside-rich salicaceous plants are able to utilize salicin for their defense. In contrast, all diabroticine beetles feeding on cucurbits seem able to sequester cucurbitacins and all studied aposematic insects feeding on plants rich in iridoid glucosides seem able to sequester them (Bowers and Puttick 1986).

Host plants can have a more indirect influence on chemical defense, which is less well understood, but certainly of prime importance. Both nutritional and microclimatic factors, as well as antiherbivore mechanisms, influence the spatial and temporal distribution of insects. The distribution of insects on their host plants will determine the amount and the nature of plant secondary compounds available to them. In some cases allelochemicals are not uniformly distributed throughout the plant, and herbivores select the plant parts that are the richest in the compound they utilize (e.g., *C. brunsvisencis* on hypericin-rich flower buds).

The location of the insects on the plant will also play a role in determining the kind of enemies and competitors they are exposed to and encounter. Unfortunately, knowledge of these indirect host plant influences remains intuitive and speculative. Detailed studies of natural communities are vital if we hope to gain a greater understanding of the rich complexity of interactions between leaf beetles and the suite of environmental factors shaping their defensive secretions.

ACKNOWLEDGMENTS

We thank Hugh Rowell, Doyle McKey, and John Smiley for many helpful suggestions that improved this chapter.
Scientific Article No. A-4470, Contribution No. 7462 of the Maryland Agricultural Experiment Station, Department of Entomology.

REFERENCES

Ahmad, S. 1983. Herbivorous Insects: Host-Seeking Behavior and Mechanisms. Academic Press, New York.

Berenbaum, M. R., and E. Miliczky. 1984. Mantids and milkweed bugs: efficiency of aposematic coloration against invertebrate predators. Amer. Midl. Nat. 111:64–68.

Bernays, E. 1982. The insect on the plant. A closer look. In J. H. Visser and A. K. Minks (eds.), Proceedings of the 5th International Symposium on Insect–Plant Relationships. Pudoc, Wageningen, The Netherlands.

Blum, M. S., J. M. Brand, J. B. Wallace, and H. F. Fales. 1972. Chemical characterization of the defensive secretion of a chrysomelid larva. Life Sci. 11:525–531.

Blum, M. S., J. B. Wallace, R. M. Duffield, J. M. Brand, H. M. Fales, and E. A. Sokoloski. 1978. Chrysomelidial in the defensive secretion of the leaf beetle, *Gastrophysa cyanea* Melsheimer. J. Chem. Ecol. 4:47–53.

Boevé, J. L., and J. M. Pasteels. 1985. Modes of defense in Nematine sawfly larvae. Efficiency against ants and birds. J. Chem. Ecol. 11:1019–1036.

Bowers, M. D. Recycling plant allelochemicals for insect defense. In J. O. Schmidt and D. E. Evans (eds.), Defense: Predator-Prey Interactions. SUNY Press, Albany, New York (in press).

Bowers, M. D., and G. M. Puttick. 1986. Fate of ingested iridoid glycosides in lepidopteran herbivores. J. Chem. Ecol. 12:169–178.

Burkot, T. R., and D. M. Benjamin. 1979. The biology and ecology of the cottonwood leaf beetle, *Chrysomela scripta,* on tissue cultured hybrid Aigeiros (*Populus × Euroamericana*) subclones in Wisconsin. Can. Ent. 111:551–556.

Chambliss, O. L., and C. M. Jones. 1966. Cucurbitacins: specific insect attractants in Cucurbitaceae. Science 153:1392–1393.

Cuénot, L. 1896. Sur la saignée réflexe et les moyens de défense de quelques insectes. Arch. Zool. Exp. 4:655–680.

Daloze, D., and J. M. Pasteels. 1979. Production of cardiac glycosides by Chrysomelid beetles and larvae. J. Chem. Ecol. 5:63–77.

Daloze, D., J. C. Braekman, and J. M. Pasteels. 1985. New polyoxygenated steroidal glucosides from *Chrysolina hyperici* (Coleoptera: Chrysomelidae). Tetrahedron Lett. 26:2311–2314.

Daloze, D., J. C. Braekman, and J. M. Pasteels. 1986. γ-L-glutamyl-L-2-amino-3(Z),5-hexadienoic acid, a toxic dipeptide from the defense glands of the Colorado beetle (*Leptinotarsa decemlineata* (Say)). Science 233:221–223.

Denno, R. F., and M. S. McClure (eds.). 1983. Variable Plants and Herbivores in Natural and Managed Systems. Academic Press, New York.

DeRoe, C., and J. M. Pasteels. 1982. Distribution of adult glands in Chrysomelids (Coleoptera: Chrysomelidae) and its significance in the evolution of defense mechanisms within the family. J. Chem. Ecol. 8:67–82.

Duffey, S. S. 1980. Sequestration of plant natural products by insects. Annu. Rev. Ent. 25:447–477.

Duffey, S. S., and G. G. E. Scudder. 1972. Cardiac glycosides in north American Asclepiadaceae, a basis for unpalatability in brightly coloured Hemiptera and Coleoptera. J. Insect Physiol. 18:63–78.

Edwards, W. R. N. 1978. Effect of salicin content on palatability of *Populus* foliage to oppossum (*Trichosurus vulpecula*). N.Z. J. Sci. 21:103–106.

Eisner, T. 1970. Chemical defense against predation in arthropoids. In E. Sondheimer and J. B. Simeone (eds.), Chemical Ecology. pp. 157–217. Academic Press, New York.

Fabre, J. H. 1891. Souvenirs Entomologiques, Vol. 4. pp. 173–190. Delagrave, Paris.

Ferguson, J. E., and R. L. Metcalf. 1985. Cucurbitacins. Plant-derived defense compounds for Diabroticites (Coleoptera: Chrysomelidae). J. Chem. Ecol. 11:311–317.

Ferguson, J. E., R. A. Metcalf, R. L. Metcalf, and A. M. Rhoades. 1983. Influence of cucurbitacin content in cotyledons of Cucurbitaceae cultivars upon feeding behavior of Diabroticina beetles (Coleoptera: Chrysomelidae). Econ. Ent. 76:47–51.

Gould, F., and A. Massey. 1985. Cucurbitacins and predation of the spotted cucumber beetle, *Diabrotica undecimpunctata howardi*. Ent. Exp. Appl. 36:273–278.

Head, R. B., W. W. Neal, and R. C. Morris. 1977. Seasonal occurrence of the cottonwood leaf beetle *Chrysomela scripta* (Fab.) and its principal insect predators in Mississippi and notes on parasites. J. Ga. Ent. Soc. 12:157–163.

Hennig, W. 1938. Übersicht über die Larven der wichtigsten Deutschen Chrysomelinen. Arb. Physiol. angew. Ent. 5:85–136.

Hinton, H. E. 1951. On a little known protective device of some chrysomelid pupae (Coleoptera). Proc. Roy. Ent. Soc. London (A) 26:67–73.

Hoffman, A. 1958. Faune de France, Coleoptères Curculionides. 62:1782–1785.

Hollande, A. C. 1909. Sur la fonction d'excretion chez les insectes salicicoles et an particulier sur l'existence de dérives salicyles. Ann. Univ. Grenoble 21:459–517.

Horn, J. M. 1986. Distribution of phenolglycosides in willow (*Salix* spp.) leaves. Proc. White Mountain Res. Stn. Symp. 1:99–105.

Howard, D. F., D. W. Phillips, T. H. Jones, and M. S. Blum. 1982a. Anthraquinones and anthrones: occurrence and defensive function in a chrysomelid beetle. Naturswissenschaften 69:91–92.

Howard, D. F., M. S. Blum, T. H. Jones, and D. W. Phillips. 1982b. Defensive adaptations of eggs and adults of *Gastrophysa cyanea* (Coleoptera: Chrysomelidae). J. Chem. Ecol. 8:453–462.

Howe, W. L., J. R. Sanborn, and A. M. Rhoades. 1976. Western corn rootworm adult and spotted cucumber beetle associations with *Cucurbita* and cucurbitacins. Environ. Ent. 5:1043–1048.

Hsiao, T. H., and G. Fraenkel. 1969. Properties of leptinotarsin, a toxic haemolymph protein from the Colorado beetle (*Leptinotarsa decemlineata:* Col., Chysomelidae). Toxicon 7:119–130.

Isman, M. B., S. S. Duffey, and G. G. E. Scudder. 1977. Cardenolide content of some leaf- and stem-feeding insects on temperate North American milkweeds (*Asclepias* spp.). Can. J. Zool. 55:1024–1028.

Jolivet, P. 1950. Les parasites, prédateurs et phoretiques des Chrysomeloidae (Coleoptera) de la faune franco-belge. Bull Inst. Roy. Sci. Nat. Belg. 26:1–39.

Kanervo, V. 1946. Studien über die naturlichen Feinde das Erlenblaatkäfers *Melasoma aenea* Linne. Ann. Zool. Soc. Zool. Bot. Fenn. Vanamo 12:1–206.

Lawton, J. H. 1986. The effect of parasitoids on phytophagous insect communities. In J. Waage and D. Greathead (eds.), Insect Parasitoids. Symposium of the Royal Entomology Society of London. pp. 265–287. Academic Press, London.

Lorenz, H., and M. Krauss. 1957. Die larval Systematik der Blattwespen (Tenthredinoidea und Megalocontoidea). Akademie-Verlag, East Berlin.

Marks, V., M. J. H. Smith, and A. C. Cunliffe. 1961. The mechanism of the antiflammatory activity of salicylate. J. Pharmacol. 13:218–223.

Matsuda, K., and F. Sugawara. 1980. Defensive secretion of chrysomelid larvae *Chrysomela vigintipunctata costella* (Marseul), *C. populi* L. and *Gastrolina depressa* Baly (Coleoptera: Chrysomelidae). Appl. Ent. Zool. 15:316–320.

Mebs, D., F. Bruning, and N. Pfaff. 1982. Preliminary studies on the chemical properties of the toxic principle from *Diamphidia nigroornata*. J. Ethnopharmacology 6:1–11.

Metcalf, R. L., R. A. Metcalf, and A. M. Rhoades. 1980. Cucurbitacins as kairomones for diabroticite beetles. Proc. Natl. Acad. Sci. USA 77:3769–3772.

Metcalf, R. L., A. M. Rhoades, R. A. Metcalf, J. E. Ferguson, E. R. Metcalf, and P. Y. Lu. 1982. Cucurbitacin contents and diabroticite feeding upon *Cucurbita* spp. Environ. Ent. 11:931–934.

Moore, B. P. 1967. Hydrogen cyanide in the defensive secretions of larval Paropsini (Coleoptera: Chrysomelidae). J. Aust. Ent. Soc. 6:36–38.

Morrow, P. A., T. E. Bellas, and T. Eisner. 1976. Eucalyptus oils in the defensive discharge of Australian sawfly larvae (Hymenoptera: Pergidae). Oecologia (Berlin) 24:193–206.

Palo, R. T. 1984. Distribution of birch (*Betula* spp.), willow (*Salix* spp.), and poplar (*Populus* spp.) secondary metabolites and their potential role as chemical defense against herbivores. J. Chem. Ecol. 10:499–520.

Parker, R. 1972. A comparison of the toxic protein in two species of *Leptinotarsa*. Toxicon 10:79–80.

Pasteels, J. M., and D. Daloze, 1977. Cardiac glycosides in the defensive secretion of Chrysomelid beetles: evidence for their production by the insects. Science 197:70–72.

Pasteels, J. M., and J. C. Gregoire. 1984. Selective predation on chemically defended chysomelid larvae. A conditioning process. J. Chem. Ecol. 10:1693–1700.

Pasteels, J. M., J. C. Braekman, D. Daloze, and R. Ottinger. 1982. Chemical defence in chrysomelid larvae and adults. Tetrahedron 38:1891–1897.

Pasteels, J. M., J. C. Gregoire, and M. Rowell-Rahier. 1983a. The chemical ecology of defense in arthropods. Annu. Rev. Ent. 28:263–289.

Pasteels, J. M., M. Rowell-Rahier, J. C. Braekman, and A. Dupont. 1983b. Salicin from host plant as precursor of salicylaldehyde in defensive secretion of Chrysomeline larvae. Physiol. Ent. 8:307–314.

Pasteels, J. M., M. Rowell-Rahier, J. C. Braekman, and D. Daloze, 1984. Chemical defenses in leaf beetles and their larvae: the ecological, evolutionary and taxonomic significance. Biochem. Syst. Ecol. 12:395–406.

Pasteels, J. M., D. Daloze, and M. Rowell-Rahier. 1986. Chemical defense in Chrysomelid eggs and neonate larvae. Physiol. Ent. 11:29–37.

Pavan, M. 1953. Stúdio sugli antibiòtici e insetticídi di orígine animàle. I. Sul princípio attívo della làrva di *Melasoma populi* L. Arch. Zool. Ital. 38:157–183.

Price, P. W. 1984. Insect Ecology, 2nd ed. Wiley, New York.

Raupp, M. J., and R. J Denno. 1983. Leaf age as a predictor of herbivore distribution and abudance. In R. F. Denno and M. S. McClure (eds.), Variable Plants and Herbivores in Natural and Managed Systems. pp. 91–124. Academic Press, New York.

Raupp, M. J., F. R. Milan, P. Barbosa, and B. Leonhardt. 1986. Methylcyclopentanoid monoterpenes mediate interactions among insect herbivores. Science 232:1408–1409.

Rees, J. C. 1969. Chemoreceptor specificity associated with choice of feeding site by the beetle *Chrysolina brunsvicensis* on its foodplant, *Hypericum hirsutum.* Ent. Exp. Appl. 12:565–583.

Renner, K. 1970. Über die ausstülpbaren Hautblasen der Larven von *Gastroidae viridula* De Geer und ihre okologische Bedeutung (Coleoptera: Chrysomelidae). Beitr. Ent. 20:527–533.

Rhem, S., P. R. Enslin, A. D. J. Meeuse, and J. H. Wessels. 1957. Bitter principles of the cucurbitaceae. VII. The distribution of bitter principles in this plant family. J. Sci. Food Agric. 8:679–686.

Risch, S. 1981. Ants as important predators of rootworm eggs in the neotropics. J. Econ. Ent. 77:88–90.

Rothschild, M. 1973. Secondary plant substances and warning coloration in insects. In H. F. van Emden (ed.), Insect/Plant Relationships. Symposium of the Royal Entomology Society of London, Vol. 6. pp. 59–83. Blackwell Scientific, Oxford.

Rowell-Rahier, M. 1984a. The presence or absence of phenolglycosides in *Salix* (Salicaceae) leaves and the level of dietary specialization of some of their herbivorous insects. Oecologia (Berlin) 62:26–30.

Rowell-Rahier, M. 1984b. The food plant preferences of *Phratora vitellinae* (Coleoptera: Chrysomelidae). A. Field observations. Oecologia (Berlin) 64:369–374.

Rowell-Rahier, M. 1984c. The food plant preferences of *Phratora vitellinae* (Coleoptera: Chrysomelinae). B. A laboratory comparison of geographically isolated populations and experiments on conditioning. Oecologia (Berlin) 64:375–380.

Rowell-Rahier, M., and J. M. Pasteels. 1982. The significance of salicin for a *Salix*-feeder, *Phratora* (*Phyllodecta*) *vitellinae*. In J. H. Visser and A. K. Minks (eds.), Proceedings of the 5th International Symposium on Insect–Plant Relationships. pp. 73–79. Pudoc, Wageningen, The Netherlands.

Rowell-Rahier, M., and J. M. Pasteels. 1986. Economics of chemical defense in Chrysomelinae. J. Chem. Ecol. 12:1189–1203.

Scudder, G. G. E., L. V. Moore, and M. B. Isman. 1986. Sequestration of cardenolides in *Oncopeltus fasciatus:* morphological and physiological adaptations. J. Chem. Ecol. 12:1171–1187.

Smereka, E. P. 1965. The life history of *Chrysomela crotchi* Brown (Coleoptera: Chrysomelidae) in northwestern Ontario. Can. Ent. 97:541–549.

Smiley, J. T. 1986. Chemical ecology of willows, leaf, beetles, and predators: adaptation along an elevation gradient in the eastern Sierra Nevada. pp. 106–113. Proceedings of the High Elevation Research Symposium.

Smiley, J. T., J. M. Horn, and N. E. Rank. 1985. Ecological effects of salicin at three trophic levels: new problems from old adaptation. Science 229:649–650.

South, R. 1948. The Caterpillars of British Moths. Frederick Warne, London.

Southwood, T. R. E. 1973. The insect/plant relationship. An evolutionary perspective. In H. F. van Emden (ed.), Insect/Plant Relationships. Symposium of the Royal Entomology Society of London, Vol. 6. pp. 3–30. Blackwell Scientific, Oxford.

Strong, D. R., J. H. Lawton, and T. R. E. Southwood. 1984. Insects on Plants. Blackwell Scientific, Oxford.

Sugawara, F., K. Matsuda, A. Kobayashi, and K. Yamashita. 1979. Defensive secretion of chysomelid larvae *Gastrophysa atrocyanea* Motschulsky and *Phaedon brassicae* Baly. J. Chem. Ecol. 5:635–641.

Tahvanainen, J., R. Julkunen-Tuttor, and R. Kettunen. 1985. Phenolic glycosides govern the food selection pattern of willow feeding leaf beetles. Oecologica (Berlin) 67:52–56.

Takizawa, H. 1976. Larvae of the genus *Gonioctena* Chevrolat (Coleoptera, Chrysomelidae): description of Japanese species and the implications of larval characters for the phylogeny. Kontyu 44:444–468.

Tallamy, D. 1985. Squash beetle feeding behavior: an adaptation against induced cucurbit defenses. Ecology 66:1574–1579.

Thieme, H. 1965. Die Phenolglykoside der Salicaceen. Pharmazie 20:688–691.

Thieme, H., and R. Benecke. 1971. Die Phenolglykoside der Salicaceen. Pharmazie 26:227–231.

Thomson, R. H. 1957. Naturally Occurring Quinones. Butterworth, London.

Vickery, M. L., and B. Vickery. 1981. Secondary Plant Metabolism, Macmillan, London.

Visser, J. H., and A. K. Minks (eds.). 1982. Proceedings of the 5th International Symposium on Insect-Plant Relationships. Pudoc, Wageningen, The Netherlands.

Wain, R. L. 1943. The secretion of salicylaldehyde by the larvae of the brassy willow beetle (*Phyllodecta vitellinae* L.). Annual Report of the Agricultural National Research Stations. pp. 108–110.

Wallace, J. B., and M. S. Blum. 1969. Refined defensive mechanisms in *Chysomela scripta*. Ann. Ent. Soc. Amer. 62:503–506.

Webster, F. M. 1913. The southern corn rootworm, or budworm. Bull. U.S. Dep. Agric. 5:1–11

Whitehead, D. R., and R. M. Duffield. 1982. An unusual specialized predator-prey association (Coleoptera, Coccinellidae, Chrysomelidae): failure of a chemical defense and possible practical application. Coleopt. Bull. 36:96–97.

CHAPTER 8

Plant Allelochemistry and Mimicry

M. D. Bowers
Harvard University
Cambridge, Massachusetts

CONTENTS

1 Introduction
2 Unpalatable insects and the chemical bases of unpalatability
  2.1 Danaine butterflies and cardenolides
  2.2 Checkerspot butterflies of the genus *Euphydryas* and iridoid glycosides
  2.3 Ithomiine butterflies
  2.4 Heliconiines and *Passiflora*
  2.5 *Eumaeus* (Lycaenidae) and cycads
  2.6 Troidine swallowtails (Papilionidae) and the Aristolochiaceae
  2.7 Zygaenidae
  2.8 Tiger moths (Arctiidae)
3 Allelochemical variation and implications for mimicry
  3.1 Variation in plant allelochemistry
    3.1.1 Milkweeds (Asclepiadaceae) and monarchs (*Danaus plexippus*)
    3.1.2 Figworts (Scrophulariaceae) and Checkerspots (*Euphydryas*)
  3.2 Variation in insect oviposition and feeding behavior
  3.3 Insect processing of plant allelochemicals
  3.4 Is there a metabolic cost to sequestering plant allelochemicals?
4 Plant allelochemistry and predator behavior
5 Plant allelochemistry and the evolution of mimicry
  5.1 The palatability spectrum and a Batesian–Müllerian continuum
  5.2 What makes a good model?

6   Future directions for research on the role of plant allelochemistry in
mimicry
References

1   INTRODUCTION

The phenomenon of insect mimicry has intrigued biologists in a multitude
of fields since it was first described by Bates in 1862. The existence of
mimetic resemblance and the implication of this resemblance were a critical part of the evidence for Darwin's theory of natural selection. In addition to being of evolutionary interest, however, mimicry systems have
proven to be rich areas of research for behaviorists, population biologists,
theoreticians, geneticists, and physiologists. The reviews published on
mimicry in the past several years have considered many of these facets
(e.g., Wickler 1968; Rettenmeyer 1970; Turner 1977, 1984; Rothschild
1981, 1985; Pasteur 1982; Gilbert 1983; Brower 1984; Huheey 1984).

A critical component of any mimicry system, whether Batesian or
Müllerian, is the existence of an unpalatable model. This unpalatability
may be due to the possession of a venomous sting as in many Hymenoptera, to the presence of urticating spines or defensive glands, or to the
presence of defensive compounds distributed throughout the body (see
Chapter 7). In the latter case, the defensive compounds may be manufac-
tured by the insect itself, for example, the production of cantharadin by
blister beetles in the family Meloidae (Capinera et al. 1985). Alternatively,
the defensive compounds may be "recycled" from those present in the
host plant. During larval and/or adult feeding these compounds are ingested and sequestered, and render the insect unpalatable to its own
predators. The interrelationships of such plant allelochemicals, the insects that use them, and mimicry comprise the topic of this chapter.

Because much of the work on unpalatability as a defense strategy of
insects (from both a chemical and an ecological perspective) and on mimicry has been done with Lepidoptera and in particular butterflies, I discuss primarily these systems. Recent classifications of mimicry (Vane-
Wright 1976, Pasteur 1982) have described as many as 40 different kinds
of mimicry. Only some of these involve unpalatability, and it is only these
that I will consider in any detail: Batesian mimicry, Müllerian mimicry,
and automimicry [or Browerian mimicry (Pasteur 1982)].

As Huheey (1984, p. 209) pointed out: "Increasingly, the traditional question, palatable or unpalatable? will be supplemented or supplanted by chemical characterization of the toxin and quantitative estimates of the amount present. Many of the puzzling cases that we are presented with today may be resolved with good data of this sort." In only a few systems are such data available, but they are increasing rapidly. Combining qualitative and quantitative chemical information, life history, host plant, and population biology data with results from controlled feeding experiments using potential predators of insect species involved in mimetic relationships, will greatly enhance our understanding of mimicry.

In considering the role of plant allelochemistry in mimicry, I develop five major themes. First, what do we know about the chemical basis of unpalatability in those systems where it is clearly based on the chemistry of the host plant? Second, what are the implications of variation in host plant chemistry, insect feeding patterns, and insect processing capabilities for unpalatability and mimicry? Third, how might plant chemistry affect the prey–predator interactions in mimicry systems? Fourth, how does incorporation of knowledge on the importance of plant allelochemistry affect our thoughts on the evolution of mimicry? Finally, what areas of research on various aspects of mimicry might be especially productive to pursue?

2 UNPALATABLE INSECTS AND THE CHEMICAL BASES OF UNPALATABILITY

In some mimicry systems the chemical basis of the unpalatability in the models is known [Table 8.1; e.g., cardenolides and monarchs and some other milkweed butterflies (Ackery and Vane-Wright 1984); iridoid glycosides and butterflies in the genus *Euphydryas* (Bowers and Puttick 1986, Stermitz et al. 1986)]. In others, however, such is not the case. For example, in one of the most spectacular mimicry systems, butterflies of the Heliconiini [the larvae of which are specialist herbivores on passionflowers, *Passiflora* (Passifloraceae)], little is known about the chemical basis of unpalatability, although recent evidence suggests that they manufacture their own chemical defense in the form of cyanogenic glycosides (Nahrstedt and Davis 1983). Even in the best-studied system, that of the monarch, *Danaus plexippus* L. (Nymphalidae), its host plants (which contain cardiac glycosides), and its mimetic relationship with the viceroy, *Limenitis archippus* (Cramer) (Nymphalidae), there are still unanswered

TABLE 8.1 Chemical Compounds Making Lepidoptera Unpalatable, Some Representative Host Plant Families Containing Them, and the Lepidopteran Groups That Use Them

Allelochemicals	Representative Plant Families Containing These Compounds	Unpalatable Lepidoptera Using These Compounds	Comment
Cardenolides	Asclepiadaceae Apocynaceae Scrophulariaceae	Danainae (Nymphalidae) Arctiidae	Sequestered during larval feeding
Iridoid glycosides	Scrophulariaceae Plantaginaceae Caprifoliaceae Bignoniaceae	*Euphydryas* (Nymphalidae)	Sequestered during larval feeding
Pyrrolizidine alkaloids	Boraginaceae Leguminosae Asteraceae Apocynaceae	Ithomiini (Nymphalidae) Danainae (Nymphalidae) Arctiidae	Ingested by adults except Arctiidae which may sequester them during larval feeding
Cycasin (azoxyglycosides)	Cycadaceae	*Eumaeus* (Lycaenidae) *Seirarctia echo* (Arctiidae)	Sequestered or metabolized from other azoxyglycosides during larval feeding
Aristolochic acids	Aristolochiaceae	*Battus, Parides* (Papilionidae)	Sequestered during larval feeding
Cyanogenic glycosides	Rosaceae Leguminosae Graminae Araceae Compositae Euphorbiaceae Passifloraceae	Heliconiinae (Nymphalidae) Acraeinae (Nymphalidae) Zygaenidae	Generally produced *de novo* by adults, but in some cases may also be sequestered during larval feeding

questions pertaining to many aspects of this interaction. Further study of systems considered to be "well known" will certainly be productive.

Within the Lepidoptera, there are several systems which have been studied with regard to the chemical basis of the unpalatability of the models and the ecology of the mimetic relationship. These include both Batesian and Müllerian mimicry systems, although what is known about the role of plant allelochemicals in these systems varies enormously. Excellent reviews of what is known about some of these systems are provided in more detail than is possible here by Ackery and Vane-Wright (1984) (danaine butterflies), Brower (1984) (unpalatable butterflies in general), Brown (1984b) (ithomiine butterflies), and Rothschild et al. (1979a) (arctiid moths), and these can be consulted for additional information. Here, I specifically consider the importance of plant allelochemistry for unpalatable Lepidoptera.

2.1   Danaine Butterflies and Cardenolides

The larvae of danaine butterflies specialize primarily on the plant families Asclepiadaceae and Apocynaceae, which are known to contain cardenolides (Fig. 8.1), although not all species within these families contain these compounds, nor are cardenolides found in all plants fed on by danaine larvae (Ackery and Vane-Wright 1984). The chemistry of those plant species that do not contain these compounds is often unknown, and they may contain other compounds that contribute to the unpalatability of their associated insects (Ackery and Vane-Wright 1984).

Although many danaids and other milkweed-feeding insects sequester cardenolides which are effective deterrents to predators (Brower 1958a,c; Berenbaum and Miliczky 1984; references in Brower 1984), the ability of different danaine species to store cardenolides varies substantially (e.g., Brower et al. 1975, Isman et al. 1977, Cohen 1985). In addition, for a single danaine species, such as *D. plexippus,* different host plant species may contain different cardenolides or none at all, so different individuals within the same species are protected to a greater or lesser degree: the phenomenon of automimicry (Brower et al. 1968, 1970) or Browerian mimicry (Pasteur 1982).

A fascinating adjunct to this story is the inclusion of another group of compounds, the pyrrolizidine alkaloids (PAs; Fig. 8.1), These substances are found in a variety of plant families, including the Boraginaceae, Faba ceae, Asteraceae, and Apocynaceae (Pliske 1975, Culvenor 1978, Edgar

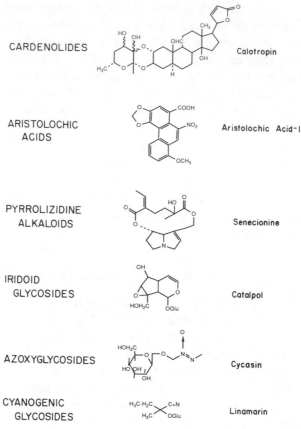

CARDENOLIDES — Calotropin

ARISTOLOCHIC ACIDS — Aristolochic Acid-I

PYRROLIZIDINE ALKALOIDS — Senecionine

IRIDOID GLYCOSIDES — Catalpol

AZOXYGLYCOSIDES — Cycasin

CYANOGENIC GLYCOSIDES — Linamarin

Figure 8.1 Groups of chemical compounds important in the unpalatability of Lepidoptera, with representative compounds.

1982), to which members of a variety of insect orders are attracted (Boppre 1986). In particular, danaine (and ithomiine; see below) butterflies visit plants containing PAs and the adults ingest these compounds, which may contribute to their unpalatability (Edgar et al. 1976a, 1979; Boppre 1978; Edgar 1982; but see Brower 1984). Although the importance of PAs for unpalatability of the danaines remains to be investigated experimentally, work on ithomiines has shown that accumulation of pyrrolizidines by the adults renders them unpalatable to a spider, *Nephila*, while adults denied access to these compounds are palatable (Brown 1984a,b), The case may be similar for danaines (Ackery and Vane-Wright 1984). Also, PAs are sequestered by some hemipterans rendering them unpalatable to lizards (McLain 1984, McLain and Shure 1985; see discussion below). Both ithomiines and danaines and some arctiid moths (Connor et al. 1981) also use pyrrolizidines as precursors for male pheromones (Schneider et al. 1975, Edgar et al. 1976b).

2.2 Checkerspot Butterflies of the Genus *Euphydryas* and Iridoid Glycosides

Butterflies of the genus *Euphydryas* specialize on plants containing iridoid glycosides (Fig. 8.2), most notably in the families Scrophulariaceae, Plantaginaceae, and Caprifoliaceae (Ehrlich et al. 1975, Bowers 1981). These butterflies are unpalatable to birds (Bowers 1980, 1981) apparently due to sequestration of iridoid glycosides (Bowers and Puttick 1986, Stermitz et al. 1986) during larval feeding. The larval host plant species may also determine the relative palatability of these butterflies to birds (Bowers 1980), due, at least in part, to the iridoid glycoside profile of the host plant (Bowers and Puttick 1986). Detailed analysis of the amounts of iridoid glycosides in *Euphydryas anicia* (Doubleday and Hewitson) (Gardner 1987, Gardner and Stermitz 1988) showed that the amount and kinds of iridoid glycosides in individual butterflies varies from almost nothing to as

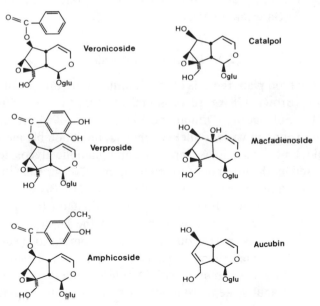

Figure 8.2   Iridoid glycosides.

much as 10% dry weight. Like *D. plexippus,* automimicry is clearly an important aspect of the ecology of these butterflies.

One member of this genus, *Euphydryas phaeton* (Drury) (Nymphalidae), appears to serve as a model for the palatable mimic, *Chlosyne harrisii* (Scudder) (Nymphalidae) (Bowers 1983b). *Chlosyne harrisii* feeds on *Aster umbellatus* Mill. (Asteraceae) and feeding experiments with birds showed that it was palatable. The underside of *C. harrisii* is very similar to that of *E. phaeton* and very unlike its close relatives. Other *Chlosyne* species may also be mimics of *Euphydryas.* Although this hypothesis has not been tested experimentally, the undersides of many *Chlosyne* species are very similar to *Euphydryas.* To make the situation even more interesting, larvae of some *Chlosyne* species, such as *C. leanira* (Felder and Felder), feed on plants such as *Castilleja* species, which contain iridoids. *Chlosyne leanira* sequesters iridoids during larval feeding and adult butterflies contain these compounds (L'Empereur, Gardner, and Stermitz, unpublished data). Thus, *Chlosyne* and *Euphydryas* species may be involved in an extensive and complex Batesian–Müllerian mimicry ring.

In addition, the larvae of *E. phaeton* and *C. harrisii* are virtually identical, probably leading to erroneous reports of *E. phaeton* feeding on asters, which it will not eat (Bowers 1983b). This is one of the few reported cases of larval Batesian mimicry.

2.3 Ithomiine Butterflies

Although the food plant records of the ithomiines are only beginning to be well known (Brown 1984b), records indicate that they are confined primarily to the Solanaceae (Drummond 1976, Brown 1984b). These plants are known to contain various allelochemicals, most notably alkaloids (but also steroidal bitter principles, saponins, and phenolic glycosides), which might by used by the ithomiines to render themselves unpalatable (Brown 1984a,b). Such a relationship was in fact suggested by Brower and Brower (1964). Recent evidence, however, indicates that, on the the contrary, ithomiines do not sequester compounds from the larval host plant, but rather the adults ingest pyrrolizidine alkaloids from various sources, and it is these that render them unpalatable, at least to spiders (Brown 1984a,b). Feeding experiments with *Nephila* spiders determined that newly emerged adults, kept from access to PAs, were palatable to the spiders, while wild-caught adults, which have presumably fed at PA sources, were rejected (Brown 1984a,b).

Although the ithomiines have been much less studied than the danaines with regard to their defensive chemistry, the two cases appear to be different. Many (although not all; see Ackery and Vane-Wright 1984) danaines use cardenolides sequestered from the larval host plant as their primary line of defense and may use PAs as the reinforcements. In ithomiines, there has been no defensive compound from the larval host plant identified in the adult butterflies, but rather, adult defense appears to depend solely on adult intake of PAs (Brown 1984a,b).

Pyrrolizidines are also important in the unpalatability of the hemipterans, *Neacoryphus bicrucis* Say (Lygaeidae) and *Lopidea instabile* (Miridae) (McLain 1984, McLain and Shure 1985). Both species were found to sequester PAs when fed on *Senecio smallii* (Britton) (Asteraceae) (which contains PAs) and were unpalatable to lizards. These two insect species appeared to be Müllerian mimics: lizards refused each after attacking and rejecting the other (McLain and Shure 1985).

2.4 Heliconiines and *Passiflora*

The heliconiine butterflies are conspicuous members of several different mimicry rings in the new world tropics, including both Batesian and Müllerian assemblages (Gilbert 1983). They specialize on many species of Passifloraceae (Gilbert 1983) and were predicted by Brower and Brower (1964) to sequester chemicals from these plants during larval feeding and use them as a defense. Feeding experiments with heliconiines (Brower et al. 1963, Brower and Brower 1964, Boyden 1976) have shown that although many of them are very unpalatable [*Heliconius erato* (L.), *H. numata* (Cramer), *H. melpomene* (L.), and *H. sara* (Fab.)], some, such as *Agraulis vanillae* (L.) and *Dryas julia* (Fab.), are much less so. The host plant relationships of the heliconiines are presently under study (Copp and Davenport 1978a,b; Smiley 1978, 1985a,b; Gilbert 1983; Smiley and Wisdom 1985) and much remains to be discovered. In the genus *Heliconius,* there appear to be species groups of *Heliconius* that preferentially attack certain species groups of *Passiflora* (Benson et al. 1975, Smiley 1985b). This relationship suggests a chemical basis for host plant specialization, but there is not as yet any such evidence (Smiley and Wisdom 1985). Thus, at this time there is no clear link between host plant range, host plant choice, host plant chemistry, and degree of unpalatability in these butterflies.

The chemistry of the Passifloraceae is complex. There are a variety of cyanogenic glycosides that occur in the family, as well as aklaloids, tannins, and flavonoids (Smiley and Wisdom 1985, Spencer and Seigler 1985a, Spencer et al. 1986). Because of the potential of cyanogenic compounds to be used as a defense by heliconiine butterflies, these compounds have been investigated.

There are three different biosynthetic pathways by which cyanogens are produced in the genus *Passiflora:* they may be cyclopentene derived, valine/isoleucine derived, or phenylalanine derived (see references in Spencer and Seigler 1985b). Most *Passiflora* species, however, contain cyclopentene-derived cyanogens (Nahrstedt and Davis 1983 and references therein). Nahrstedt and Davis (1981, 1983) found the cyanogenic glycosides, linamarin and lotaustralin, which are valine/isoleucine derived, in several species of heliconiines as well as some species of African acraeines. Because linamarin and lotaustralin were presumably not present in the host plants, Nahrstedt and Davis (1981, 1983) suggested that these compounds are produced by the butterflies themselves rather than being sequestered during larval feeding. However, some *Passiflora* species, such as *Passiflora lutea,* contain both cyclopentene-derived and valine/isoleucine-drieved cyanogens (Spencer and Seigler 1985b), suggesting that these valine/isoleucine-derived cyanogens, such as linamarin and lotaustralin, may also be sequestered by *Heliconius* if they feed on a *Passiflora* species that contains those compounds. The ability both to biosynthesize and sequester linamarin and lotaustralin was recently found in larvae of a zygaenid moth, a lepidopteran also known to produce cyanide (Nahrstedt and Davis 1986); such may be the case with *Heliconius* species. There are a variety of questions remaining to be answered in this fascinating complex of butterflies.

2.5   *Eumaeus* (Lycaenidae) and Cycads

The host plants of butterflies in the lycaenid genus *Eumaeus* are cycads of the genera *Cycas* and *Xamia* (Rawson 1961, DeVries 1976). These plants contain a group of unusual nitrogen-containing compounds, the azoxyglycosides (Fig. 8.1), which are carcinogenic and mutagenic and may cause severe probelms with livestock poisoning (Hopper 1978). The *Eumaeus* species are notable among the lycaenids for the warning coloration of larvae and adults. The larvae are red with yellow tubercles and are gregarious. The pupae are also warningly colored red-orange and are often found conspicuously clumped (DeVries 1977), in contrast to other lepidopteran pupae, which are solitary and cryptic (Brower 1984). In general, the adults are black with iridescent blue-green on the wings and a bright red abdomen, except for *E. childrenae* (Gray), which does not have a red abdomen. The adults of *Eumaeus atala* (Poey) contain cycasin (Rothschild et al. 1986, Bowers and Larin 1988), a relatively common azoxyglycoside. Adult *E. atala* butterflies and pure cycasin are deterrent to ants (Fig. 8.3): 25% sucrose solution containing 0.10% cycasin or ground *E. atala* butterfly received significantly fewer ant visits than a control of 25% sucrose. The butterflies are also unpalatable to birds (Bowers and Farley, unpublished data). Three out of four gray jays (*Perisoreus canadensis* (L.): Corvidae) tasted but did not eat, and rejected, *E. atala* adults. The fourth bird ate one *E. atala* and subsequently refused all others (Bowers and Farley, unpublished data).

Although there has been no experimental work done, there are several species of butterflies and a moth that appear to be mimics of *Eumaeus* species. For example, *Hades noctula* Westwood (Lycaenidae) appears to be a very close mimic of *Eumaeus* species. This species has the red spot on the underside of the forewing next to the thorax instead of on the hindwing next to the abdomen as do *Eumaeus,* but otherwise it is very similar. Another fascinating mimic is a newly described species of moth in the genus *Castnia* (Castniidae) (Miller, unpublished data). The Castniidae are often brightly colored, day-flying moths and many species are Batesian mimics. The mimic has the iridescent wings typical of *Eumaeus,* as well as the red abdomen and red spots on the ventral hindwing next to the abdomen typical of most *Eumaeus,* and was collected in an area where *Eumaeus* were flying (J. Miller, personal communication). Although work is just beginning on this fascinating complex (Rothschild et al. 1986, Bowers and Larin 1988), it should certainly prove to be a fruitful research area.

2.6   Troidine Swallowtails (Papilionidae) and the Aristolochiaceae

The *Aristolochia*-feeding swallowtails include the genera *Battus, Parides,* and *Troides.* Feeding trials with various species in these genera have shown that they are quite unpalatable to birds [e.g., *Battus philenor* L. (Brower 1958b), *Parides anchises* (L.) and *P. neophilus* (Hub.) (Brower

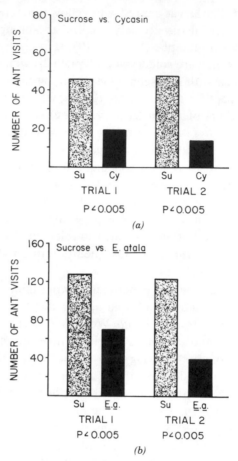

Figure 8.3 Responses of *Camponotus abdominalis floridanus* ants to (a) solutions of 25% sucrose versus cycasin in 25% sucrose (0.10%) and (b) solutions of 25% sucrose versus ground *Emaeus atala* in 25% sucrose. Chi-square analyses showed that the ants significantly preferred the pure sucrose solution in all trials. (Data from Bowers and Larin 1988.)

and Brower 1964)]. In addition, *Battus* and *Parides* species serve as models in a variety of mimicry complexes (Bates 1862, Poulton 1898, Brower 1958b, West 1985).

A search for the chemical basis of unpalatability in these butterflies led to host plants in the Aristolochiaceae. Many of these plant species contain aristolochic acids (Fig. 8.1), but tests of various species of *Aristolochia-*

feeding swallowtails for such compounds have given equivocal results (von Euw et al. 1968; Rothschild et al. 1970, 1972). Some swallowtails apparently do contain aristolochic acids [e.g., *Ornithoptera priamus* (L.), *Battus philenor, B. polydamas* (L.), *Pachliopta aristolochiae* (Fab.) and *Zerynthia polyxena* Schif.], while others do not [*Troides aeacus* Felder, three other species of *Troides,* and *Papilio hector* (L.)] (op. cit.). These variable results of chemical analyses, coupled with the unpalatability of this group of swallowtails in feeding tests, suggest that there may be other chemicals responsible for their unpalatability (Brower 1984). Further research on these insects and their host plants is certainly needed.

2.7 Zygaenidae

Certain moth species of the family Zygaenidae are well-known components of Müllerian mimicry complexes in Europe (Turner 1971, Sbordoni et al. 1979). It had been thought that their chemical defense was not derived from the host plant but rather from the autogenous production of cyanide (Jones et al. 1962). The sources of cyanide in 39 species of Zygaenidae, feeding on plant species in six different families (Fabaceae, Lamiaceae, Apiaceae, Rosaceae, Celastraceae, Polygonaceae), were found to be the autogenously produced cyanogenic glycosides, linamarin and lotaustralin (Davis and Nahrstedt 1979, 1982). However, recently, Nahrstedt and Davis (1986) found that larvae of one species of zygaenid, *Zygaena trifolii,* were able to sequester linamarin and lotaustralin. They fed larvae unlabeled or [aglycone ^{14}C]-labeled linamarin and lotaustralin on leaves of an acyanogenic strain of the food plant, *Lotus corniculatus* (L.) (Fabaceae), and found that larvae retained 20 to 45% of what they were fed. Thus, these insects appear to be able to sequester as well as biosynthesize their defense compounds.

2.8 Tiger Moths (Arctiidae)

Adults of the Arctiidae are usually aposematically colored black and yellow, black and red, or black and white. In contrast to most other unpalatable insects that sequester plant allelochemicals, larvae of these moths are relatively polyphagous and have been recorded feeding from a very diverse array of plant families (Rothschild et al. 1979b). Although at least some species of arctiids contain toxins that do not appear to be host plant derived (Rothschild et al. 1979a,b), larvae of some species of these moths also sequester pyrrolizidine alkaloids from some host plant species (Rothschild et al. 1979b) and cardenolides from others (Rothschild and Aplin 1971). Larvae of another arctiid species, *Seirarctia echo* Abbot and Smith, sequester cycasin from their cycad host plants in the genera *Cycas* and *Xamia* (Teas et al. 1966, Teas 1967). Pyrrolizidines are also important in producing the sex pheromones of some arctiids, notably *Utetheisa bella* (L.), and may also be important cues for female mate selection (Connor et al. 1981).

A variety of plant allelochemicals are implicated in the unpalatability of many lepidopteran species. Yet it is clear from this brief review that there is still much to be learned about the chemical basis of unpalatability in the Lepidoptera, as well as in insects in general. From what is known, however, it is apparent that plant allelochemistry is a critical component of any mimetic relationship where the defense of the unpalatable model (and mimic) is derived from the host plant.

3 ALLELOCHEMICAL VARIATION AND IMPLICATIONS FOR MIMICRY

The amount of defensive chemicals present in unpalatable insects may determine their degree of protection from predators and therefore their effectiveness as models in mimetic relationships. Variation in host plant chemistry, insect oviposition and feeding behavior, and insect processing of plant allelochemicals may all potentially affect the amounts of these compounds in the insect. Although these factors are considered independently below, they will act in concert to produce the defensive chemical profile of an individual insect or a particular population of unpalatable insects. The relative importance of these factors may vary substantially among different populations of the same species, but they may all play a role.

3.1 Variation in Plant Allelochemistry

Recent advances in chemical techniques for the isolation, identification, and quantification of plant allelochemicals have revealed the existence of what is to many researchers a perhaps startling amount of both qualitative and quantitative variation in the chemical makeup of plant individuals, populations, and species (e.g., Dolinger et al. 1973; McKey 1979; Brower et al. 1982; Louda and Rodman 1983a,b; Roby and Stermitz 1984a,b; Harris et al. 1986). The patterns and amounts of allelochemicals present may also change through the season (Mooney and Chu 1974, Nelson et al. 1981, Johnson et al. 1984; see also references in Denno and McClure 1983), and the conditions under which the plant is growing (Cooper-Driver et al. 1977, McKey et al. 1978, Lincoln and Mooney 1984).

For unpalatable insect species which sequester allelochemicals to use in their own defense and which may serve as models in mimicry systems this amount of variation could have large effects on their degree of unpalatability (i.e., how well defended they are). Indeed, such is the case in the only system where this has been at all well studied, that of cardenolides in *Asclepias* and other milkweeds and associated herbivores (see below). This variation in degree of noxiousness or unpalatability might then affect associated mimicry complexes. For example, a less unpalatable model that has fed on a plant low in the compounds responsible for its unpalatability will theoretically be able to support fewer associated mimics (Brower et al. 1970), and might in fact be considered relatively palatable by certain predators (Fink and Brower 1981, Fink et al. 1983, Bowers et al. 1985).

Variation in palatability within populations of a single species was first described by Brower et al. (1967) for the monarch, *D. plexippus,* and called automimicry (Brower et al. 1967) or Browerian mimicry (Pasteur 1982). This has also been described for *Euphydryas* butterflies (Bowers 1980, 1981) and will undoubtedly be true for other species of unpalatable insects as well. In both these cases, the chemical makeup of the host plant was the primary determinant of the allelochemical profile and relative unpalatability of the individual insect (Roeske et al. 1976, Bowers 1980, Bowers and Puttick 1986).

3.1.1 Milkweeds (*Asclepiadaceae*) and Monarchs (*Danaus plexippus*)

The ecological chemistry of the monarch, *D. plexippus,* and its host plants in North America provides a spectacular example of how variation in plant allelochemistry can and does affect the defensive chemistry of a principal herbivore (Roeske et al. 1976; Brower et al. 1982, 1984; Brower 1984; Seiber et al. 1986). Females of *D. plexippus* oviposit on and larvae feed on a wide variety of *Asclepias* species in North America (Roeske et al. 1976). The cardenolide profiles of these different species vary in both the kinds and amounts of the compounds present (Roeske et al. 1976). In addition, cardenolide profiles may vary with the age of a leaf, the part of the plant, and the individual plant (Nelson et al. 1981, Brower et al. 1982, Brower 1984). Despite this variation, the profiles of different *Asclepias* species are diagnostic enough that adult Monarch butterflies can be "fingerprinted" by the cardenolide profile of the larval host plant (Brower et al. 1982, 1984; Brower 1984; Seiber et al. 1986).

The specific cardenolides and the amounts of those cardenolides in an individual butterfly determine its relative unpalatability and emetic potential (Brower et al. 1968, Roeske et al. 1976, Brower 1984). This variation can encompass three orders of magnitude: adults reared from larvae collected from seven species of *Asclepias* in California contained amounts of cardenolides ranging from unmeasurable to over 1200 μg per butterfly (Brower 1984). The emetic potency of these butterflies ranged from none to enough to make 26 bluejays vomit (those latter butterflies packed quite a wallop!). In addition to their toxic qualities (i.e., effects on heart rate and inducing emesis), cardenolides are quite bitter. Thus, high concentrations of cardenolides may make a butterfly taste bad as well as causing it to be emetic [see Brower (1984) for a discussion of the difference between toxicity and distastefulness].

3.1.2 Figworts (Scrophulariaceae) and Checkerspots (Euphydryas)

Although checkerspot butterflies in the genus *Euphydryas* feed on species in several plant families that contain iridoid glycosides, members of the Scrophulariaceae are the predominant host plants (Ehrlich et al. 1975, Bowers 1981). Within this family, the plant genera with species used as hosts by these butterflies include *Aureolaria, Chelone, Scrophularia, Diplacus, Pedicularis, Orthocarpus, Penstemon, Castilleja,* and *Besseya.* The iridoid chemistry of some species in the latter three genera has been investigated by Stermitz and coworkers (Roby and Stermitz 1984a,b; McCoy and Stermitz 1983; Stermitz et al. 1986; Harris et al. 1986; Gardner 1987). In those plant species that have been investigated, there is substantial variation between individuals and within populations in the type of iridoid glycosides present and in their relative and absolute concentrations. For example, analysis of *Castilleja sulphurea* Rydb. (Scrophulariaceae) populations showed that some have as many as eight iridoid glycosides (Harris et al. 1986). The total as well as relative amounts of these different iridoid glycosides may vary from one population to another: one population was found to contain one iridoid glycoside, catalpol, almost exclusively (95%); in contrast to other populations sampled, which contained catalpol in amounts ranging from 21 to 46% of the total iridoid glycoside content (Harris et al. 1986).

Chemical analyses of the adult butterflies of two species of *Euphydryas, E. phaeton* and *E. anicia,* have shown somewhat different results from those of research on *D. plexippus* and cardenolides. Although the iridoid glycoside profile of the host plant clearly affects that of the butterflies which fed as larvae on that plant species (Bowers and Puttick 1986, Stermitz et al. 1986, Gardner 1987), only some of the available iridoid glycosides are sequestered (i.e., catalpol, aucubin, and macfadienoside; Fig. 8.2; see also Bowers and Puttick 1986; Stermitz et al. 1986; Gardner and Stermitz 1988). In addition, esters of catalpol (Fig. 8.2) appear to be metabolized to catalpol by the larvae (Gardner 1987, Gardner and Stermitz 1988). Catalpol appears to be of particular importance in the unpalatability of *Euphydryas* (Bowers 1980, Bowers and Puttick 1986); thus, larvae feeding on individual plants high in catalpol and/or catalpol esters will produce more unpalatable, and potentially more fit, butterflies.

3.2   Variation in Insect Oviposition and Feeding Behavior

Because plant species, populations, and individuals may vary in their allelochemical composition, the feeding behavior of an individual insect or insect population may be a critical determinant of the relative unpalatability of those individuals. In species in which all, or most, of larval development is completed on a single host plant, the oviposition behavior of the female may also be extremely important in determining the degree of unpalatability of her offspring. In two systems of unpalatable insects and their host plants, enough is known about the population biology and host plant chemistry to examine some of the implications of such variation in insect oviposition patterns and feeding behavior: *Danaus plexippus* and some related danaids and their milkweed host plants containing cardenolides, and *Euphydryas* species and their iridoid glycoside–containing host plants.

Danaus plexippus is known to use a variety of *Asclepias* species as host plants (Roeske et al. 1976). Oviposition choice tests with *D. plexippus* among three potential asclepiad host plants, *Asclepias currassivica* L., *Gomphocarpus fruticosus* (L.), and *Calotropis gigantea* (Dryand.) (Asclepiadaceae), showed that females preferred to oviposit on *A. currassavica,* the plant species lowest in cardenolide content but which produced the most emetic butterflies (Dixon et al. 1978). However, it was also noted (Dixon et al. 1978, p. 443) that one "aberrant" female concentrated on *C. gigantea*, "one of the least favored danaid food plants." This female may not have been aberrant at all, but rather, an example of individual variation. Such variation among individual females may be an important component of larval success (Singer 1984), as well as, potentially, relative unpalatability.

A study in Florida, more relevant to natural populations of *D. plexippus*, showed that oviposition patterns and larval survival were unrelated to the cardenolide content of the host plant, *Asclepias humistrata* Walt. (Cohen and Brower 1982). Cardenolide concentration of the host plant was not important in larval growth rate of *D. plexippus* (Erickson 1973), but leaf nitrogen and water content were, as in other insects (Scriber and Slansky 1981). However, cardenolides do not appear to be oviposition or larval feeding stimulants for *D. plexippus* (Dixon et al. 1978) and larvae will feed on asclepiad plants that appear to lack cardenolides (Brower et al. 1968). Factors such as presence of other eggs or larvae, plant age (Dixon et al. 1978), and plant size (Cohen and Brower 1982) appear to be more important than cardenolide content in determining oviposition patterns in *D. plexippus*.

Euphydryas and *Junonia coenia* butterflies and their iridoid-containing host plants provide interesting comparisons and contrasts with monarchs and asclepiads. Most notably, in both these taxa, iridoids are larval feeding stimulants (Bowers 1983a, 1984) and they are oviposition stimulants for *J. coenia* (Pereyra and Bowers 1988). Different populations within a single species of *Euphydryas* may vary from monophagous to oligophagous in their patterns of host plant use. Field studies of oviposition behavior in *E. editha* (Boisduval) (Singer 1971, 1982, 1983; Rausher et al. 1982), *E. gillettii* (Barnes) (Williams 1981), and *E. phaeton* (Stamp 1982) have shown that the females are very discriminating about where they will lay their eggs. In *J. coenia*, females can discriminate and prefer to oviposit on agar disks highest in iridoids (Pereyra and Bowers 1988). Such experiments remain to be done with *Euphydryas*. The iridoid glycoside content of different host plant species (and probably individuals) may vary substantially (Harris et al. 1986), and this may, in turn, affect the iridoid glycoside content of adult butterflies (Bowers and Puttick 1986, Stermitz et al. 1986, Gardner 1987). Thus, discrimination among individual plants on the basis of their iridoid content, on the part of *Euphydryas* females, might be important in determining the unpalatability of their offspring.

Oviposition choices by *Euphydryas* females may be less important in determining the iridoid glycoside content of offspring than in *D. plexippus* because of the life history of *Euphydryas*. These species are univoltine with diapause occurring when larvae are half-grown (usually as fourth instars). Although early, prediapause larvae feed on the plant on which the female oviposited, they may move if the host plant is consumed (Holdren and Ehrlich 1982). More important, however, when the larvae emerge from diapause in the spring, host plants are usually quite small and larvae may wander quite far in search of appropriate food. At this time, they may encounter a variety of iridoid glycoside-containing host plants and may choose to feed or to continue searching. Because these postdiapause larvae ingest many times more food (and iridoid glycosides) than early instars, the behavior of these postdiapause larvae may be the most critical determinant of the unpalatability of the adult butterfly. A few host plant species, however, are large shrubs, such as *Lonicera involucrata* Rich. (Banks) (Caprifoliaceae), used by *E. gillettii,* and *Penstemon breviflorus* Lindl. (Scrophulariaceae), used by some populations of *E. chalcedona*. In these, oviposition choice may indeed be important in determining the degree of adult defense.

3.3   Insect Processing of Plant Allelochemicals

The ability of an insect to regulate metabolically the defensive compounds ingested from the host plant may be another factor affecting the relative unpalatability of an individual. Comparative studies on different species of danaids, including *D. plexippus, D. gilippus* (Cramer), and *D. chrysippus* (L.), have shown that *D. plexippus* appears to be the most efficient at sequestering cardenolides (Brower et al. 1975, Rothschild et al. 1975, Cohen 1985). Yet both *D. gilippus* and *D. plexippus* sequester the same individual cardenolides from their host plant, although there may be substantial quantitative differences in the amounts of these cardenolides in the butterflies (Cohen 1985). In addition, *D. plexippus* larvae appear to regulate the amount of cardenolide that they sequester (Brower et al. 1982, 1984; Cohen 1985; Seiber et al. 1986). If the plant has a low cardenolide concentration, the insect is able to incorporate a higher percentage of cardenolide than occurs in the plant. Alternatively, if the plant is very high in cardenolides, the insect eliminates the excess (Seiber et al. 1980, 1986; Brower et al. 1982, 1984).

Cohen (1985) discussed the implications of his findings on relative cardenolide concentrations for understanding the monarch/queen-viceroy mimetic relationship. In Florida, the viceroy, *Limenitis archippus,* mimics the queen instead of the monarch (J. Brower 1958a,c), yet the queen is a less toxic model. Cohen suggests that this may be due to the more sedentary population biology of *D. gilippus,* which is not migratory as is *D. plexippus,* and thus would be a more predictable and available model than *D. plexippus.* Alternatively, he proposes that pyrrolizidine alkaloids, which are used as pheromone precursors by male queens but not monarchs (Meinwald et al. 1969, Pliske and Eisner 1969), may be a more important component of the chemical defense of queens than of monarchs. However, monarchs will also store pyrrolizidines (Edgar et al. 1976b, Rothschild and Edgar 1978), and these compounds have recently been found in wild-collected monarchs from Mexico (Kelley et al. 1986). Study of this relationship will certainly provide some fascinating data relevant to the role of plant allelochemicals in the evolution of mimicry.

Butterflies of the genus *Euphydryas* have a very different pattern of sequestration of iridoid glycosides, their defensive compounds. Although at the present time only *E. anicia* and *E. phaeton* have been examined in any detail, chemical studies of the other species are under way. Preliminary evidence from three of the other four *Euphydryas* species, *E. chalcedona, E. editha,* and *E. gillettii* coupled with the more complete data from *E. anicia* and *E. phaeton,* indicate that only certain iridoid glycosides are sequestered (Stermitz et al. 1986; Gardner 1987; Gardner and Stermitz 1988; Bowers, Belovsky, Janzen, Seewald, L'Empereur, Williams, unpublished). Specifically, the more polar iridoid glycosides, catalpol, aucubin, and macfadienoside (Fig. 8.2), have been found in butterflies, despite an array of as many as nine iridoids in the host plant (Stermitz et al. 1986). In addition, preliminary evidence suggests that larvae metabolize various esters of catalpol (Fig. 8.2), which may be in relatively high amounts in some host plants (Stermitz et al. 1986; Gardner 1987, Gardner and Stermitz 1988), and hydrolyze them to catalpol (Gardner and Stermitz 1988). Interestingly, catalpol appears to be particularly important in the unpalatability of *E. phaeton* (Bowers and Puttick 1986). However, the critical experiment to causally link catalpol concentration in butterflies with degree of unpalatability, that is, feeding the pure compound to potential predators, has not yet been done, but this is an obvious area for the focus of future research on this system.

Because certain individual cardenolides, iridoid glycosides, or other defensive chemicals may be particularly effective as deterrents to potential predators, the degree of protection of an unpalatable insect may be enhanced by the ability either to selectively sequester those compounds or to metabolize compounds that might be less effective as defenses to those that are more effective (as in the case of catalpol esters metabolized to catalpol). Whether this ability is due to specific selection by certain predators or whether it is a function of the chemical properties of the compound (Duffey 1980) has yet to be determined, even in a system as well studied as that of insects feeding on milkweeds.

3.4 Is There a Metabolic Cost to Sequestering Plant Allelochemicals?

Analogous to an examination of the cost and benefit to plants of producing allelochemicals potentially used for defense (Chew and Rodman 1979), examination of unpalatable insects for possible costs of storing defensive compounds is extremely complex. Data suggesting that there is no cost to sequestering cardenolides include those of Smith (1978), who reared larvae of *D. chrysippus* on milkweed species with and without cardenolides, and found that larvae grew significantly faster on cardenolide-containing species. In addition, Erickson (1973) found that *D. plexippus* larvae grew equally well on four species of *Asclepias,* ranging from one very high in cardenolides to two very low.

In contrast to these results, Cohen (1985) found a negative correlation between cardenolide concentration and various size and weight parameters in *D. plexippus*. However, neither he nor Seiber et al. (1980) found any direct evidence of a causal connection between cardenolide differences and weight differences of larvae. There is, however, metabolism of cardenolides during larval and pupal development (Nishio 1980, Seiber et al. 1980, Brower et al. 1982) that may exact a metabolic cost that is reflected in the negative relationship of adult size and cardenolide concentration. Cohen (1985) found no such negative correlation between cardenolide concentration and body size in queen butterflies, *D. gillippus,* which also were generally lower in cardenolide concentration than monarchs.

Analysis of iridoid concentration of *Euphydryas anicia* males and females from a single population in Colorado showed that there was no correlation between iridoid glycoside content and weight of males (Fig. 8.4), but that there was a positive correlation between female weight and iridoid glycoside content: females that weighed more contained more iridoid glycosides (Fig. 8.4). These data do not support the hypothesis that there is a cost to tolerating and/or sequestering such allelochemicals, at

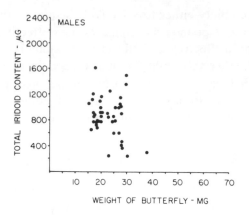

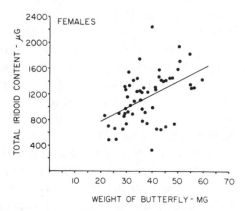

Figure 8.4   Iridoid glycoside content of *E. anicia* males ($N = 43$) and females ($N = 57$) from a single population (Red Hill, Colorado). Males: $R^2 = 0.01$, $Y = 916X - 3.94$; females: $R^2 = 0.20$, $Y = 368.69X + 20.75$. No line drawn for males because the correlation coefficient is so low. (Data from D. Gardner 1987.)

least for adapted species. More detailed analyses of other unpalatable species and populations are necessary before the potential and actual costs of sequestering plant allelochemicals can be assessed accurately.

4   PLANT ALLELOCHEMISTRY AND PREDATOR BEHAVIOR

Predators serve as the primary selective agents in the evolution of mimicry. In view of this, there have been a variety of both laboratory (J.

Brower 1958a,b,c; Brower et al. 1963; Brower and Brower 1964; Bowers 1980, 1981; Jarvi et al. 1981a; Wourms and Wasserman 1985a,b) and field (Brower et al. 1964; Benson 1972; Boyden 1976; Jeffords et al. 1979, 1980; Brown 1984a; Vasconcellos-Neto and Lewinsohn 1984) experiments that have addressed different aspects of this relationship. These include experimental and theoretical assessments of the importance of individual versus kin selection in the evolution of unpalatability and aposematic coloration (Harvey and Paxton 1981; Jarvi et al. 1981a,b; Harvey et al. 1982), the effect of aposematic coloration and/or relative unpalatability on predators (Sexton 1960; Coppinger 1969, 1970; O'Donald and Pilecki 1970; Pilecki and O'Donald 1971; Jeffords et al. 1979, 1980; Smith 1980), and the effectiveness of mimetic resemblance (Brower 1958a,b,c; Schmidt 1960; Brower et al. 1971; Boyden 1976; Bowers 1983b).

Variation in palatability among individuals of a single unpalatable species seems inevitable in species that recycle plant allelochemicals to use in their own defense, for the variety of reasons discussed above. This suggests that populations of most unpalatable species exist as a palatability spectrum, and therefore predators may be exposed to a range of palatabilities in the model as well as to the palatable mimic.

Several characteristics of potential predators are important to consider when evaluating their role in the evolution of mimicry. First, they can be extremely discriminating. For example, spiders can discriminate individual ithomiine butterflies with and without pyrrolizidines and eat those that do not contain those compounds, while cutting those with PAs out of their web (Brown 1984a,b). In addition, black-backed orioles (*Icterus abeillei* Lesson) eat less of individual monarch butterflies high in cardenolide concentration than of those that have low concentrations. They attack and kill the monarchs without regard to cardenolide content, but ingest significantly less of those high in cardenolides (Fink and Brower 1981). Under appropriate experimental conditions, birds are able to discern relatively minute differences in form and coloration of potential prey items (Pietrewicz and Kamil 1977). But how much time do they have to be discriminating under natural conditions? In nature, where the time available for feeding may be limited or energetic requirements may determine what prey are taken (Davies 1977), such discrimination may not be possible. Then, imperfect resemblance may be enough to protect mimics (Brower et al. 1971; Jeffords et al. 1979, 1980).

Second, the availability of alternative prey may be important in determining whether models and/or mimics are eaten. Butterflies, in general, do not appear to be preferred prey of birds (Sargent, unpublished), although there are innumerable references to birds eating butterflies (references in Bowers et al. 1985; Wourms and Wasserman 1985c). Availability of alternative prey may thus be an important component of the effectiveness of mimicry (Holling 1965, Dill 1975, Matthews 1977). If alternative prey are abundant, a mimic may not have to be as good as when alternative prey are scarce.

Third, there may be interspecific as well as intraspecific variation in the response of predators to unpalatable models as well as mimics. For example, different bird species are known to have very different tolerance for cardenolides (Fink and Brower 1981, Brower 1984); and some eat monarch butterflies with impunity (Brower 1984). There is also substantial variation among individuals within a species in how they respond to unpalatable insects of model-mimic pairs (e.g., Brower 1958a,b,c; Brower and Brower 1964; Bowers 1980, 1981). Some individuals may refuse to attack an unpalatable insect after tasting only one, while others may get sick more than once before learning (e.g., Bowers 1980).

5   PLANT ALLELOCHEMISTRY AND THE EVOLUTION OF MIMICRY

5.1   The Palatability Spectrum and a Batesian–Müllerian Continuum

Variation in allelochemical content of host plant species coupled with population differences in feeding behavior, and the consequent variation in degree of unpalatability, make the existence of a continuum between Batesian and Müllerian mimicry inevitable. Although Fisher (1930) did not have the chemical data currently available for plants and insects, he realized this as well (p. 167):

> When no question of degree is introduced into the discussion nothing is clearer than the distinction between the Batesian and the Müllerian factors. If however, we take into consideration that butterflies may exist in all degrees of palatability, and that avoidance or acceptance by the predator must depend greatly upon its appetite, there is some danger that the distinctness of the evolutionary tendencies pointed out by these two authors may be lost in the complexity of actual biological facts.

As pointed out in many discussions, both pro and con, regarding the existence of a Batesian–Müllerian continuum, the predictions generated from these two concepts are in some cases diametrically opposed (Fisher

1930; Huheey 1976, 1980; Benson 1977; Sheppard and Turner 1977; Sbordoni et al. 1979). For example, in Batesian mimicry, the predator (or dupe) is deceived into mistaking a palatable mimic for its noxious, unpalatable model; while in Müllerian mimicry, there is no deception because both "models" are unpalatable. However, those cases where the defensive chemicals have been analyzed have shown that even within an unpalatable species, individuals can range from palatable (containing little or no defensive chemicals) to extremely unpalatable. Therefore, within a mimicry ring (whether Batesian or Müllerian) the palatability of the various species (or individuals within a species) involved may change depending on the larval host plant, and range from a truly Batesian relationship where the mimic is completely palatable, to the extreme of a Müllerian one, where the mimic is as unpalatable as the model. In unpalatable insect species that are dependent on the larval host plant for their defensive chemicals, this relationship is dynamic and continuously changing.

A return to the often-cited but still revealing relationship of the "unpalatable" monarch and the mimicking viceroy provides an example. Evidence from Jane Brower's initial feeding experiments with these species (1958a,c) indicated that the Viceroy was somewhat unpalatable, suggesting that these two species may exist as Müllerian mimics rather than Batesian ones. However, rendering such judgments is difficult, because what is tasty to one predator may be noxious to another. It is hypothetically possible that the viceroy may be less preferred than the most palatable monarchs. As pointed out by several investigators of mimicry, each mimetic relationship must be investigated individually (Rothschild 1981, Brower 1984).

5.2   What Makes a Good Model?

Because there are so many aspects of the model–mimic–predator relationship that can vary (e.g., host plant chemistry, insect feeding behavior, predator behavior, model and mimic abundance), it is difficult to predict what optimal characteristic(s) would be associated with the best model or mimic. For models, however, one major trait would seem to be chemical constancy, that is, little variation in chemical makeup, and thus little variation in degree of unpalatability. This would make the monarch a relatively poor model because there is so much variation in the amount of cardenolides contained by different individuals. Because the monarch must support its own automimetic complex, it may be difficult for it to support a suite of other mimics. This may be why the monarch has only one very good mimic, the viceroy.

The other major butterfly mimetic complex in North America is the pipevine swallowtail *Battus philenor* (Papilionidae) and its associated mimics. *Battus philenor* has several mimics, none of which are very good: *Limenitis arthemis astyanax* (Drury) (Nymphalidae), the females of *Speyeria diana* (Cramer) (Nymphalidae), the females of *Papilio polyxenes* Fab. and *P. troilus* L. (Papilionidae), and the dark-form females of *P. glaucus*. The day-flying saturniid moth, *Callosamia promethea* (Drury) (Saturniidae), also appears to be a mimic of *B. philenor* (Jeffords et al. 1979). This is a much more extensive complex than that supported by the monarch, and most of the mimics are very poor. Feeding experiments with birds have shown that they can discriminate the mimics from the model in some cases (J. Brower 1958b). It may be that *B. philenor* is a more predictably unpalatable model than *D. plexippus* and thus can support a more extensive and less visually similar mimetic complex. Quantitative chemical analysis of *B. philenor* and its host plants in the genus *Aristolochia* would certainly be a fruitful area of research.

6 FUTURE DIRECTIONS FOR RESEARCH ON THE ROLE OF PLANT ALLELOCHEMISTRY IN MIMICRY

There are many productive and exciting avenues of research on the role of plant allelochemistry in mimicry. Although several of these have been pointed out or alluded to throughout this chapter, I will specifically discuss some of them here. There are two major foci that I see being particularly productive in future research in this area.

The first is to continue to study plant allelochemicals with particular attention to determination of both qualitative and quantitative variation in the plants, combined with studies of the fate of these compounds in insects feeding on those plants. This area of research should also include comparative studies of how different insect species process plant allelochemicals and what impact this has on unpalatability and potentially on mimicry. For example, Isman et al. (1977) analyzed milkweed-feeding insects from four orders (Coleoptera, Hemiptera, Orthoptera, and Lepidoptera) for their cardenolide content and found substantial variation among individuals and species feeding on the same species of *Asclepias*. Because the dynamics of cardenolide processing have been well studied in danaine butterflies and lygaeid bugs (Scudder and Duffey 1972), comparative studies with those other species might be extremely productive.

The second is to combine phylogenetic and systematic studies of unpalatable taxa with research into their chemistry, behavior, and physiology to try to understand the evolution of unpalatability and its concomitant adaptations. Phylogenetic studies can provide the evolutionary background on which could be superimposed data on host plant range, host plant chemistry, insect processing capabilities, and insect behavior to try to understand the evolution of these features. For example, phylogenetic analyses of danaine taxa (e.g., Ackery and Vane-Wright 1984) coupled with physiological and chemical data on their ability to sequester and process cardenolides might be used to document the evolution of unpalatability in this group. Such interdisciplinary, comparative approaches will prove to be extremely rewarding.

There are other areas relating to the interrelationship of plant allelochemistry and mimicry that are intriguing and understudied. An example is the individual variation found among particular species of predators and the importance of this variation in the effectiveness of mimetic resemblance. Most studies concentrate on overall patterns of predator behavior and not individual variation (but see Brower 1958a,b,c; Brower et al. 1971; Bowers 1980; Gibson 1980), which might also be an important component of mimetic effectiveness. Also, there are very few documented instances of larval mimicry. Even when adults exhibit very similar color patterns and behavior, the larvae are often very different. Batesian mimicry appears to be very rare among larvae, perhaps because larvae of model-mimic sets of adult species are not often found in similar habitats, and are on very different host plant species. Thus, the set of predators to which the larvae are exposed might be very different, making it difficult for the predators to act as selective agents promoting similarity of the larvae. One exception to this is the unpalatable *Euphydryas phaeton* and its mimic, *Chlosyne harrisii,* discussed previously. The larvae of these two species are virtually identical, both being bright orange with black spines (Bowers 1983b).

Another example is two species of sawflies, *Tenthredo grandis* (Norton) and *Macrophya nigra* (Norton) (Tenthredinidae), the larvae of which both feed on *Chelone glabra* L. (Scrophulariaceae), a plant containing iridoid glycosides (Bowers 1981, Bowers and Puttick 1986). Feeding experiments with birds did not distinguish between larvae of the two species because the differences are minute, but showed that one or both species were unpalatable (Bowers 1980). The larvae of these two species are both white with black markings and quite aposematic, yet the adults are quite different. Additional feeding experiments coupled with chemical analyses are needed to sort out this mimetic relationship.

A third example of larval mimicry occurs in at least two, and probably other species of geometrid larvae feeding on plant species that contain iridoid glycosides (Stermitz et al. 1988). Poole (1970) described the aposematic black and white coloration of *Meris alticola* Hulst and *Neoterpes graefiaria* (Hulst) (Geometridae), which both feed on species of *Penstemon,* a genus known to contain iridoid glycosides (Jensen et al. 1975), as convergence in larval coloration probably related to the wandering habit of the last instar. This coloration, he suggests (p. 294), is disruptive and serves to hide them in the grass. He states that the "possibility of distastefulness of the larvae and Müllerian mimicry between the two species cannot be ruled out, but [does not] think it is likely." Stermitz et al. (1988) suggested that, on the contrary, larvae of these two species are Müllerian mimics and are both protected by containing iridoid glycosides, compounds which would render them unpalatable.

Although mimicry has fascinated biologists for well over a hundred years, there remain intriguing problems associated with its many facets. Many of these are associated with the dynamics of chemical defense in unpalatable insects and the implications of variation in chemical defense for both theoretical and ecological investigations of mimicry. Recognition of the role that plant chemistry may play in mimetic relationships, the development and use of quantitative chemical techniques, and an understanding of phylogenetic relationships can be combined to help in understanding the ecology and evolution of unpalatability and the associated mimicry complexes.

ACKNOWLEDGMENTS

This review was prepared while the author was on sabbatical in the Chemistry Department at Colorado State University, supported by NSF grant RII-8503816. The support of the Chemistry Department in particular and Colorado State University in general are gratefully acknowledged. I appreciate comments by Pedro Barbosa, Lincoln Brower, Deborah Letourneau, and Frank Stermitz on earlier drafts of the manuscipt. Special thanks go to Dale Gardner and Frank Stermitz for allowing me to use unpublished data, to Kevin Spencer for discussions about cyanogenic glycosides in the Passifloraceae, and to Larry Gilbert for encouraging me to write this review.

REFERENCES

Ackery, P. R., and R. I. Vane-Wright. 1984. Milkweed Butterflies. Cornell University Press, Ithaca, N.Y.

Bates, H. W. 1862. Contributions to an insect fauna of the Amazon valley. Trans. Linn. Soc. London 23:495–566.

Benson, W. W. 1972. Natural selection for Müllerian mimicry in *Heliconius erato* in Costa Rica. Science 176:936–938.

Benson, W. W. 1977. On the supposed spectrum between Batesian and Müllerian mimicry. Evolution 31:454–455.

Benson, W. W., K. S. Brown, and L. E. Gilbert. 1975. Coevolution of plants and herbivores: passion flower butterflies. Evolution 29:659–680.

Berenbaum, M., and E. Miliczky. 1984. Mantids and milkweed bugs: efficacy of aposematic coloration against invertebrate predators. Amer. Midl. Nat. 111:64–68.

Boppre, M. 1978. Chemical communication, plant relationships, and mimicry in the evolution of danaid butterflies. Ent. Exp. Appl. 24:264–277.

Boppre, M. 1986. Insects pharmacophagously utilizing defensive plant chemicals (pyrrolizidine alkaloids). Naturwissenschaften 73:17–26.

Bowers, M. D. 1980. Unpalatability as a defense strategy of *Euphydryas phaeton* (Lepidoptera: Nymphalidae). Evolution 34:586–600.

Bowers, M. D. 1981. Unpalatability as a defense strategy of western checkerspot butterflies (*Euphydryas*, Nymphalidae). Evolution 35:367–375.

Bowers, M. D. 1983a. Iridoid glycosides and larval host plant specificity in checkerspot butterflies (*Euphydryas:* Nymphalidae). J. Chem. Ecol. 9:475–493.

Bowers, M. D. 1983b. Mimicry in north American checkerspot butterflies: *Euphydryas phaeton* and *Chlosyne harrisii* (Nymphalidae). Ecol. Ent. 8:1–8.

Bowers, M. D. 1984. Iridoid glycosides and host-plant specificity in larvae of the buckeye butterfly, *Junonia coenia* (Nymphalidae). J. Chem. Ecol. 10:1567–1577.

Bowers, M. D., and Z. Larin. 1988. Acquired chemical defense in the lycaenid butterfly, *Eumaeus atala*. J. Chem. Ecol. (in review).

Bowers, M. D., and G. M. Puttick. 1986. The fate of ingested iridoid glycosides in lepidopteran herbivores. J. Chem. Ecol. 12:169–178.

Bowers, M. D., I. L. Brown, and D. Wheye. 1985. Bird predation as a selective agent in a butterfly population. Evolution 39:93–103.

Boyden, T. C. 1976. Butterfly palatability and mimicry: experiments with *Ameiva* lizards. Evolution 30:73–81.

Brower, J. V. Z. 1958a. Experimental studies of mimicry in some North American butterflies. I. The monarch *Danaus plexippus* and viceroy *Limenitis archippus*. Evolution 12:32–47.

Brower, J. V. Z. 1958b. Experimental studies of mimicry in some North American butterflies. II. *Battus philenor* and *Papilio troilus, P. polyxenes* and *P. glaucus*. Evolution 12:123–136.

Brower, J. V. Z. 1958c. Experimental studies of mimicry in some North American butterflies. III. *Danaus gilippus berenice* and *Limenitis archippus floridensis*. Evolution 12:273–285.

Brower, L. P. 1984. Chemical defense in butterflies. In R. I. Vane-Wright and P. R. Ackery (eds.), The Biology of Butterflies. Symposium of the Royal Entomology Society of London, Vol. 11. pp. 109–134. Academic Press, London.

Brower, L. P., and J. V. Z. Brower. 1964. Birds, butterflies, and plant poisons: a study in ecological chemistry. Zoologica 49:137–159.

Brower, L. P., J. V. Z. Brower, and C. T. Collins. 1963. Experimental studies of mimicry. 7. Relative palatability and Müllerian mimicry among neotropical butterflies of the subfamily Heliconiinae. Zoologica 48:65–83.

Brower, L. P., J. V. Z. Brower, F. G. Stiles, H. J. Croze, and A. S. Hower. 1964. Mimicry: differential advantage of color patterns in the natural environment. Science 144:183–185.

Brower, L. P., J. V. Z. Brower, and J. M. Corvino. 1967. Plant poisons in a terrestrial food chain. Proc. Natl. Acad. Sci. 57:893–898.

Brower, L. P., W. N. Ryerson, L. L. Coppinger, and S. C. Glazier. 1968. Ecological chemistry and the palatability spectrum. Science 161:1349–1351.

Brower, L. P., F. H. Pough, and H. R. Meck. 1970. Theoretical investigations of automimicry. I. Single trial learning. Proc. Natl. Acad. Sci. 66:1059–1066.

Brower, L. P., J. Alcock, and J. V. Z. Brower. 1971. Avian feeding behaviour and the selective advantage of incipient mimicry. In E. R. Creed (ed.), Ecological Genetics and Evolution: Essays in Honour of E. B. Ford. pp. 261–274. Oxford University Press, Oxford.

Brower, L. P., M. Edmunds, and C. M. Moffett. 1975. Cardenolide content and palatability of a population of *Danaus chrysippus* butterflies from West Africa. J. Ent. (A) 49:183–196.

Brower, L. P., J. N. Seiber, C. J. Nelson, S. P. Lynch, and P. M. Tuskes. 1982. Plant determined variation in the cardenolide content, thin-layer chromatography profiles, and emetic potency of Monarch butterflies, *Danaus plexippus*, reared on the milkweed, *Asclepias eriocarpa* in California. J. Chem. Ecol. 8:579–633.

Brower, L. P., J. N. Seiber, C. J. Nelson, S. P. Lynch, and M. M. Holland. 1984. Plant-determined variation in the cardenolide content, thin-layer chromatography profiles, and emetic potency of monarch butterflies, *Danaus plexippus*, reared on the milkweed *Asclepias speciosa* in California. J. Chem. Ecol. 10:601–639.

Brown, K. S. 1984a. Adult-obtained pyrrolizidine alkaloids defend ithomiine butterflies against a spider predator. Nature 309:707–709.

Brown, K. S. 1984b. Chemical ecology of dehydropyrrolizidine alkaloids in adult ithomiinae (Lepidoptera: Nymphalidae). Rev. Bras. Biol. 44:435–460.

Capinera, J. L., D. R. Gardener, and F. R. Stermitz. 1985. Cantharidin levels in blister beetles (Coleoptera: Meloidae) associated with alfalfa in Colorado. J. Econ. Ent. 78:1052–1055.

Chew, F. S., and J. E. Rodman. 1979. Plant resources for chemical defense. In G. A. Rosenthal and D. H. Janzen (eds.), Herbivores: Their Interaction with Plant Secondary Metabolites. pp. 271–306. Academic Press, New York.

Cohen, J. A. 1985. Differences and similarities in carenolide contents of queen and monarch butterflies in Florida and their ecological and evolutionary consequences. J. Chem. Ecol. 11:85–103.

Cohen, J. A., and L. P. Brower. 1982. Oviposition and larval success of wild monarch butterflies in relation to host plant size and cardenolide concentration. J. Kans. Ent. Soc. 55:343–348.

Connor, W. E., T. Eisner, R. K. van der Meer, A. Guerrero, and J. Meinwald. 1981. Precopulatory sexual interaction in an arctiid moth (*Utetheisa ornatrix*): role of a pheromone derived from dietary alkaloids. Behav. Ecol. Sociobiol. 9:227–235.

Cooper-Driver, G. A., S. Finch, T. Swain, and E. Bernays. 1977. Seasonal variation in secondary plant compounds in relation to the palatability of *Pteridium aquilinum*. Biochem. Syst. Ecol. 5:211–218.

Copp, N. H., and D. Davenport. 1978a. *Agraulis* and *Passiflora*. 1. Control of specificity. Biol. Bull. 155:98–112.

Copp, N. H., and D. Davenport. 1978b. *Agraulis* and *Passiflora*. 2. Behavior and sensory modalites. Biol. Bull. 155:113–124.

Coppinger, R. P. 1969. The effect of experience and novelty on avian feeding behaviour with reference to the evolution of warning coloration in butterflies. I. Reactions of wild-caught adult blue jays to novel insects. Behaviour 35:45–60.

Coppinger, R. P. 1970. The effect of experience and novelty on avian feeding behaviour with reference to the evolution of warning coloration in butteflies. II. Reactions of naive birds to novel insects. Amer. Nat. 104:323–335.

Culvenor, C. C. J. 1978. Pyrrolizidine alkaloids, occurrence and systematic importance in angiosperms. Bot. Not. 131:473–486.

Davies, N. B. 1977. Prey selection and the search strategy of the spotted flycatcher (*Muscicapa striata*): a field study on optimal foraging. Anim. Behav. 24:1016–1033.

Davis, R. H., and A. Nahrstedt. 1979. Linamarin and lotaustralin as the source of cyanide in *Zygaena filipendulae* L. (Lepidoptera). Comp. Biochem. Physiol. 64B:395–397.

Davis, R. H., and A. Nahrstedt. 1982. Occurrence and variation of the cyanogenic glucosides linamarin and lotaustralin in species of the Zygaenidae (Insecta: Lepidoptera). Comp. Biochem. Physiol. 71B:329–332.

Denno, R. F., and M. S. McClure. 1983. Variable Plants and Herbivores in Natural and Managed Systems. Academic Press, New York.

De Vries, P. J. 1976. Notes on the behavior of *Eumaeus minyas* in Costa Rica. Brenesia 8:103.

De Vries, P. J. 1977. *Eumaeus minyas,* an aposematic lycaenid butterfly. Brenesia 10:269–270.

Dill, L. M. 1975. Calculated risk-taking by predators as a factor in Batesian mimicry. Can. J. Zool. 53:1614–1621.

Dixon, C. A., J. M. Erickson, D. N. Kellett, and M. Rothschild. 1978. Some adaptations between *Danaus plexippus* and its food plant, with notes on *Danaus chrysippus* and *Euploea core.* J. Zool. 185:437–467.

Dolinger, P. M., P. R. Ehrlich, W. L. Fitch, and D. E. Breedlove. 1973. Alkaloid and predation patterns in Colorado lupine populations. Oecologia (Berlin) 13:191–204.

Drummond, B. A. 1976. Comparative ecology and mimetic relationships of ithomiine butterflies in eastern Equador. Ph.D. thesis, University of Florida, Gainesville, Fla. 361pp.

Duffey, S. S. 1980. Sequestration of plant natural products by insects. Annu. Rev. Ent. 25:447–477.

Edgar, J. A. 1982. Pyrrolizidine alkaloids sequestered by Solomon Island Danaine butterflies. The feeding preferences of the Danainae and Ithomiinae. J. Zool. London 196:385–399.

Edgar, J. A., C. C. J. Culvenor, and T. E. Pliske. 1976a. Isolation of a lactone, structurally related to the esterifying acids of pyrrolizidine alkaloids, from the costal fringes of male Ithomiinae. J. Chem. Ecol. 2:263–270.

Edgar, J. A., P. A. Cockrum, and J. L. Frahn. 1976b. Pyrrolizidine alkaloids in *Danaus plexippus* L. and *Danaus chrysippus* L. Experientia 32:1535–1537.

Edgar, J. A., M. Boppre, and D. Schneider. 1979. Pyrrolizidine aklaloid storage in African and Australian danaid butterflies. Experientia 35:1447–1448.

Ehrlich, P. R., R. R. White, M. C. Singer, S. W. McKechnie, and L. E. Gilbert. 1975. Checkerspot butterflies: a historical perspective. Science 188:221–228.

Erickson, J. M. 1973. The utilization of various *Asclepias* species by larvae of the monarch butterfly, *Danaus plexippus*. Psyche 80:230–244.

Fink, L. S., and L. P. Brower. 1981. Birds can overcome the cardenolide defense of monarch butterflies in Mexico. Nature 291:67–70.

Fink, L. S., L. P. Brower, R. B. Wade, and P. R. Spitzer. 1983. Overwintering monarch butterflies as food for insectivorous birds in Mexico. Biotropica 15:151–153.

Fisher, R. A. 1930. The Genetical Theory of Natural Selection. Clarendon Press, Oxford.

Gardner, D. R. 1987. Iridoid chemistry of some *Castilleja* and *Besseya* plants and their hosted checkerspot butterflies. Ph.D. Thesis, Colorado State University, Fort Collins, Colorado.

Gardner, D. R. and F. R. Stermitz. 1988. Hostplant utilization and iridoid glycoside sequestration by *Euphydryas anicia* individuals and populations. J. Chem. Ecol. (in press).

Gibson, D. O. 1980. The role of escape in mimicry and polymorphism. I. The response of captive birds to artificial prey. Biol. J. Linn. Soc. 14:201–214.

Gilbert, L. E. 1983. Coevolution and mimicry. In D. J. Futuyma and M. S. Slatkin (eds.), Coevolution. pp. 263–281. Sinauer Associates, Sunderland, Mass.

Harris, G. H., F. R. Stermitz, and W. Jing. 1986. Iridoids and alkaloids from *Castilleja* (Scrophulariaceae) host plants for *Platyptilia pica* (Lepidoptera: Pterophoridae): Rhexifoline content of *P. pica*. Biochem. Syst. Ecol. 14:499–504.

Harvey, P. H., and P. J. Paxton. 1981. The evolution of aposematic coloration. Oikos 37:391–393.

Harvey, P. H., J. J. Bull, M. Pemberton, and R. J. Paxton. 1982. The evolution of aposematic coloration in distasteful prey: a family model. Amer. Nat. 119:710–719.

Holdren, C. E., and P. R. Ehrlich. 1982. Ecological determinants of food plant choice in the checkerspot butterfly, *Euphydryas editha* in Colorado. Oecologia (Berlin) 52:417–423.

Holling, C. S. 1965. The functional response of predators to prey density and its role in mimicry and population regulation. Mem. Ent. Soc. Can. 45:1–60.

Hooper, P. T. 1978. Cycad poisoning in Australia-etiology and pathology. In R. F. Keeler, K. R. van Kampen, and L. F. James (eds.), Effects of Poisonous Plants on Livestock. pp. 337–347. Academic Press, New York.

Huheey, J. E. 1976. Studies of warning coloration and mimicry. VII. Evolutionary consequences of a Batesian–Müllerian spectrum: a model for Müllerian mimicry. Evolution 30:86–93.

Huheey, J. E. 1980. Bastesian and Müllerian mimicry: semantic and substantive differences of opinion. Evolution 34:1212–1215.

Huheey, J. E. 1984. Warning coloration and mimicry. In W. J. Bell and R. T. Cardé (eds.), Chemical Ecology of Insects. pp. 257–297. Chapman & Hall, London.

Isman, M. B., S. S. Duffey, and G. G. E. Scudder. 1977. Cardenolide content of some leaf- and stem-feeding insects on temperate North American milkweeds (*Asclepias* spp.). Can. J. Zool. 55:1024–1028.

Jarvi, T., B. Sillen-Tullberg, and C. Wiklund, 1981a. The cost of being aposematic. An experimental study of the predation on larvae of *Papilio machaon* by the great tit *Parus major*. Oikos 36:267–272.

Jarvi, T., B. Sillen-Tullberg, and C. Wiklund. 1981b. Individual versus kin selection for aposematic coloration: a reply to Harvey and Paxton. Oikos 37:393–395.

Jeffords, M. R., J. G. Sternburg, and G. P. Waldbauer. 1979. Batesian mimicry: field demonstration of the survival value of pipevine swallowtail and monarch color patterns. Evolution 33:275–286.

Jeffords, M. R., J. G. Sternburg, and G. P. Waldbauer. 1980. Determination of the time of day at which diurnal moths painted to resemble butterflies are attacked by birds. Evolution 34:1205–1211.

Jensen, S. R., B. J. Nielsen, and R. Dahlgren, 1975. Iridoid compounds, their occurrence and systematic importance in the angiosperms. Bot. Not. 128:148–180.

Johnson, N. D., C. C. Chu, P. R. Ehrlich, and H. A. Mooney. 1984. The seasonal dynamics of leaf resin, nitrogen, and herbivore damage in *Eriodictyon californicum* and their parallels in *Diplacus aurantiacus*. Oecologia 61:398–402.

Jones, D. A., J. Parsons, and M. Rothschild. 1962. Release of hydrocyanic acid from crushed tissues of all stages in the life-cycle of species of the Zygaeninae (Lepidoptera). Nature 193:52–53.

Kelley, R. B., J. N. Seiber, A. B. Jones, and H. J. Segall. 1986. Isolation and structural identification of pyrrolizidine alkaloids in overwintering monarchs (poster abstract). American Chemical Society, Rocky Mountain regional meeting program, Denver, Colo. (unpublished).

Lincoln, D. E., and H. A. Mooney. 1984. Herbivory on *Diplacus aurantiacus* shrubs in sun and shade. Oecologia (Berlin) 64:173–178.

Louda, S. M., and J. E. Rodman. 1983a. Ecological patterns in the glucosinolate content of a native mustard, *Cardamine cordifolia* in the Rocky Mountains. J. Chem. Ecol. 9:397–421.

Louda, S. M., and J. E. Rodman. 1983b. Concentration of glucosinolates in relation to habitat and insect herbivory for the native crucifer *Cardamine cordifolia*. Biochem. Syst. Ecol. 11:199–207.

Matthews, E. G. 1977. Signal-based frequency-dependent defense strategies and the evolution of mimicry. Amer. Nat. 111:213–222.

McCoy, J. W., and F. R. Stermitz. 1983. Alkaloids from *Castilleja miniata* and *Penstemon whippleanus,* two host species for the plume moth. J. Nat. Prod. 46:902–905.

McKey, D. 1979. The distribution of secondary compounds within plants. In G. A. Rosenthal and D. H. Janzen (eds.), Herbivores: Their Interaction with Secondary Plant Metabolites. pp. 55–133. Academic Press, New York.

McKey, D., P. G. Waterman, C. N. Mbi, G. N. Gartlan, and T. T. Struhsaker. 1978. Phenolic content of vegetation in two African rainforests: ecological implications. Science 202:61–64.

McLain, D. K. 1984. Coevolution: Müllerian mimicry between a plant bug (Miridae) and a seed bug (Lygaeidae) and the relationship between host plant choice and unpalatability. Oikos 43:143–148.

McLain, D. K., and D. J. Shure. 1985. Host plant toxins and unpalatability of *Neacoryphus bicrucis* (Hemiptera: Lygaeidae) Ecol. Ent. 10:291–298.

Meinwald, J., Y. C. Meinwald, and P. H. Mazzochi. 1969. Sex pheromone in the queen butterfly: chemistry. Science 164:1174–1175.

Mooney, H. A., and C. Chu. 1974. Seasonal carbon allocation in *Heteromeles arbutifolia,* a California evergreen shrub. Oecologia (Berlin) 14:295–306.

Nahrstedt, A., and R. H. Davis. 1981. The occurrence of the cyanoglucosides, linamarin and lotaustralin, in *Acraea* and *Heliconius* butterflies. Comp. Biochem. Physiol. 68B:575–577.

Nahrstedt, A., and R. H. Davis. 1983. Occurrence, variation and biosynthesis of the cyanogenic glucosides linamarin and lotaustralin in species of the Heliconiini (Insecta: Lepidoptera). Comp. Biochem. Physiol. 75B:65–73.

Nahrstedt, A., and R. H. Davis. 1986. Uptake of linamarin and lotaustralin from their food plant by larvae of *Zygaena trifolii.* Phytochemistry 25:2299–2302.

Nelson, C. J., J. N. Seiber, and L. P. Brower. 1981. Seasonal and intraplant variation of cardenolide content in the California milkweed, *Asclepias eriocarpa,* and the implications for plant defense. J. Chem. Ecol. 7:981–1010.

Nishio, S. 1980. The fates and adaptive significance of cardenolides sequestered by larvae of *Danaus plexippus* (L.) and *Cycnia inopinatus* (Hy. Edwards). Ph.D. thesis. University Microfilms, University of Georgia, Athens, Ga.

O'Donald, P., and C. Pilecki. 1970. Polymorphic mimicry and natural selection. Evolution 24:395–401.

Pasteur, G. 1982. A classificatory review of mimicry systems. Annu. Rev. Ecol. Syst. 13:169–199.

Pereyra, P. and M. D. Bowers. 1988. Iridoid glycosides as oviposition stimulants for the Buckeye, *Junonia coenia* (Nymphalidae). J. Chem. Ecol. (in press).

Pietrewicz, A. T., and A. C. Kamil. 1977. Visual detection of cryptic prey by blue jays (*Cyanocitta cristata*). Science 195:580–582.

Pilecki, C., and P. O'Donald. 1971. The effects of predation on artifical mimetic polymorphisms with perfect and imperfect mimics at varying frequencies. Evolution 25:365–370.

Pliske, T. E. 1975. Attraction of Lepidoptera to plants containing pyrrolizidine alkaloids. Environ. Ent. 4:455–473.

Pliske, T. E., and T. E. Eisner. 1969. Sex pheromone of the queen butterfly: biology. Science 164:1170–1172.

Poole, R. W. 1970. Convergent evolution in the larvae of two *Penstemon*-feeding geometrids (Lepidoptera: Geometridae). J. Kans. Ent. Soc. 43:292–297.

Poulton, E. B. 1898. Natural selection: the cause of mimetic resemblance and common warning colours. Linn. Soc. J. Zool. 26:558–612.

Rausher, M. D., D. A. MacKay, and M. C. Singer. 1982. Pre- and post-alighting host discrimination by *Euphydryas editha:* the behavioral mechanisms causing clumped distribution of egg clusters. Anim. Behav. 29:1220–1228.

Rawson, G. W. 1961. The recent rediscovery of *Eumaeus atala* (Lycaenidae) in southern Florida. J. Lepid. Soc. 15:237–244.

Rettenmeyer, C. W. 1970. Insect mimicry. Annu. Rev. Ent. 15:43–74.

Roby, M. R., and F. R. Stermitz. 1984a. Pyrrolizidine and pyridine monoterpene alkaloids from two *Castilleja* plant hosts of the plume moth, *Platyptilia pica*. J. Nat. Prod. 47:846–853.

Roby, M. R., and F. R. Stermitz. 1984b. Penstemonoside and other iridoids from *Castilleja rhexifolia:* conversion of penstemonoside to the pyridine monoterpene alkaloid rhexifoline. J. Nat. Prod. 47:853–859.

Roeske, C. N., J. S. Seiber, L. P. Brower, and C. M. Moffitt. 1976. Milkweed cardenolides and their comparative processing by monarch butterflies (*Danaus plexippus*). In J. W. Wallace and R. L. Mansell (eds.), Biochemical Interactions between Plants and Insects, Vol. 10, Recent Advances in Phytochemistry. pp. 93–167. Plenum Press, New York.

Rothschild, M. 1981. Mimicry, butterflies and plants. Symb. Bot. Ups. 22:82–99.

Rothschild, M. 1985. British aposematic Lepidoptera. In J. Heath and A. M. Emmet (eds.), The Moths and Butterflies of Great Britain and Ireland, pp. 9–62. Harley Books, Essex, England.

Rothschild, M., and R. Aplin. 1971. Toxins in tiger moths (Arctiidae: Lepidoptera). In A. S. Tahori (ed.), Pesticide Chemistry, Vol. 3, Chemical Releasers in Insects. pp. 177–182. Gordon and Breach, London.

Rothschild, M., and J. A. Edgar. 1978. Pyrrolizidine alkaloids from *Senecio vulgaris* sequestered by *Danaus plexippus*. J. Zool. London 185:347–349.

Rothschild, M., R. T. Reichstein, J. von Euw, R. T. Aplin, and R. R. M. Harman. 1970. Toxic Lepidoptera. Toxicon 8:293–299.

Rothschild, M., J. von Euw, and T. Reichstein. 1972. Aristolichic acids stored by *Zerynthia polyxena*. Insect Biochem. 2:334–343.

Rothschild, M., J. von Euw, T. Reichstein, D. A. Smith, and J. Pierre. 1975. Cardenolide storage in *Danaus chrysippus* with additional notes on *D. plexippus*. Proc. Roy. Soc. London Ser. B 190:1–31.

Rothschild, M., R. T. Aplin, P. A. Cockrum, J. A. Edgar, P. Fairweather, and R. Lees. 1979a. Pyrrolizidine alkaloids in arctiid moths (Lep.) with a discussion on host plant relationships and the role of these secondary plant substances in the Arctiidae. Biol. J. Linn. Soc. 12:305–326.

Rothschild, M., H. Keutmann, N. J. Lane, J. Parsons, W. Prince, and L. S. Swales. 1979b. A study of the mode of action and composition of a toxin from the female abdomen and eggs of *Arctia caja:* an electrophysiological, ultrastructural and biochemical analysis. Toxicon 17:285–306.

Rothschild, M., R. J. Nash, and E. A. Bell. 1986. Cycasin in the endangered butterfly *Eumaeus atala florida*. Phytochemistry 25:1853–1854.

Sbordoni, V., L. Bullini, G. Scarpelli, S. Forestiero, and M. Rampini. 1979. Mimicry in the burnet moth *Zygaena ephialtes:* population studies and evidence of a Batesian-Müllerian situation. Ecol. Ent. 4:83–93.

Schmidt, R. S. 1960. Predator behavior and the perfection of incipient mimetic resemblances. Behaviour 16:110–148.

Schneider, D., M. Boppre, H. Schneider, W. R. Thompson, C. J. Boriack, R. L. Petty, and J. Meinwald. 1975. A pheromone precursor and its uptake in male *Danaus* butterflies. J. Comp. Physiol. 97:245–256.

Scriber, J. M., and F. Slansky. 1981. The nutritional ecology of immature insects. Annu. Rev. Ent. 26:183–211.

Scudder, G. G. E., and S. S. Duffey. 1972. Cardiac glycosides in the Lygaeinae (Hemiptera: Lygaeidae). Can. J. Zool. 50:35–42.

Seiber, J. N., P. M. Tuskes, L. P. Brower, and C. J. Nelson. 1980. Pharmacodynamics of some individual milkweed cardenolides fed to the larvae of the monarch butterfly (*Danaus plexippus*). J. Chem. Ecol. 6:321–339.

Seiber, J. N., L. P. Brower, S. M. Lee, M. M. McChesney, H. T. A. Cheung, D. J. Nelson, and T. R. Watson. 1986. The cardenolide connection between overwintering monarch butterflies from Mexico and their larval food plant, *Asclepias syriaca*. J. Chem. Ecol. 12:1157–1170.

Sexton, O. J. 1960. Experimental studies of artificial Batesian mimics. Behaviour 15:244–252.

Sheppard, P. M., and J. R. G. Turner. 1977. The existence of Müllerian mimicry. Evolution 31:452–453.

Singer, M. C. 1971. Evolution of food plant preferences in the butterfly. *Euphydryas editha.* Evolution 25:383–389.

Singer, M. C. 1982. Quantification of host specificity by manipulation of oviposition behavior in the butterfly. *Euphydryas editha.* Oecologia (Berlin) 52:224–229.

Singer, M. C. 1983. Determinants of multiple host use in a phytophagous insect population. Evolution 37:389–403.

Singer, M. C. 1984. Butterfly-host plant relationships: host quality, adult choice and larval success. In R. I. Vane-Wright and P. R. Ackery (eds.), The Biology of Butterflies. Symposium of the Royal Entomology Society of London, Vol. 11. pp. 81–87. Academic Press, London.

Smiley, J. T. 1978. Plant chemistry and the evolution of host specificity: new evidence from *Heliconius* and *Passiflora.* Science 201:745–747.

Smiley, J. T. 1985a. *Heliconius* caterpillar mortality during establishment of plants with and without attending ants. Ecology 66:845–849.

Smiley, J. T. 1985b. Are chemical barriers necessary for evolution of butterfly-plant associations? Oecologia (Berlin) 65:580–583.

Smiley, J. T., and C. S. Wisdom. 1985. Determinants of growth rate on chemically heterogeneous host plants by specialist insects. Biochem. Syst. Ecol. 13:305–312.

Smith, D. A. S. 1978. The effect of cardiac glycoside storage on growth rate and adult size in the butterfly *Danaus chrysippus* (L.) Experientia 34:845–846.

Smith, S. M. 1980. Responses of naive temperate birds to warning coloration. Amer. Midl. Nat. 103:346–352.

Spencer, K. S., and D. S. Seigler. 1985a. Passibiflorin, eripassibiflorin and passitrifasciatin: novel cyclopentenoid cyanogenic glycosides from *Passiflora.* Phytochemistry 24:981–896.

Spencer, K. C., and D. S. Seigler. 1985b. Co-occurrence of valine/isoleucine-derived and cyclopentenoid cyanogens in a *Passiflora* species. Biochem. Syst. Ecol. 13:303–304.

Spencer, K. C., D. S. Seigler, and A. Nahrstedt. 1986. Linamarin, lotaustralin, linustatin, and neolinustatin from *Passiflora* species. Phytochemistry 25:645–649.

Stamp, N. E. 1982. Selection of oviposition sites by the Baltimore checkerspot, *Euphydryas phaeton* (Nymphalidae). J. Lepid. Soc. 36:290–302.

Stermitz, F. R., D. R. Gardner, F. J. Odendaal, and P. R. Ehrlich. 1986. *Euphydryas anicia* utilization of iridoid glycosides from *Castilleja* and *Besseya* (Scrophulariaceae). J. Chem. Ecol. 12:1456–1468.

Stermitz, F. R., D. R. Gardner, and N. McFarland. 1988. Iridoid glycoside sequestration by two aposematic *Penstemon*-feeding geometrid larvae. J. Chem. Ecol. 14:435–441

Teas, H. J. 1967. Cycasin synthesis in *Seirarctia echo* (Lepidoptera) larvae fed methylazoxymethanol. Biochem. Biophys. Res. Commun. 26:686–690.

Teas, H. J., J. G. Dyson, and B. R. Whisenant. 1966. Cycasin metabolism in *Seirarctia echo* Abbot and Smith (Lepidoptera: Arctiidae). J. Ga. Ent. Soc. 1:21–22.

Turner, J. R. G. 1971. Studies of Müllerian mimicry and its evolution in burnet moths and heliconid butterflies. In E. R. Creed (ed.), Ecological Genetics and Evolution: Essays in Honour of E. B. Ford. pp. 224–260. Oxford University Press, Oxford.

Turner, J. R. G. 1977. Butterfly mimicry: the genetical evolution of an adaptation. Evol. Biol. 10:163–206.

Turner, J. R. G. 1984. Mimicry: the palatability spectrum and its consequences. In R. I. Vane-Wright and P. R. Ackery (eds.), The Biology of Butterflies. Symposium of the Royal Entomology Society of London, Vol. 11. pp. 141–161. Academic Press, London.

Vane-Wright, R. I. 1976. A unified classification of mimetic resemblances. Biol. J. Linn. Soc. 8:25–56.

Vasconcellos-Neto, J., and T. M. Lewinsohn. 1984. Discrimination and release of unpalatable butterflies by *Nephila clavipes*, a neotropical orb-weaving spider. Ecol. Ent. 9:337–344.

Von Euw, J., T. Reichstein, and M. Rothschild. 1968. Aristolochic acid in the swallowtail butterfly *Pachlioptera aristolochiae*. Isr. J. Chem. 6:659–670.

West, D. A. 1985. The Batesian polymorphic butterfly, *Eurytides lysithous*. Natl. Geogr. Res. Rep. 21:501–506.

Wickler, W. 1968. Mimicry in Plants and Animals (translated from original German by R. D. Martin). McGraw-Hill, New York.

Williams, E. H. 1981. Thermal influences on oviposition in the montane butterfly, *Euphydryas gillettii* (Nymphalidae). Oecologia (Berlin) 50:342–346.

Wourms, M. K., and F. E. Wasserman. 1985a. Butterfly wing markings are more advantageous during handling than during the initial strike of an avian predator. Evolution 39:845–851.

Wourms, M. K., and F. E. Wasserman. 1985b. Prey choice by blue jays based on movement patterns of artificial prey. Can. J. Zool. 63:781–784.

Wourms, M. K., and F. E. Wasserman. 1985c. Bird predation on Lepidoptera and the reliability of beak-marks in determining predation pressure. J. Lepid Soc. 39:239–261.

CHAPTER 9

Potential Role of Plant Allelochemicals in the Development of Insecticide Resistance

L. B. Brattsten
Rutgers University
New Brunswick, New Jersey

CONTENTS

1 Introduction
2 Resistance and induction: similarities and differences
    2.1 Resistance development
3 Resistance mechanisms
    3.1 Behavioral resistance mechanisms
    3.2 Physiological resistance mechanisms
    3.3 Biochemical resistance mechanisms
        3.3.1 Target site resistance mechanisms
        3.3.2 Metabolic resistance mechanisms
        3.3.3 Importance of metabolic resistance mechanisms
4 Induction of insecticide-metabolizing enzymes
5 Does induction help insects develop resistance?
6 Summary
    References

1 INTRODUCTION

It has been routine to assume that resistance is a classical case of pure Darwinian selection (O'Brien 1967, Corbett et al. 1984) of preexisting

313

genes conferring resistance through mutations that occurred at least hundreds or thousands of years ago. Indeed, there are examples of species that were resistant to synthetic insecticides from the very beginning of their application. One such case is the redbanded leafroller, *Argyrotaenia velutiana* (Walker), which was always unusually resistant to DDT (Glass and Chapman 1952). This is ascribed to a metabolic resistance mechanism, DDT dehydrochlorination, which is catalyzed by a glutathione transferase (see later), an enzyme that may be important in eliminating plant phenolics.

It is clear that plants can also influence the toxicity of insecticides to herbivorous insects indirectly. This can occur via several major mechanisms: (1) plant allelochemicals can induce higher activities of the insecticide-detoxifying enzymes in the insects or inhibit these enzymes; (2) the nutrient content in the plant may limit the energy available to the insects to perform detoxification reactions, many of which depend on endogenous high-energy intermediates such as NADPH and phosphorylated nucleosides; (3) the diversity and variability in composition and concentration of plant allelochemicals (as influenced by plant variety, growth condition, plant part, and season, as well as defensive biosynthesis and/or translocation) may impose a corresponding phenotypic diversity and flexibility of detoxifying capabilities in the insects; and (4) plant allelochemicals may interfere with DNA and cause mutations which may necessitate a high degree of genotypic variability, as well, in the insects. Of these possibilities, none is well understood or has even been investigated extensively. However, induction by plant allelochemicals of insect enzyme activities, a clear manifestation of biochemical phenotypic plasticity, has been documented in several instances.

I will therefore discuss the possibility that possession of metabolic defenses, inducible by plant allelochemicals, can help insects develop permanent resistance (which is an expression of the genotypic diversity in resistance mechanisms to synthetic insecticides). This idea can best be evaluated in the framework of a discussion of the mechanistic aspects of resistance and induction.

2   RESISTANCE AND INDUCTION: SIMILARITIES AND DIFFERENCES

Resistance and induction affect insecticide toxicity in similar ways but are very different phenomena. The World Health Organization (Brown 1958) defines resistance as "the development of an ability in a strain of insects to tolerate doses of toxicants which would prove lethal to a majority of individuals in a normal population of the same species." The definition employs the term "toxicant" and can therefore apply to toxic phytochemicals as well as synthetic insecticides. Host races of many insects may differ from each other genetically through acquired and expressed resistance mechanisms to allelochemicals in the plants they have specialized to feed on (Knerer and Atwood 1973, Scriber 1986). The definition above does not discriminate between resistance and tolerance. Indeed, there are no mechanistic differences. The only real difference is that the selection pressure from a synthetic insecticide, almost always a fast-acting nerve poison with high acute toxicity, is probably always greater, more focused on one single defense mechanism, and operates in a situation that otherwise, largely favors insect growth and reproduction. Resistance to synthetic insecticides can therefore develop in a dramatically shorter time than resistance to plant allelochemicals (Brattsten et al. 1986a). Furthermore, this definition emphasizes that resistance is a phenomenon that belongs to a population (strain) of insects, implying that it is a heritable characteristic of the population. This is different from induction, which is a strictly temporary phenomenon, persisting only as long as the inducing chemical is present in sufficient concentration in the tissues of an individual insect.

Induction is not heritable, although the ability to be induced may be. In mice, certain strains have a hereditary receptor deficiency and cannot be induced (Nebert 1980). No receptor protein has been demonstrated yet in insects (Denison et al. 1985), however, induction may occur via some different mechanism. Induction has been shown in herbivorous and omnivorous insects, in generalist and specialist herbivores, in adult and immature life stages, and in insects with relatively low as well as relatively high initial enzyme activities (see Brattsten 1987b). No systematic study has been undertaken to determine the patterns or conditions required for induction to occur. Maybe all insects and life stages are inducible at all times. The data at hand indicate, however, that the degree of induction varies between species and life stages.

Table 9.1 shows the similarities between resistance in the tobacco budworm, *Heliothis virescens* (Fabr.), and induction in the southern armyworm, *Spodoptera eridania* (Cramer). The budworm larvae represent two different populations, one originally collected in Texas and the other one in Arizona. The Texas population has a metabolic resistance to methomyl, which is reflected in the higher rate of N-demethylation of p-chloro N-

TABLE 9.1   Similarities between Resistance and Induction

a. Resistance in *H. virescens*

Strain	Cytochrome P-450 Activity (NCH$_3$)[a]	Methomyl Toxicity LC$_{50}$ (ppm; 48hr)
AZ	3.55	36
TX	7.15	412

b. Induction in *S. eridania*[b]

Diet	Cytochrome P-450 Activity (AE)[c]	Carbaryl Toxicity LD$_{50}$ (μg/g; 24hr)
Control	2.84	30
0.2% Pentamethylbenzene	8.91	350

[a] Measured as *p*-chloro *N*-methylaniline N-demethylation.
[b] Data from Brattsten and Wilkinson (1973).
[c] Measured as aldrin epoxidation.

methylaniline in their midgut tissues than in the larvae of the Arizona population. The southern armyworm larvae were siblings separated into two groups, one of which was fed a control diet, whereas the other group fed on a diet containing 0.2% pentamethylbenzene for 3 days. In this case the higher rate of epoxidation of aldrin correlated well with the lower susceptibility to carbaryl in the induced sibling group. The differences are that the two tobacco budworm populations have permanently different metabolic activities and responses to the insecticide. However groups of southern armyworms from the same population have identical metabolic activities and reponses to the insecticide only as long as they are fed identical diets.

2.1   Resistance Development

Resistance depends on one or more genes that are permanently expressed. The frequency of those genes in a population depends on the intensity of a selection pressure (insecticide exposure) and on many biological factors characteristic of the insect species. Resistance development has two distinct components and it is important to distinguish between them because they depend on entirely different sets of factors. The first prerequisite for resistance to occur is the *de novo* genesis of a resistance mechanism in the genome of an individual insect. This could have happened a million years ago and it can happen in the present generation; it matters not when it happens, but for a population to have a chance to develop resistance, a change in the DNA of at least one individual has to occur. This is true of biochemical and physiological resistance mechanisms and may also be true for behavioral resistance, depending on the extent to which molecular changes in receptors and corresponding reprogramming of central circuits are involved. If a large component of learning is involved, genetic changes may not be important. The second component is the spread of that resistance mechanism throughout a population.

It has never been possible to develop a resistant strain in the laboratory by treating insects with sublethal insecticide doses (Crow 1957). Therefore, presumably, the insecticide itself does not change the insect by any direct action in laboratory cultures. This does not preclude the possible occurrence of sublethal genetic effects in field populations, where the gene pool is more heterogeneous and where several chemicals may act simultaneously at varying and indeterminable concentrations under a variety of microclimatic and microbiotic conditions (Brattsten et al. 1986a). Compared to selection of preexisting, resistant individuals, it is also likely that resistance mechanisms arise with a certain low frequency in all herbivorous insect populations as a result of contemporaneous, spontaneous, or environmentally induced mutations. The term *mutation* is used throughout this chapter in a broad sense to denote any hereditary gene rearrangement, including a single base-pair substitution affecting a single amino acid, or large recombinations or amplifications of DNA sequences, cytoplasmic genes, sex-linked genes, or polyploidy. A spontaneous frequency of mutations leading to resistance was estimated to be smaller than 1 in 3 million in a laboratory culture of the spider mite, *Tetranychus urticae* Koch (Oppenoorth 1985). This frequency occurred in a highly homogeneous gene pool in a milieu kept at optimum, uniform, and constant environmental conditions and completely free from extraneous chemicals. Many insecticides are known to be genotoxic and produce DNA rearrangements in a variety of organisms, including insects (Fishbein 1976). These effects have usually been demonstrated at rather high doses of pure material but not as they occur in spray formulations; it would be difficult to pinpoint a relationship between a resistance mutation and any factor causing it in a field situation. Other than insecticides, many plant allelochemicals are known to interfere with DNA, for example, the linear furanocoumarins (Pearlman et al. 1985) and the pyrrolizidine alkaloids (Bull et al. 1968).

3 RESISTANCE MECHANISMS

Metabolic resistance to plant allelochemicals or synthetic insecticides is only one mechanism out of many possible ones (Brattsten and Ahmad 1986). There are also behavioral resistance mechanisms. Physiological processes can be modified for resistance, and specific target sites can change without loss of function but so that toxicants can no longer interact with them. These and other mechanisms are discussed below.

3.1 Behavioral Resistance Mechanisms

In nature, herbivorous insects have two major ways of dealing with detrimental, nonnutritive chemicals in their food plants: adaptations in feeding behavior and biochemical adaptations. Two beetle species feeding on plants containing cucurbitacins illustrate this.

The spotted cucumber beetle, *Diabrotica undecimpunctata howardi* Barber, and other diabroticite beetles are strongly attracted to cucurbitacins, which are oxygenated tetracyclic triterpenes and acutely toxic to mice at LD_{50} values of 0.8 to 1.2 mg/kg (David and Vallance 1955). These beetles ingest large doses without any acutely ill effects. Metcalf et al. (1980) estimated the LD_{50} for the spotted cucumber beetle to greatly exceed 2000 mg/kg. In contrast, the squash beetle, *Epilachna borealis* (Fabr.), although confined to feed on cucurbitacin-containing plants, avoids them by cutting a trench through most of the leaf tissue so that the cucurbitacins cannot be translocated to the site of feeding, behind the trench (Carroll and Hoffman 1980, Tallamy 1985, 1986). The trenching behavior is facultative. The squash beetles do not trench when feeding on squash plants with a low cucurbitacin content, but trench "furiously" on plants with a high level of cucurbitacins. Nevertheless, they die on the latter kind of squash plant.

Apparently, the squash beetle relies to a great extent on behavioral defenses, whereas the spotted cucumber beetle obviously has very effective biochemical defenses. The spotted cucumber beetle sequesters cucurbitacins (Ferguson and Metcalf 1985) and probably has an insensitive target site. It does metabolize a low percent of the ingested dose but excretes most of it unchanged (Ferguson et al. 1985). Energetically, it would be enormously expensive, perhaps impossible, for this beetle to rely on metabolic detoxification only. Metabolic defenses in the squash beetle have not been investigated; however, the squash beetle may have a low detoxifying capacity sufficient only to allow it to forgo trenching on plants with a low cucurbitacin content. The squash beetle and other herbivorous coccinellids are thought to have derived, relatively recently, from carnivorous ancestors. In general, carnivores would not need as highly developed metabolic defenses as herbivores because their food is largely free of potentially toxic foreign compounds.

Behavioral adaptations are very important and widely documented in nature but play a minor role in resistance to synthetic insecticides. There is, in fact, not a single example of behavioral resistance, that is, some behavior that allows an insect to consume a plant despite the presence on it or in it of a synthetic insecticide. However, many insects show the necessary prerequisite for the potential development of behavioral resistance: they avoid insecticide-treated surfaces (Georghiou 1972, Gould 1984, Lockwood et al. 1984, Pluthero and Singh 1984). Many naturally occurring poisons are irritants, but a toxicant is not necessarily always a feeding deterrent or irritant. For example, black swallowtail, *Papilio polyxenes* (Fabr.), larvae prefer furanocoumarin-containing plants (Berenbaum 1981); the baltimore, *Euphydryas phaeton* (Drury), and other checkerspot caterpillars prefer iridoid glycoside-containing plants (Bowers and Puttick 1986); and southern armyworm larvae prefer cyanogenic foliage (Brattsten et al. 1983). There are many similar examples and most involve a strong element of some biochemical defense. There are also examples, other than the squash beetle, where the defense appears to be predominantly behavioral (see Tallamy 1986). Today, several of the synthetic insecticides have irritant or antifeedant properties at very low, nontoxic concentrations. It does not appear to have been recorded whether the avoidance behavior to these insecticides was present from the beginning of their use or developed after some period of exposure (attention tended to be focused on acutely toxic effects). The avoidance behavior may be a first step in the evolution of behavioral resistance, perhaps in the form of different host plant choices or feeding behaviors on the treated plant, whereby the toxic surface is bypassed. After four decades of intensive use of synthetic insecticides frequently accompanied by resistance, no single case of host shift or host race formation attributable to the insecticide exposure has been documented, nor has any shift in the feeding behavior of a phytophagous insect. If behavior is important, the potential for such shifts would be highest in those insects which may have developed avoidance. These insects may, indeed, constitute important cases for investigating host race formation as well as the importance of behavioral resistance to fast-acting nerve poisons with high acute toxicity.

3.2 Physiological Resistance Mechanisms

Physiological resistance mechanisms include cases where modification in a dynamic, precisely coordinated process helps insects survive exposure to toxicants. These resistance mechanisms are rare and insignificant for synthetic insecticides but abundant in natural insect–plant associations. This may reflect the short time that synthetic insecticides have been used. Examples are the extra-rapid excretion of nicotine in the tobacco hornworm, *Manduca sexta* (L.) larvae (Self et al. 1964) and xanthotoxin in black swallowtail larvae (Ivie et al. 1983). These species have adapted to utilize a few plants containing these and biosynthetically related compounds and over, presumably, a long period of time have developed modification(s) in an essential physiological process. Whereas very rapid metabolism of xanthotoxin is also a resistance factor in the black swallowtail larvae, tobacco hornworm larvae excrete most of the ingested nicotine unmetabolized. Another example is the sequestration of cardenolides into dorsolateral epidermal spaces in the large milkweed bug, *Oncopeltus fasciatus* (Dallas) (Scudder and Meredith 1982, Moore and Scudder 1985). Physiological resistance mechanisms, in particular different forms of sequestration, often or perhaps always appear combined with other types of resistance mechanisms.

Physiological resistance mechanisms to synthetic insecticides include a reduced rate of penetration through the integument and binding to proteins. Reduced cuticular penetration has been shown in houseflies, *Musca domestica* L. (Hoyer and Plapp 1968, Sawicki 1970) and is thought to arise from a heavier deposition of phospholipids and other compounds in the cuticle of resistant flies (Patil and Guthrie 1979). The major effect of this mechanism is to allow more time for detoxifying enzymes to remove the incoming dose so that a lethal concentration does not accumulate inside the flies.

Binding of insecticides to proteins allows for their transport, within the insect, to sites where metabolism may occur or to target sites. If the release of the insecticide is slow, binding to protein can serve as an effective protective mechanism. An isoenzymic form, E4, of carboxylesterase in the green peach aphid, *Myzus persicae* (Sulzer), hydrolyzes organophosphate, carbamate, and pyrethroid insecticides and also binds these compounds (Devonshire and Moores 1982). The E4 esterase occurs in molar amounts corresponding to a toxic dose of insecticide in highly resistant clones of the aphid. Abdallah et al. (1973) suggest fat body storage as a resistance mechanism to DDT and methyl parathion in the Egyptian armyworm, *Spodoptera littoralis* (Boisduval). In the southern and fall armyworm, *S. eridania* (Cramer) and *S. frugiperda* (J. E. Smith), respectively, the apparent kinetic characteristics of carboxylesterases in midgut and fatbody tissues are different (Table 9.2), and suggestive of a storage function for the fat body carboxylesterase. Hydrolytic activities are high in the midgut tissue accompanied by apparent K_m values, indicating low affinity or loose binding of the substrate to the enzyme active site. In contrast, in the fat bodies, the hydrolytic activity is very low and the apparent K_m values are also very low, indicating high affinity for the substrates. This would make fat body esterases ideally suited for storing lipophilic compounds containing an ester bond. Lipophorin (Chino et al. 1981) and arylphorin (Telfer et al. 1983) are hemolymph proteins, found in lepidopterans and orthopterans, with high binding affinity for insecticides of intermediate polarity, such as dieldrin (Haunerland and Bowers 1986).

3.3 Biochemical Resistance Mechanisms

The large milkweed bug also has Na^+, K^+-ATPases that are less sensitive to inhibition by cardenolides than the ATPases of herbivores not adapted to feeding on cardenolide-containing plants (Moore and Scudder 1986). This resistance mechanism involves interaction with one single macromolecule which has undergone some modification to reduce its affinity for binding cardenolides. It is an interaction that can be characterized in biochemical terms and is thus a biochemical resistance mechanism. It is a case of target site resistance. Most cases of resistance to synthetic insecticides depend on one or both of two kinds of biochemical resistance mechanisms: target site resistance and metabolic resistance.

Whereas physiological resistance mechanisms may require changes in several or many different genes to accomplish a modified process that still serves its vital function, biochemical resistance mechanisms may require only one single gene-dependent molecular change.

A critical change in a protein molecule such as an enzyme or a receptor could require as little as a change in one single amino acid (i e , only a single base pair would be involved). This could provide for very rapid and

TABLE 9.2 Apparent Kinetic Characteristics of Carboxyesterase Activities from Midgut and Fat Body Tissues of the Southern and Fall Armyworm Larvae[a]

Activity	1-Naphthyl Acetate		2-Naphthyl Butyrate	
	K_m	V_{max}	K_m	V_{max}
Southern Armyworm				
Gut, soluble	77.53 ± 6.68 (5)	19.78 ± 0.83 (5)	120.70 ± 11.42 (5)	26.53 ± 2.37 (6)
Gut, microsomal	78.73 ± 11.63 (3)	25.53 ± 3.74 (3)	45.84 ± 6.42 (3)	21.29 ± 4.18 (3)
Fat body, soluble	10.39 ± 1.12 (6)	0.76 ± 0.11 (6)	9.01 ± 0.77 (6)	1.32 ± 0.19 (6)
Fat body, microsomal	5.33 ± 0.73 (3)	2.59 ± 0.51 (4)	9.45 ± 2.01 (3)	6.39 ± 0.64 (3)
Fall Armyworm				
Gut, soluble	204.60 ± 30.82 (4)	48.47 ± 4.33 (4)	220.50 ± 13.04 (4)	48.41 ± 1.23 (4)
Gut, microsomal	22.73 ± 1.92 (4)	8.25 ± 0.69 (4)	71.42 ± 12.93 (3)	23.03 ± 1.01 (3)
Fat body, soluble	26.67 ± 1.12 (4)	5.90 ± 0.59 (4)	8.28 ± 0.91 (5)	3.04 ± 0.55 (5)
Fat body, microsomal	13.57 ± 1.19 (3)	1.71 ± 0.19 (3)	7.87 ± 0.26 (3)	1.78 ± 0.15 (3)

[a] Data are mean ± S.D. of (*n*) experiments; $K_m = \mu M$; $V_{max} = \mu mol/min$ per mg protein.

effective adaptations to toxicants and probably has, in both agricultural and nonagricultural ecosystems.

3.3.1 Target Site Resistance Mechanisms

Two kinds of target site resistance mechanisms are particularly important in resistance to synthetic insecticides: acetylcholinesterases (which are abnormally insensitive to inhibition by organophosphates and carbamates) and sodium gates (which do not bind DDT and pyrethroids). The nature of the molecular changes in these two target sites that protect them from interference by the insecticides is unclear. In the case of acetylcholinesterase-based resistance, first discovered in 1964 (Smissaert 1964), the most likely mechanism is a small change in the active site. Enzymes from insensitive insects have binding affinities to the insecticides several orders of magnitude lower than enzyme from sensitive insects. This is sometimes accompanied by a reduced reaction rate with the natural substrate, acetylcholine [see Oppenoorth and Welling (1976) and Oppenoorth (1985) for details]. The sodium gate–based resistance, also known as "kdr" (knock-down resistance), was first discovered in 1951 (Busvine 1951). It may be due to a molecular modification in the sodium gate protein (Soderlund et al. 1983a), a change in the number of sodium gate molecules (Jackson et al. 1984), or a change in the associated membrane lipids (Chialiang and Devonshire 1982).

Changes in target sites are extremely risky. Undoubtedly, many insects die in the process of evolving resistance based on a modification in a target site. Only a fraction of mutations of those that actually occur and affect a certain target site would produce a change that protects it against the insecticide and at the same time allows for its continued function. However, once acquired, target site resistance is very effective in protecting the population from all kinds of insecticides which would normally interfere with that particular target site. Also, this is a "built-in" defense that does not require the extra energy expenditure necessitated by metabolic resistance mechanisms. If a target site resistance does not impose any impairment of normal function nor is associated with genetic fitness modifiers (Roush and Plapp 1982, McKenzie and Purvis 1984), it could persist in a population forever and thus provide the opportunity for sequestration and the possibility for the insects to take advantage of the sequestered toxicant for their own defensive purposes as undoubtedly occurs in nature (Berenbaum 1986; Chapters 7 and 8). Houseflies with sufficiently high body burdens of DDT to kill susceptible flies have been reported (Perry and Hoskins 1950).

3.3.2 Metabolic Resistance Mechanisms

A group of enzymes comprising cytochrome P-450 (E.C. 1.14.14.1), esterases (mainly E.C. 3.1.1.1 and 3.1.1.2), epoxide hydrolases (E.C. 3.3.2.3), glutathione transferases (E.C. 2.5.1.18), and other conjugating enzymes, probably mainly phenol β-glucosyl transferases (E.C. 2.4.1.35, also called UDP glucosyl transferases) are prominent in constituting a major metabolic defense in insects and other organisms against lipophilic foreign compounds that may be toxic. The biochemical details of these enzymes in insects are described in several reviews (Wilkinson and Brattsten 1972; Dauterman 1976, 1985; Nakatsugawa and Morelli 1976; Yang 1976; Brattsten 1979; Hodgson 1983, 1985; Hodgson and Kulkarni 1983; Wilkinson 1984; Agosin 1985; Ahmad et al. 1986). Only the general properties of the enzymes are pointed out here.

The basic role of foreign compound–metabolizing enzymes is to produce polar, excretable metabolites. In this process, compounds usually lose their toxicity or biological activity, but not always. Cytochrome P-450 regularly produces highly reactive epoxide metabolites which can destroy cellular macromolecules, including the cytochrome itself. Epoxide metabolites are rendered harmless either spontaneously, if they are aromatic, or by reactions catalyzed by epoxide hydrolases and glutathione transferases. This emphasizes the importance of the concerted action of these enzymes in producing the final excretable metabolite.

The four major enzymes have several characteristics in common. They are all specialized to accept rather highly lipophilic molecules as substrates without any other stringent structural requirements. This low substrate specificity, together with the occurrence of several isoenzymic forms with broadly overlapping substrate preferences, facilitates the function of these enzymes as a general-purpose metabolic defense system. The binding to the active site occurs mainly through lipophilic and electronic interactions. This diversity in ability to bind many structurally different molecules is at the expense of high catalytic activity. The lipophilic binding is not very precise, and consequently cytochrome P-450, glutathione transferases, and epoxide hydrolases have activities six to seven orders of magnitude lower than enzymes in the glycolytic pathway or some other part of the intermediary metabolism. The latter enzymes, as a rule, have very strict structural requirements of their substrates. However, the low catalytic rates of the foreign compound–metabolizing enzymes is to some extent compensated for by their presence in cells of certain tissues in high concentrations. For instance, 10% of all soluble proteins in normal rat liver cells can be glutathione transferases (DePierre and Morgenstern 1983) and up to 20% of all membrane-bound proteins in liver cells from induced rats can be cytochrome P-450 (Wilkinson 1984). The esterases regularly have about 1000-fold higher activities than those of the other enzymes. They have a higher degree of substrate selectivity in binding only compounds containing ester bonds and may also participate in nutrient metabolism and the turnover of endogenous substrates.

The foreign compound–metabolizing enzymes are supported with regard to their synthesis, catalytic energy requirements, and subcellular localization by normal intermediary metabolic pathways but, with exceptions, do not participate in the metabolism of endogenous substances. Certain specialized forms of cytochrome P-450, often found in mitrochondrial membranes, participate in the metabolism of insect hormones such as ecdysone 20-monooxygenase. In an environment completely free from lipophilic foreign compounds, these enzymes are not needed. Such an environment is the mammalian uterus, and thus fetal mammals have no detectable activities. It takes several weeks after birth until cytochrome P-450 activity is fully developed (Jondorf et al. 1959). Similarly, this activity is not detectable in insect pupae and eggs except those that imbibe water during embryogenesis (Wilkinson and Brattsten 1972; B. T. Walton, personal communication).

Another important characteristic of these enzymes is their ability to respond rapidly to foreign chemicals which act as inducers (see Section 4). Due to their specialization for lipophilic foreign compounds, changes in the genes coding for these enzymes are not nearly as risky as changes in genes directing the synthesis of a target site: no vital function is at risk. Therefore, metabolic adaptations to toxicants are the most facile defense mechanism available to insects. The genetic changes required may be quite minor. A single base-pair substitution could produce an isoenzyme with increased affinity for a certain kind of toxicant, or amplification of a gene could very substantially increase the amount of an enzyme important in detoxifying a consistently encountered insecticide. Since mutations are random events, there is a likelihood of a certain frequency of "wrong" mutations that either impair or do not affect enzyme function.

However, because the organism does not depend on these enzymes for any critical life function, at least in the absence of toxicants, little or no harm may be done.

3.3.3 Importance of Metabolic Resistance Mechanisms

The facility with which genetic adaptations can occur in metabolic mechanisms leads to the idea that metabolism is a primary defense and may facilitate resistance development simply by allowing insects to survive long enough to acquire other resistance mechanisms. The data in Table 9.3 support this idea. Metabolic resistance was established before target site resistance against DDT, the organophosphorous and carbamate insecticides, and probably most other synthetic insecticides as well.

The use of DDT started in 1942, and in 1949, after seven years, metabolic resistance was documented in houseflies (Ferguson and Kearns 1949). This was due to an enzyme, DDT-dehydrochlorinase, which converts DDT to DDE (Fig. 9.1). In 1984, it was identified as an isoenzymic form of the glutathione transferases (GST) (Clark and Shamaan 1984). DDT resistance had been reported even earlier. The first cases were reported as early as 1946 (Brown 1971), only 4 years after DDT was patented as an insecticide. The resistance mechanism in these very early cases is not known. The ideas at the time were that resistant flies had larger fat deposits in which to store DDT, a mechanism that is now considered to account for only minor effects. Target site resistance to DDT had appeared and was reported only two years later, in 1951, also in houseflies (Busvine 1951).

The major metabolic detoxification of DDT in many insects species and populations (strains) is by DDT-dehydrochlorinase–catalyzed conversion to DDE. In several of the early cases of housefly resistance to DDT there were no differences in the production of DDE between susceptible and

TABLE 9.3   Rate Of Development of Resistance to Three Major Insecticides

Insecticide (Year of Patent)	Metabolic Resistance[a]	Target Site Resistance[a]
DDT (1942)	1949 (7)	1951 (9)
Methyl parathion (1944)	1958 (14)	1964 (20)
Carbaryl (1956)	1961 (5)	1971 (15)

[a] Numbers in parentheses are the number of years since patent that a metabolic or target site resistance case was published.

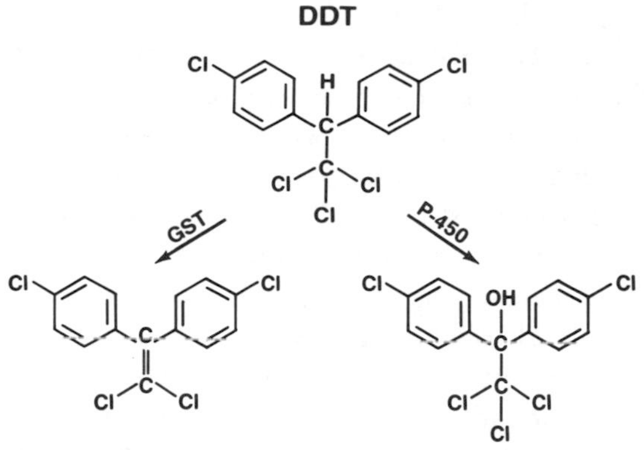

Figure 9.1 Major metabolites of DDT produced by DDT-dehydrochlorinase (GST) and cytochrome P-450.

resistance strains implying that there might have been cases in which target site resistance was not preceded by metabolic resistance. However, in addition to dehydrochlorination, DDT is also detoxified by cytochrome P-450–catalyzed conversion to dicofol (Fig. 9.1).

Cytochrome P-450 was first discovered in pig liver (Garfinkel 1958) and rat liver (Klingenberg 1958) seven years after the appearance of DDT target site resistance. The enzyme was not characterized until 1964 (Omura and Sato 1964) and its existence in insects was also recognized only in the early and mid-1960s. *In vivo* work with cytochrome P-450–specific synergists, compounds that enhance insecticide toxicity by inhibiting the enzyme catalyzing its major detoxifying reaction (Metcalf 1967), indicated its existence in house flies (a DDT-resistant strain) (Eldefrawi et al. 1960). The availability of *in vitro* methods allowed its identification in a trypanosome-vectoring bug, *Triatoma infestans* Klug. (Agosin et al. 1961), and in houseflies (Schonbrod et al. 1965). It is thus possible or even likely that some early cases of metabolic resistance to DDT due to cytochrome P-450–catalyzed conversion to dicofol went unnoticed. To be sure, the synergistic effect of the methylenedioxyphenyl compounds was known already in 1938 (Weed 1938), but these useful inhibitors of cytochrome P-450 were for a long time regarded exclusively as pyrethrin synergists. In 1953 they were reported to have only slight, if any activity as DDT synergists in the large milkweed bug, which was not DDT resistant anyway. This was apparently one of the rare attempts to synergize

DDT with methylenedioxyphenyl compounds. Nearly all attempts at synergizing DDT toxicity were done with the purpose of inhibiting DDT dehydrochlorination and used very many structural analogs of DDT (Metcalf 1955); although an experiment without success had been done with piperonyl butoxide as a DDT synergist to house flies while these were still DDT susceptible (Lindquist et al. 1947). This lack of success is not surprising, in retrospect, because it is characteristic of synergists not to be very effective in enhancing insecticide toxicity in susceptible populations. In other words, it is characteristic of susceptible insects not to have high detoxifying enzyme activities that can be inhibited by synergists.

It is remarkable and interesting how much longer after their introduction it took for target site resistance to develop to the organophosphothionates (20 years) than to DDT (9 years). Methyl parathion and other organophosphothionates undergo complex metabolic transformations (Fig. 9.2). They can be attacked in several different sites by three of the major enzymes. Carboxylesterases with phosphatase activity (Oppenoorth 1985) hydrolyze them at the phosphoester bonds, often at the one next to the bulkiest substituent (Lewis and Sawicki 1971, Welling et al. 1971). This results in detoxification. Glutathione transferases can replace any of the substituents with the cysteinyl sulfhydryl group of glutathione (Oppenoorth et al. 1977, Motoyama and Dauterman 1980). This also results in detoxification. Cytochrome P-450 can catalyze dealkylations of substi-

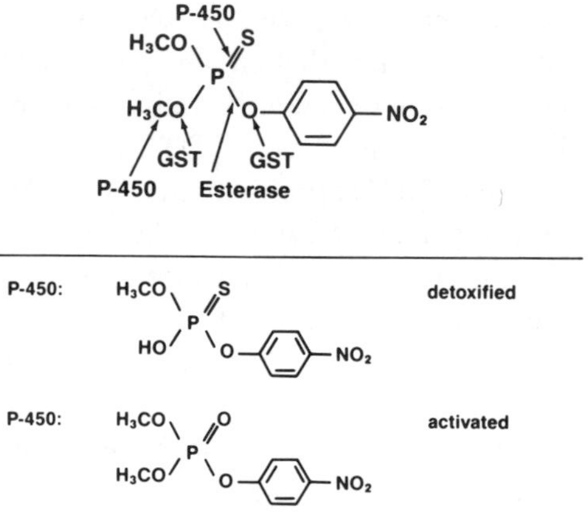

Figure 9.2 Metabolic attack on methyl parathion by cytochrome P-450, glutathione transferase, and esterase enzymes; cytochrome P-450 produces both activated and detoxified metabolites.

tuents, including propyl and isopropyl groups (Nakatsugawa and Morelli 1976, Kulkarni and Hodgson 1984), resulting in detoxification. In addition, cytochrome P-450 catalyzes the oxidative desulfuration of the phospho-sulfur double bond. This reaction is thought to be analogous to an epoxidation of a carbon–carbon double bond (Nakatsugawa and Morelli 1976) and results in activation of the insecticide to a potent acetylcholinesterase inhibitor.

The insecticidal selection pressure is spread out over three different mechanisms, one of which, cytochrome P-450, not only detoxifies but also activates the compound. No experiments have been done to evaluate the possibility that this may be related to delayed metabolic resistance development, although one of the ideas in current resistance management theory is that resistance may be delayed by not focusing the selection pressure on any single potential resistance mechanism (Georghiou 1980, 1983). However, rapid metabolic, environmental, and photodegradation of the organophosphothionates may sufficiently relieve the selection pressure between applications to preserve a susceptible gene pool. In contrast, DDT has very high environmental stability. It is also possible that some target sites have more molecular flexibility than others. The organophosphates inhibit enzymes, implying that two processes must be disrupted, the binding of the normal substrate and the subsequent catalytic event. DDT interferes with a receptorlike protein, the sodium gate, which functions by changes in structural configuration in response to a change in the membrane potential. Only one event, the configuration change, needs to be disrupted, possibly allowing a higher degree of structural diversity without loss of function.

Carbamate metabolism can be catalyzed by both cytochrome P-450 and esterase. These compounds used to be only rarely hydrolyzed in insects. Recent intensive selection pressures by photostable pyrethroids may have made carbamate ester hydrolysis more common. However, cytochrome P-450, possibly in different isoenzymic forms, oxidizes carbamate molecules in several different sites. Some examples are shown in Fig. 9.3. All carbamate oxidation products are detoxified, just as all metabolic changes of the DDT molecule lead to inactivation. This is also true of the pyrethroids. The short time required for metabolic carbamate resistance to develop, compared to the organophosphothionates, could thus be related to the selection pressure being focused on one potential resistance mechanism, cytochrome P-450. The environmental stability of the carbamates is comparable to that of the organophosphate insecticides. In addi-

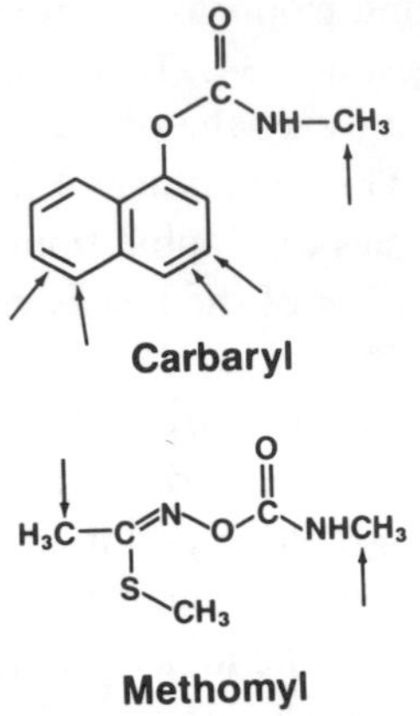

Carbaryl

Methomyl

Figure 9.3 Metabolic attack by cytochrome P-450 on the carbamate insecticides carbaryl and methomyl; all metabolites are less toxic than the parent molecules.

tion, a previous history of insecticide exposure could have influenced the rate of this process. If DDT or any other chlorinated hydrocarbon, organophosphate, or other insecticide had already selected for populations with permanently expressed high cytochrome P-450 activity, carbamates would have reduced toxicity to these populations also, because they undergo detoxification by the same enzyme. Likewise, insect populations with target site resistance to organophosphates may be cross-resistant to carbamates because both groups of compounds interfere with the same target site. However, organophosphate target site resistance is not always accompanied by carbamate target site resistance, indicating that several different target site resistance mechanisms occur (Oppenoorth 1985).

In the case of the photostable pyrethroids (introduced around 1978), detoxified by cytochrome P-450 and carboxylesterases (Casida and Ruzo 1980, Soderlund et al. 1983b), both metabolic and target site resistance occurred in field populations very shortly after their introduction (Gammon 1980, Liu et al. 1981). In addition, resistance had been observed in laboratory insect cultures even before their release. This situation is complicated, largely by previous exposure to other insecticides, which selected for metabolic resistance mechanisms, and by exposure to DDT, which selected for target site resistance (Miller et al. 1979). Pyrethroids and DDT both interfere with the sodium gates (Farnham 1977). Thus, past experience strongly implies the fundamental importance of metabolic resistance in providing an opportunity for the development of target site resistance.

Just as there are insect species with preexisting natural resistance to synthetic insecticides (e.g., the redbanded leafroller), there are also those that remain susceptible despite intensive exposure. The boll weevil, *Anthonomus grandis grandis* Boheman, is still effectively controlled by organophosphate insecticides after some 30 years of intensive use. In the case of the boll weevil, insecticide applications are directed at the adult beetles rather than at the damaging life stage (the larvae) because it is inaccessible, feeding inside the plant tissues, and because the use of systemic insecticides can lead to boll abscission. Azinphosmethyl (Fig. 9.4) is an organophosphothionate widely used in the southern United States. The LD_{50} for adult boll weevils is 2.8 μg per gram of body weight. In contrast, another cotton pest, the tobacco budworm, readily develops resistance to the organophosphates. A carbamate- and pyrethroid-resistant population of the tobacco budworm suffered zero mortality from an enormous dose of 11 mg per gram of body weight applied topically. In contrast, the 48-hour LD_{50} for a susceptible strain was 207 μg per gram of body weight (Brattsten 1987a), indicating more than a 53-fold decrease in toxicity to the resistant population.

A comparison of the metabolic capacities in the susceptible tobacco budworm population with those in boll weevils should indicate whether originally existing metabolic defenses help in the development of metabolic and other types of resistance. Azinphosmethyl (Fig. 9.4) can be

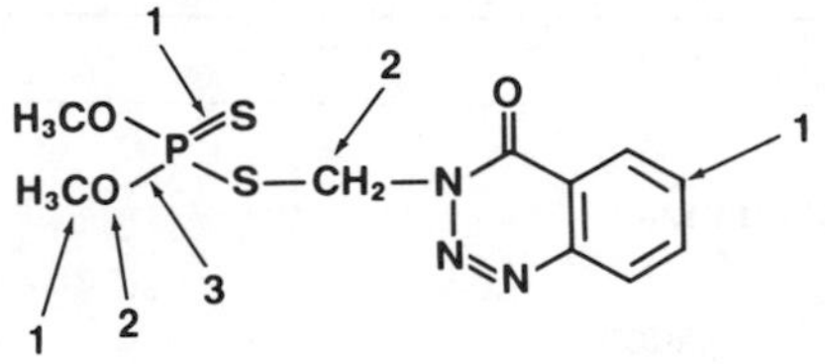

1. Cytochrome P-450

2. Glutathione transferase

3. Esterase

Figure 9.4 Metabolic attack on azinphosmethyl. Oxidation of the benzotriazin moiety is putative.

metabolized by glutathione transferases (Motoyama and Dauterman 1980), cytochrome P-450 (Motoyama and Dauterman 1980, Levine and Murphy 1977), and probably by esterases (Oppenoorth 1985). The data in Table 9.4 are a comparison of the activities of the three enzymes toward model substrates. Specific activities were measured in midgut and fat body tissue from fifth instar tobacco budworms and in 4 to 5-day-old adult boll weevils. The specific activities were converted to approximate total activities per insect based on the protein fraction used for the measurements (Brattsten 1987a). This underrepresents total activities in the tobacco budworm larvae since activities in other tissues (e.g., nerve, Malpighian tubules, integument and testes) were not included, but probably represents boll weevil activities fairly well since no endogenous inhibition was noticed despite the mixing of tissues during enzyme preparation (Brattsten 1987a). The data indicate up to 46-fold higher activities in the susceptible tobacco budworm larvae compared to those in adult boll weevils.

The data in Table 9.4 do not rule out the possibility that resistance is purely a selection of preadapted individuals. They, in fact, reveal nothing about the mechanisms of resistance development, and being averages of many pooled insects, they do not reveal the individual variation, which may, however, be equally high in both species. This comparison may

TABLE 9.4   Ratios of Organophosphate-Metabolizing Enzyme Activities Between Tobacco Budworm Larvae and Adult Boll Weevils

Activity Measured[a]	Ratio
Cytochrome P-450	16.3
N-Demethylation	13.9
Epoxidation	46.6
CDNB conjugation	20.4
Microsomal esterase	36.6
Soluble esterase	13.6

Source: Brattsten (1987a).

[a] Cytochrome P-450 was quantified as the carbon monoxide complex. N-demethylation was measured with *p*-chloro *N*-methylaniline and epoxidation with aldrin as substrates. CDNB (1-chloro-2, 4-dinitrobenzene) conjugation measures glutathione transferase activity. Microsomal and soluble esterases were measured with 1-naphthyl acetate as substrate.

unfairly favor the tobacco budworm by the use of last instars as enzyme sources, whereas in a field, a mixture of instars may be present, the earlier ones with lower activities. The younger larvae would be killed more often by insecticides than the fifth instars. This is a selection process but one that does not necessarily result in resistance.

4   INDUCTION OF INSECTICIDE-METABOLIZING ENZYMES

The most important difference between induction and resistance, as pointed out earlier (Table 9.1), is the strictly temporary, nonhereditary nature of induction compared to resistance. Induction may be an energy-saving device. Instead of maintaining high levels of detoxifying enzymes at all times, it allows their rapid increase by new synthesis when needed. Enzyme proteins are constantly broken down and resynthesized; the half-life of a cytochrome P-450 molecule is about 22 hours (Remmer 1972). The caloric expense of protein biosynthesis may account for a significant portion of the available energy in a small, short-lived organism such as an insect (Brattsten 1979). Southern armyworm larvae, fed carrot foliage continuously throughout the sixth (last) stadium, had higher cytochrome P-450 activities and attained lower maximal body weights than did larvae fed kidney bean foliage throughout the last stadium (Brattsten et al. 1984). This could have resulted from energy spent to biosynthesize extra enzymes or from lower nutrient or water content in carrot leaves, or both. When larvae were fed alternatingly on carrot foliage (the first day after the molt) and on bean foliage, they had higher enzyme activities and higher body weights than those of larvae fed continuously on bean foliage. The southern armyworm may be able to compensate for unsuitable food quality by increasing its food intake (not measured in this work) when suitable foliage is available (Scriber 1979, 1981). Effects on the energy balance of the insect may be obscured in experiments where the inducer is administered within an optimized artificial diet provided *ad libitum*.

Induction of cytochrome P-450 in insects has been studied as long as the enzyme has been known to occur in them. For almost a decade the increases in activity observed were only minor. Morello (1964) demonstrated a small (12%) increase in tolerance to DDT in *Triatoma infestans* that was accompanied by increased production of a dicofol-like material, a cytochrome P-450 metabolite. Similar small increases were observed in

DDT-treated house flies (Gil et al. 1968). Less than twofold increases in metabolic rates of several insecticides in house flies were also reported (Plapp and Casida 1970). Brattsten and Wilkinson (1973) reported a substantial, threefold increase in cytochrome P-450 activity in an insect which was accompanied by an 11-fold decreased toxicity of carbaryl (Table 9.1).

Induction in insects was, and still is almost always, studied only in regard to insecticide resistance. Its role in insect–plant interactions has only been implied. A widely occurring plant allelochemical, (+)-α-pinene, induced cytochrome P-450 in southern armyworm larvae and, at the same time, reduced the toxicity of another allelochemical, nicotine, to larvae (Brattsten et al. 1977). The toxicity of the mint monoterpene pulegone was reduced to southern armyworm larvae fed a diet containing coumarin (Gunderson et al. 1985). Coumarin does not significantly change the activity of cytochrome P-450 in the southern armyworm (Brattsten et al. 1984) but does increase its glutathione transferase activity sevenfold (Gunderson et al. 1986). In the latter case, induction of cytochrome P-450 activities does not reduce the toxicity of pulegone, although induction of another enzyme does. Both examples indicate that having inducible enzymes, although not necessarily always cytochrome P-450, is advantageous to the insect.

Induction of cytochrome P-450 and glutathione transferase activities by allelochemicals and plants, when eaten by insects, and accompanying changes in insecticide toxicity, are now well-established phenomena. Most of this work has been done with herbivorous lepidopteran larvae, excellent starting material for enzyme activity measurements. Tables 9.5 and 9.6 show induction in the variegated cutworm, *Peridroma saucia* (Hubner), and alfalfa looper, *Autographa californica* (Speyer), by pepper-

TABLE 9.5   Induction of Cytochrome P-450 in the Variegated Cutworm and Effect on Carbaryl Toxicity

Diet	P-450	Epoxidation	24-hr Mortality from 0.1% Carbaryl (%)
Bean leaves	1×	1×	90
Peppermint leaves	1.9×	32.2×	20
Menthol, 0.1%	4.6×	24.7×	
α-Pinene, 0.2%	6.5×	23.8×	

Source: Based on Yu et al. (1979).

TABLE 9.6   Effect of Peppermint on Cytochrome P-450 and Insecticide Toxicities in Variegated Cutworm and Alfalfa Looper Larvae

Insect Diet	P-450	Epoxidation	24-hr Mortality (%)		
			Acephate 0.1%	*Methomyl* 0.1%	*Malathion* 0.5%
		Variegated Cutworm			
Snap beans	1×	1×	100	100	90
Peppermint	4.2×	6.6×	60	20	20
			Acephate 0.15%	*Methomyl* 0.3%	*Carbaryl* 0.5%
		Alfalfa Looper			
Alfalfa	1×	1×	100	85	45
Peppermint	2.3×	7.8×	90	70	20

Sources: Variegated cutworm data based on Berry et al. (1980); alfalfa looper data based on Farnham et al. (1981).

mint leaves and monoterpenes which occur in leaves, and their effect on the toxicities of several insecticides (Yu et al. 1979, Berry et al. 1980, Farnham et al. 1981). Carbaryl and methomyl are detoxified by cytochrome P-450 (Fig. 9.3). Both contain hydrolyzable bonds but are rarely hydrolyzed by insect esterases. Acephate and malathion are detoxified by cytochrome P-450, esterases, and glutathione transferases (Fig. 9.5). Malathion can, in addition, be activated by a cytochrome P-450-catalyzed conversion of the $P = S$ group to $P = 0$. The data show that there are differences in the response to plant allelochemicals between the species. Even though cytochrome P-450 content and aldrin epoxidation activity were induced in both the variegated cutworm larvae and alfalfa loopers, insecticide toxicities were more drastically reduced for the variegated cutworm larvae. It is possible that in addition to the form of the cytochrome which epoxidizes aldrin and was induced and measured in both species, another form (which more effectively detoxified these insecticides) was induced in the variegated cutworm, but not measured. This is a difficult problem and can only be solved by the purification and individual

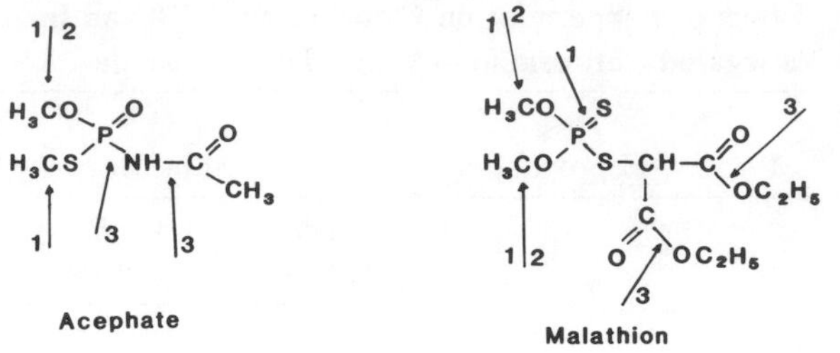

Acephate

Malathion

1. Cytochrome P-450
2. Glutathione transferase
3. Esterase

Figure 9.5 Metabolic attack on the organophosphorous insecticides acephate and malathion.

assay of each isoenzymic form with the insecticide in question. A somewhat clearer picture can be obtained by using several different model substrates in this sort of study (Brattsten et al. 1986b, Yu 1986).

Induction by plants or plant chemicals of cytochrome P-450 has been demonstrated with very few insect species other than Lepidoptera. Japanese beetle, *Popillia japonica* Newman, larvae feeding on mixed roots of two or three grass species had higher activity than that of larvae feeding on roots of *Poa annua* only (Ahmad 1983). A similar effect was demonstrated with adults: beetles fed on broccoli had higher activity (N-demethylation of *p*-chloro *N*-methylaniline) than did phlox-fed beetles, and beetles fed a mixture of phlox, sassafras, and broccoli foliage also had higher activity. This is also, perhaps, an example of compensatory food intake (when suitable or preferred foliage is available) that provides enough energy both for growth and to sustain high detoxifying activities, as noted for the southern armyworm, mentioned before.

Glutathione transferase activity is also induced by plants and plant chemicals. Parsnip, parsley, mustard, turnip, and other plants induced the activity, measured as 1,2-dichloro-4-nitrobenzene (DCNB) conjugation, in fall armyworm larvae (Yu 1982, 1984). The inducing allelochemical in parsnip plants is the linear furanocoumarin, xanthotoxin. Data in Table 9.7 show induction of this activity in fall armyworm larvae and its effect on the toxicity of three organophosphorous insecticides, a very good illustration of the importance of glutathione transferase in organophosphate detoxification.

TABLE 9.7 Effect of Cowpea Foliage on Glutathione Transferase Activity and Organophosphorous Insecticide Toxicity in the Fall Armyworm

Diet	DCNB Conjugation	LC_{50} (24 hr ppm)		
		Diazinon	Methamidophos	Methyl parathion
Soybean leaves	1×	191.6	53.6	242.8
Cowpea leaves	7.7×	414.8	132.3	641.6

Source: Based on data in Yu (1982).

Several insecticides, including chlorinated hydrocarbons, carbamates, and nicotine, are inducers of these enzymes in mammals. No insecticide has been shown to induce its own metabolism or that of any other compound in insects. Highly toxic insecticides would reach a lethal concentration in insect tissues before they could cause induction. It is, however, a fact, based on experimental evidence, that plant allelochemicals induce insecticide-metabolizing enzymes in insects.

5 DOES INDUCTION HELP INSECTS DEVELOP RESISTANCE?

Terriere (1983, 1984), Hodgson (1983), Oppenoorth (1985), and others have concluded that there is no obvious relationship between induction and resistance. These conclusions rest squarely on the assumption that resistance arises exclusively by selection of individuals already possessing metabolic resistance as a function of genetic variability that may be associated with defense against host plant allelochemicals within a population. There may exist within-population variation in detoxifying capabilities that escape detection because, except for carboxylesterases, there are no good methods yet for measuring activities in individual insects. However, as long as no other dominating selection pressure operated on the pre-DDT insect populations, even the long-term selection pressure exerted by plant allelochemicals (although not nearly as intensive as that by synthetic insecticides) would have diminished the genetic variability of their enzyme activities. It is also likely that low frequencies of spontaneous and environmentally induced mutations cause changes in genetically expressed enzyme activities in herbivorous insects. Such changes also are undetected because the mutations need not affect the viability of the insect, and the changes may also disappear in the measurement of average activities in pooled insect tissues.

Induction of metabolic defenses may be important in allowing a susceptible individual, in which a resistance mechanism is present but unexpressed, to survive (e.g., in cases of a new mutation or a single copy of a recessive resistance allele or an unsuitable chromosomal environment). It may be expressed in the next generation only if the insect survives and reproduces. The insecticide selection pressure could prevent this, unless the insect possessed inducible detoxifying enzymes. There is no evidence that cases like this occur, but there is no evidence against it either. Figure 9.6 outlines the possibilities. Any contribution, by induction, to resistance development would be important only in the very early stages of the process.

Induction of metabolic defenses could certainly help increase the frequency of a recessive resistance gene if it helps heterozygous and therefore susceptible individuals to survive and reproduce, even though simultaneously, susceptible noncarriers would also be helped. In the next few

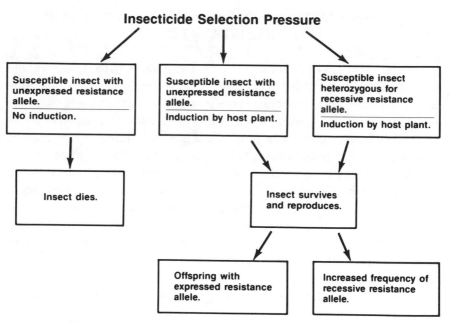

Figure 9.6 Role of induction in resistance development.

generations the frequency of heterozygotes may remain the same, but their total number would increase and therefore raise the probability of an increasing frequency of homozygous and therefore resistant individuals over subsequent generations. Subsequent selection pressure would probably increase, probably to the point where induction would no longer help susceptible carriers or noncarriers. Thus, induction is likely to be important only in the very early stages of resistance development. This may have happened in kdr housefly populations. Although the inducers would not have been plant allelochemicals, they could have been livestock feed additives or metabolites thereof.

Results from induction studies in insects show that insect populations with low cytochrome P-450 activities are equally or sometimes more inducible than species or populations with high activities, at least in the very few species that have been used. For instance, the rate of parathion metabolism increased 3.9-fold in a susceptible housefly strain fed phenobarbital, but only 1.7-fold and 1.3-fold in two different resistant strains treated the same way (Plapp and Wang 1983). Control variegated cutworm larvae had low aldrin epoxidation activity (0.15 nmol/min per milligram of protein) (Yu et al. 1979) compared to control southern armyworm larvae (3.60 nmol/min per milligram of protein) (Brattsten et al. 1984). This activity was induced 23.8-fold by 0.2% dietary (+)-α-pinene in the variegated cutworm, and 2.3-fold by the same treatment in the southern armyworm. In other cases the degree of induction is about equal in insects with low and high original activities.

Cytochrome P-450 and other insecticide-detoxifying enzymes are inducible to some extent in most herbivorous insects examined. Only one case of induction in a specialist herbivore, the velvetbean caterpillar, *Anticarsia gemmatalis* Hubner, has been reported (Christian and Yu 1985). This paucity of data may be due to difficulties in getting specialists to feed on diets containing inducing chemicals. If a specialist with low activities were less inducible than a generalist with high activities, it may have trouble developing resistance to insecticides.

The potential role of inducibility in the development of resistance was assessed by comparing it in the susceptible boll weevil and the tobacco budworm with a documented ability to develop resistance. Of five cotton allelochemicals added to their diet—(+)-α-pinene, caryophyllene, gossypol, umbelliferone, and scopoletin all (except gossypol) induced cytochrome P-450 activities significantly ($P < 0.001$; Brattsten

1987b): more in the tobacco budworm larvae than in boll weevil adults. Glutathione transferase activities were induced to similar extents in both exposed life stages except by $(+)$-α-pinene, which induced this activity more in the adult boll weevils than in the tobacco budworms. Esterase activities responded similarly in the two species. The soluble esterase activity was unaffected by all dietary allelochemicals except scopoletin, which moderately increased the activity in the tobacco budworm. The microsomal esterase activity was either unaffected or suppressed by these allelochemicals. The data in this very limited study provide some support for the idea that induction helps insects develop resistance, and certainly show that the potential exists (Brattsten 1987b).

6 SUMMARY

Insecticide resistance is a function of permanently expressed, heritable mechanisms allowing insects to circumvent the lethal effects of insecticides. The major resistance mechanisms are metabolic resistance due to high activities of cytochrome P-450, glutathione transferases and esterases, and modified target sites that have a reduced affinity for binding the insecticides but still function in their critical roles for the insect's survival.

Metabolic resistance is probably the first kind of resistance to develop upon exposure to any synthetic insecticide and may provide opportunities for target site and other kinds of resistance to develop, as long as the population is exposed to biochemically related insecticides.

Induction causes a temporary, nonhereditary increase in metabolic defenses and can be effected by a large variety of foreign compounds. Allelochemicals in the food plants of herbivorous insects are known inducers and can change the toxicity of insecticides.

There is a real possibility that induction helps susceptible insects with an unexpressed resistance mechanism survive and reproduce so that the frequency of the resistance mechanism increases in the next generation(s).

ACKNOWLEDGMENTS

I thank K. L. Strickler, P. J. C. Kuiper, J. L. Frazier, and F. Gould for constructive critiques of the manuscript and the editors for the opportunity to participate.

REFERENCES

Abdallah, M., H. Zazon, M. Kandil, and M. H. Balba. 1973. Storage of insecticides in the fat body of *Spodoptera littoralis* (Boisd.) as a possible mechanism of resistance. Experientia 29:318–319.

Agosin, M. 1985. Role of microsomal oxidation in insecticide degradation. In G. A. Kerkut and L. I. Gilbert (eds.), Comprehensive Insect Physiology, Biochemistry, and Pharmacology, Vol. 12. pp. 647–712. Pergamon Press, Oxford.

Agosin, M., D. Michael, R. Miskus, S. Nagasawa, and W. W. Hoskins. 1961. A new DDT-metabolizing enzyme in the German cockroach. J. Econ. Ent. 54:340–342.

Ahmad, S. 1983. Mixed-function oxidase activity in a generalist herbivore in relation to its biology, food plants, and feeding history. Ecology 64:235–243.

Ahmad, S., L. B. Brattsten, C. A. Mullin, and S. J. Yu. 1986. Enzymes involved in the metabolism of plant allelochemicals. In L. B. Brattsten and S. Ahmad (eds.), Molecular Aspects of Insect Plant Associations. pp. 73–151. Plenum, New York.

Berenbaum, M. 1981. Patterns of furanocoumarin distribution and insect herbivory in the Umbelliferae: plant chemistry and community structure. Ecology 62:1254–1266.

Berenbaum, M. R. 1986. Target site insensitivity in insect-plant interactions. In L. B. Brattsten and S. Ahmad (eds.), Molecular Aspects of Insect-Plant Associations. pp. 257–271. Plenum, New York.

Berry, R. E., S. J. Yu, and L. C. Terriere. 1980. Influence of host plants on insecticide metabolism and management of variegated cutworm. J. Econ. Ent. 73:771–774.

Bowers, M. D., and G. M. Puttick. 1986. Fate of ingested iridoid glycosides in lepidopteran herbivores. J. Chem. Ecol. 12:169–178.

Brattsten, L. B. 1979. Biochemical defense mechanisms in herbivores against plant allelochemicals. In G. A. Rosenthal and D. H. Janzen (eds.), Herbivores: Their Interaction with Secondary Plant Metabolites. pp. 199–270. Academic Press, New York.

Brattsten, L. B. 1987a. Metabolic insecticide defenses in the boll weevil compared to those in a resistance-prone species. Pestic. Biochem. Physiol. 27:1–12.

Brattsten, L. B. 1987b. Inducibility of metabolic insecticide defenses in boll weevils and tobacco budworm caterpillars. Pestic. Biochem. Physiol. 27:13–23.

Brattsten, L. B., and S. Ahmad. 1986. Molecular Aspects of Insect-Plant Associations. Plenum, New York.

Brattsten, L. B., and C. F. Wilkinson. 1973. Induction of microsomal enzymes in the southern armyworm (*Prodenia eridania*). Pestic. Biochem. Physiol. 3:393–407.

Brattsten, L. B., C. F. Wilkinson, and T. Eisner. 1977. Herbivore-plant interactions: mixed-function oxidases and secondary plant substances. Science 196:1349–1352.

Brattsten, L. B., J. H. Samuelian, K. Y. Long, S. A. Kincaid, and C. K. Evans. 1983. Cyanide as a feeding stimulant for the southern armyworm, *Spodoptera eridania*. Ecol. Ent. 8:125–132.

Brattsten, L. B., C. K. Evans, S. Bonetti, and L. H. Zalkow. 1984. Induction by carrot allelochemicals of insecticide-metabolizing enzymes in the southern armyworm (*Spodoptera eridania*). Comp. Biochem. Physiol. 77C:29–37.

Brattsten, L. B., C. W. Holyoke, Jr., J. R. Leeper, and K. F. Raffa. 1986a. Insecticide resistance: challenge to pest management and basic research. Science 231:1255–1260.

Brattsten, L. B., C. A. Gunderson, J. T. Fleming, and K. N. Nikbahkt. 1986b. Temperature and diet modulate cytochrome P-450 activities in southern armyworm, *Spodoptera eridania* (Cramer), caterpillars. Pestic. Biochem. Physiol. 25:346–347.

Brown, A. W. A. 1958. Laboratory studies on behavioristic resistance of *Anopheles albimanus* in Panama. Bull. WHO 19:1053–1061.

Brown, A. W. A. 1971. Pest resistance to pesticides. In R. White-Stevens (ed.), Pesticides in the Environment, Vol. 1, Part 2. pp. 457–552. Marcel Dekker, New York.

Bull, L. B., C. C. J. Culvenor, and A. J. Dick. 1968. The Pyrrolizidine Alkaloids. Wiley, New York.

Busvine, J. R. 1951. Mechanism of resistance to insecticides in house flies. Nature 168:193–195.

Carroll, C. R., and C. H. Hoffman. 1980. Chemical feeding deterrent mobilized in response to insect herbivore and counter adaptation by *Epilachna tredecimnotata*. Science 209:414–416.

Casida, J. E., and L. O. Ruzo. 1980. Metabolic chemistry of pyrethroid insecticides. Pestic. Sci. 11:257–269.

Chialiang, C., and A. L. Devonshire. 1982. Changes in membrane phospholipids identified by Arrhenius plots of acetylcholinesterase and associated with pyrethroid resistance (kdr) in house flies (*Musca domestica*). Pestic. Sci. 13:156–160.

Chino, H., R. G. H. Downer, G. R. Wyatt, and L. I. Gilbert. 1981. Lipophorin, a major class of lipoproteins of insect hemolymph. Insect Biochem. 11:491.

Christian, M. F., and S. J. Yu. 1986. Cytochrome P-450-dependent monoxygenases in the velvetbean caterpillar, *Anticarsia gemmatalis* Hubner. Comp. Biochem. Pysiol. 83C:23–27.

Clark, A. G., and N. A. Shamaan. 1984. Evidence that DDT-dehydrochlorinase from the house fly is a glutathione *S*-transferase. Pestic. Biochem. Physiol. 22:249–261.

Corbett, J. R., K. Wright, and A. C. Baille. 1984. The Biochemical Mode of Action of Pesticides, 2nd. ed. Academic Press, New York.

Crow, J. F. 1957. Genetics of insect resistance to chemicals. Annu. Rev. Ent. 2:227–246.

Dauterman, W. C. 1976. Extramicrosomal metabolism of insecticides. In C. F. Wilkinson (ed.), Insecticide Biochemistry and Physiology. pp. 149–176. Plenum, New York.

Dauterman, W. C. 1985. Insect metabolism: extramicrosomal. In G. S. Kerkut and L. I. Gilbert (eds.), Comprehensive Insect Physiology, Biochemistry and Pharmacology, Vol. 12. pp. 713–730. Pergamon Press, Oxford.

David, A., and D. K. Vallance. 1955. Bitter principles of Cucurbitaceae. J. Pharm. Pharmacol. 7:295–296.

Denison, M. S., J. W. Hamilton, and C. F. Wilkinson. 1985. Comparative studies of aryl hydrocarbon hydroxylase and the Ah receptor in non-mammalian species. Comp. Biochem. Physiol. 80C:319–324.

DePierre, J., and R. Morgenstern. 1983. Comparison of the distribution of microsomal and cytosolic glutathione *S*-transferase activities in different organs of the rat. Biochem. Pharmacol. 32:721–723.

Devonshire, A. L., and G. D. Moores. 1982. A carboxyesterase with broad substrate specificity causes organophosphorous, carbamate and pyrethroid resistance in peach-potato aphids (*Myzus persicae*). Pestic. Biochem. Physiol. 18:235–246.

Eldefrawi, M. E., R. Miskus, and V. Sutcher. 1960. Methylene-dioxyphenyl derivatives as synergists for carbamate insecticides in susceptible, DDT-and parathion-resistant house flies. J. Econ. Ent. 53:231–234.

Farnham, A. W. 1977. Genetics of resistance in house flies (*Musca domestica* L.) to pyrethroids. Knock-down resistance. Pestic. Sci. 8:631–636.

Farnham, D. E., R. E. Berry, S. J. Yu, and L. C. Terriere. 1981. Aldrin epoxidase activity and cytochrome P-450 content of microsomes prepared from alfalfa and cabbage looper larvae fed various plant diets. Pestic. Biochem. Physiol. 15:158–165.

Ferguson, W., and C. W. Kearns. 1949. The metabolism of DDT in the large milkweed bug. J. Econ. Ent. 42:810–817.

Ferguson, J. E., and R. L. Metcalf. 1985. Cucurbitacins: plant-derived defense compounds for diabroticites (Coleoptera: Chrysomelidae). J. Chem. Ecol. 11:311–318.

Ferguson, J. E., R. L. Metcalf, and D. C. Fischer. 1985. Disposition and fate of cucurbitacin B in five species of diabroticites. J. Chem. Ecol. 11:1307–1321.

Fishbein, L. 1976. Teratogenic, mutagenic, and carcinogenic effects of insecticides. In C. F. Wilkinson (ed.), Insecticide Biochemistry and Physiology. pp. 555–603. Plenum, New York.

Gammon, D. W. 1980. Pyrethroid resistance in a strain of *Spodoptera littoralis* is correlated with decreased sensitivity of CNS in vitro. Pestic. Biochem. Physiol. 13:53–62.

Garfinkel, D. 1958. Studies on pig liver microsomes. I. Enzymic and pigment composition of different microsomal fractions. Arch. Biochem. Biophys. 77:493–509.

Georghiou, G. P. 1972. The evolution of resistance to pesticides. Annu. Rev. Ecol. Syst. 3:133–168.

Georghiou, G. P. 1980. Insecticide resistance and prospects for its management. Residue Rev. 76:131–145.

Georghiou, G. P. 1983. Management of resistance in arthropods. In G. P. Georghiou and T. Saito (eds.), Pest Resistance to Pesticides. pp. 769–792. Plenum, New York.

Gil, L. D., B. C. Fine, M. L. Dinamarca, I. Balazs, J. R. Busvine, and M. Agosin. 1968. Biochemical studies on insecticide resistance in *Musca domestica*. Ent. Exp. Appl. 11:17–29.

Glass, E. H., and P. J. Chapman. 1952. The redbanded leaf roller and its control. New York State Agricultural Experiment Station Bulletin 755. New York State Agricultural Experiment Station, Geneva, N.Y.

Gould, F. 1984. Role of behavior in the evolution of insect adaptation to insecticides and resistant host plants. Bull Ent. Soc. Amer. 30:34–41.

Gunderson, C. A., J. H. Samuelian, C. K. Evans, and L. B. Brattsten. 1985. Effects of the mint monoterpene pulegone on *Spodoptera eridania* (Lepidoptera: Noctuidae). Environ. Ent. 14:859–863.

Gunderson, C. A., L. B. Brattsten, and J. T. Fleming. 1986. Biochemical defense against pulegone in southern and fall armyworm larvae. Pestic. Biochem. Physiol. 26:238–249.

Haunerland, N. H., and W. S. Bowers. 1986. Binding of insecticides to lipophorin and arylphorin, two hemolymph proteins of *Heliothis zea*. Arch. Insect Biochem. Physiol. 3:87–96.

Hodgson, E. 1983. The significance of cytochrome P-450 in insects. Insect Biochem. 13:237–246.

Hodgson, E. 1985. Microsomal monooxygenases. In G. A. Kerkut and L. I. Gilbert (eds.), Comprehensive Insect Physiology, Biochemistry, and Pharmacology, Vol. 11. pp. 206–321. Pergamon Press, Elmsford, N.Y.

Hodgson, E., and A. P. Kulkarni. 1983. Characterization of cytochrome P-450 in studies of insecticide resistance. In G. P. Georghiou and T. Saito (eds.), Pest Resistance to Pesticides. pp. 207–228. Plenum, New York.

Hoyer, R. F., and F. W. Plapp. 1968. Insecticide resistance in the house fly; identification of a gene that confers resistance to organotin insecticides and acts as an intensifier of parathion resistance. J. Econ. Ent. 61:1269–1276.

Ivie, G. W., D. Bull, R. Beier, N. Pryor, and E. Oertli. 1983. Metabolic detoxification: mechanism of insect resistance to plant psoralens. Science 221:374–376.

Jackson, F. R., S. D. Wilson, and L. M. Hall. 1984. Two types of mutants affecting voltage-sensitive sodium channels in *Drosophila melanogaster*. Nature 308:189–191.

Jondorf, W. R., R. P. Maickel, and B. B. Brodie. 1959. Inability of newborn mice and guinea pigs to metabolize drugs. Biochem. Pharmacol. 1:352–354.

Klingenberg, M. 1958. Pigments of rat liver microsomes. Arch. Biochem. Biophys. 75:376–386.

Knerer, G., and C. E. Atwood. 1973. Diprionid sawflies: polymorphism and speciation. Science 179:1090–1099.

Kulkarni, A. P., and E. Hodgson. 1984. The metabolism of insecticides: the role of monooxygenase enzymes. Annu. Rev. Pharmacol. Toxicol. 24:19–42.

Levine, B. S., and S. D. Murphy. 1977. Effect of piperonyl butoxide on the metabolism of dimethyl and diethyl phosphorothionate insecticides. Toxicol. Appl. Pharmacol. 40:393–406.

Lewis, J. B., and R. M. Sawicki. 1971. Characterization of the resistance mechanism to diazinon, parathion, and diazoxon in the organophosphorous-resistant SKA strain of house flies. Pestic. Biochem. Physiol. 1:275–285.

Lindquist, A. W., A. H. Madden, and H. G. Wilson. 1947. Pre-treating house flies with synergists before applying pyrethrum sprays. J. Econ. Ent. 40:426–427.

Liu, M. Y., Y. J. Tzeng, and C. N. Sun. 1981. Diamond-back moth resistance to several synthetic pyrethroids. J. Econ. Ent. 74:393–396.

Lockwood, J. A., T. C. Sparks, and R. N. Story. 1984. Evolution of insect resistance to insecticides: a reevaluation of the roles of physiology and behavior. Bull. Ent. Soc. Amer. 30:41–51.

McKenzie, J. A., and A. Purvis. 1984. Chromosomal localization of fitness modifiers of diazinon resistance genoypes of *Lucilia cuprina*. Heredity 53:625–634.

Metcalf, R. L. 1955. Organic Insecticides, Their Chemistry and Mode of Action. Interscience, New York.

Metcalf, R. L. 1967. Mode of action of insecticide synergists. Annu. Rev. Ent. 12:229–256.

Metcalf, R. L., R. A. Metcalf, and A. M. Rhodes. 1980. Cucurbitacins as kairomones for diabroticite beetles. Proc. Natl. Acad. Sci. USA 77:3769–3772.

Miller, T. A., J. M. Kennedy, and C. Collins. 1979. CNS insensitivity to pyrethroids in the resistant kdr strain of house flies. Pestic. Biochem. Physiol. 12:224–230.

Moore, L. V., and G. G. E. Scudder. 1985. Selective sequestration of milkweed (*Asclepias* sp.) cardenolides in *Oncopeltus fasciatus* (Dallas) (Hemiptera: Lygaeidae). J. Chem. Ecol. 11:667–687.

Moore, L. V., and G. G. E. Scudder. 1986. Ouabain-resistant Na,K-ATPases and cardenolide tolerance in the large milkweed bug, *Oncopeltus fasciatus*. J. Insect Physiol. 32:27–33.

Morello, A. 1964. Role of DDT-hydroxylation in resistance. Nature 203:785–786.

Motoyama, N., and W. C. Dauterman, 1980. Glutathione *S*-transferases: their role in the metabolism of organophosphorous insecticides. Rev. Biochem. Toxicol. 2:49–69.

Nakatsugawa, T., and M. A. Morelli. 1976. Microsomal oxidation and insecticide metabolism. In C. F. Wilkinson (ed.), Insecticide Biochemistry and Physiology. pp. 61–114. Plenum, New York.

Nebert, D. W. 1980. The Ah locus. Genetic differences in toxic and tumorigenic response to foreign compounds. In M. J. Coon, A. H. Conney, R. W. Estabrook, H. V. Gelboin, J. R. Gillette, and P. J. O'Brien (eds.), Microsomes, Drug Oxidations, and Chemical Carcinogenesis. pp. 801–812. Academic Press, New York.

O'Brien, R. D. 1967. Insecticides, Action and Metabolism. Academic. Press, New York.

Omura, T., and R. Sato. 1964. The carbon monoxide-binding pigment of liver microsomes. J. Biol. Chem. 239:2370–2378.

Oppenoorth, F. J. 1985. Biochemistry and genetics of insecticide resistance. In G. A. Kerkut and L. I. Gilbert (eds.), Comprehensive Insect Physiology, Biochemistry and Pharmacology, Vol. 12. pp. 731–773. Pergamon Press, Oxford.

Oppenoorth, F. J., and W. Welling. 1976. Biochemistry and physiology of resistance. In C. F. Wilkinson (ed.), Insecticide Biochemistry and Physiology. pp. 507–551. Plenum, New York.

Oppenoorth, F. J., and H. R. Smissaert, W. Welling, L. J. T. vander Pas, and K. T. Hitman. 1977. Insensitive acetylcholinesterase, high glutathione S-transferase, and hydrolytic activity as resistance factors in a tetrachlorvinphos-resistant strain of house fly. Pestic. Biochem. Physiol. 7:34–47.

Patil, V. L., and F. E. Guthrie. 1979. Cuticular lipids of two resistant and a susceptible strain of house flies. Pestic. Sci. 10:399–406.

Pearlman, D. A., S. R. Holbrook, D. Pirkle, and S. H. Kim. 1985. Molecular models for DNA damage by photoreaction. Science 227:1304–1308.

Perry, A. and W. Hoskins. 1950. Detoxification of DDT by resistant house flies and inhibition of this process by piperonyl cyclonene. Science 111:600–601.

Plapp, F. W., and J. E. Casida. 1970. Induction by DDT and dieldrin of insecticide metabolism by house fly enzymes. J. Econ. Ent. 63:1091–1092.

Plapp, F. W., Jr., and T. C. Wang. 1983. Genetic origins of insecticide resistance. In G. P. Georghiou and T. Saito (eds.), Pest Resistance to Pesticides. pp. 47–70. Plenum, New York.

Pluthero, F. G., and R. S. Singh. 1984. Insect behavioral responses to toxins: practical and evolutionary considerations. Can. Ent. 116:57–68.

Remmer, H. 1972. Induction of drug metabolizing enzyme system in the liver. Eur. J. Clin. Pharmacol. 5:116–136.

Roush, R. T., and F. W. Plapp, Jr. 1982. Effects of insecticide resistance on biotic potential of the house fly (Diptera:Muscidae). J. Econ. Ent. 75:708–713.

Sawicki, R. M. 1970. Interaction between the factor delaying penetration of insecticides and the deethylating mechanism of resistance in organophosphorous-resistance house flies. Pestic. Sci. 1:84–87.

Schonbrod, R. D., W. W. Philleo, and L. C. Terriere. 1965. Microsomal oxidases in the house fly: A survey of fourteen strains. Life Sci. 7:681–688.

Scriber, J. M. 1979. Post-ingestive utilization of plant biomass and nitrogen by Lepidoptera: legume feeding by the southern armyworm, *Spodoptera eridania*. Oecologia (Berlin) 34:143–155.

Scriber, J. M. 1981. Sequential diets, metabolic costs, and growth of *Spodoptera eridania* (Lepidoptera: Noctuidae) feeding upon dill, lima bean, and cabbage. Oecologia (Berlin) 51:175–180.

Scriber, J. M. 1986. Allelochemicals and alimentary ecology: Heterosis in a hybrid zone. In L. B. Brattsten and S. Ahmad (eds.), Molecular Aspects of Insect–Plant Associations. pp. 43–72, Plenum, New York.

Scudder, G. G. E., and J. Meredith. 1982. Morphological basis of cardiac glycoside sequestration by *Oncopeltus fasciatus* (Dallas) (Hemiptera: Lygaeidae). Zoomorphology (Berlin) 99:87–101.

Self, L. S., F. Guthrie, and E. Hodgson. 1964. Adaptation of tobacco hornworms to the ingestion of nicotine. J. Insect Physiol. 12:224–230.

Smissaert, H. R. 1964. Cholinesterase inhibition in spider mites susceptible and resistant to organophosphate. Science 143:129–131.

Soderlund, D. M., S. M. Ghiasuddin, and D. W. Helmuth. 1983a. Receptor-like stereospecific binding of pyrethroid insecticide to mouse brain membranes. Life Sci. 33:261–267.

Soderlund, D. M., J. R. Sanborn, and P. W. Lee. 1983b. Metabolism of pyrethrins and pyrethroids in insects. In D. W. Hutson and T. R. Roberts (eds.), Progress in Pesticide Biochemistry and Toxicology, Vol. 3. pp. 401–435. Wiley, New York.

Tallamy, D. W. 1985. Squash beetle feeding behavior: an adaptation against induced cucurbit defenses. Ecology 66:1574–1579.

Tallamy, D. W. 1986. Behavioral adaptation in insects to plant allelochemicals. In L. B. Brattsten and S. Ahmad (eds.), Molecular Aspects of Insect-Plant Associations, pp. 273–300. Plenum, New York.

Telfer, W. S., P. S. Keim, and J. H. Law. 1983. Arylphorin, a new protein from *Hyalophora cecropia:* comparison with calliphorin and manducin. Insect Biochem. 13:601–613.

Terriere, L. C. 1983. Enzyme induction, gene amplification and insect resistance to insecticides. In G. P. Georghiou and T. Saito (eds.), Pest Resistance to Pesticides. pp. 265–297. Plenum, New York.

Terriere, L. C. 1984. Induction of detoxication enzymes in insects. Annu. Rev. Ent. 29:71–88.

Weed, A. 1938. New insecticide compound. Soap (Sanit. Prod. Sect.) 14:133, 135.

Welling, W., P. Blackmeer, G. J. Vink, and S. Voerman. 1971. In vitro hydrolysis of paraoxon by parathion resistant house flies. Pestic. Biochem. Physiol. 1:61–70.

Wilkinson, C. F. 1984. Metabolism and selective toxicity in an environmental context. In J. Caldwell and G. D. Paulson (eds.), Foreign Compound Metabolism. pp. 133–147. Taylor & Francis, New York.

Wilkinson, C. F., and L. B. Brattsten. 1972. Microsomal drug metabolizing enzymes in insects. Drug Metab. Rev. 1:153–228.

Yang, R. S. H. 1976. Enzymatic conjugation and insecticide metabolism. In C. F. Wilkinson (ed.), Insecticide Biochemistry and Physiology. pp. 177–225. Plenum, New York.

Yu, S. J. 1982. Host plant induction of glutathione *S*-transferase in the fall armyworm. Pestic. Biochem. Physiol. 18:101–106.

Yu, S. J. 1984. Interactions of allelochemicals with detoxication enzymes of insecticide-susceptible and resistant fall armyworms. Pestic. Biochem. Physiol. 22:60–68.

Yu, S. J. 1986. Consequences of induction of foreign compound-metabolizing enzymes in insects. In L. B. Brattsten and S. Ahmad (eds.), Molecular Aspects of Insect-Plant Associations. pp. 153–174. Plenum, New York.

Yu, S. J., R. E. Berry, and L. C. Terriere. 1979. Host plant stimulation of detoxifying enzymes in a phytophagous insect. Pestic. Biochem. Physiol. 12:280–284.

Species Index

Abies balsamea, 107
Acacia spp., 173
Acalymna vittatum, 237, 245–246, 252, 259
Acarus siro, 130
Acremonium coenophialum, 112
Acromyrmex octospinosus, 135, 140
Aedes aegypti, 142
Aerobacter aergenes, 142
Aesculus californica, 19
Agaricus bisporus, 137
Agraulis vanillae, 281
Agrotis ypsilon, 106
Allium spp., 137
Alnus spp., 255
Alternaria:
  cucumerina, 129
  solani, 129
Alysia manducator, 127
Amaranthus, spp., 69
Amylostereum spp., 145
  laevigatum, 145
Anatis ocellata, 15, 69
Anisochrysa prasina, 255

Anthonomus grandis, 105, 106, 331
Anticarsia gemmatalis, 339
Apanteles, see *Cotesia*
Apanteles flavipes, 74
Aphaereta minuta, 148
Aphelenchus avenae, 131, 132
Aphidoletes aphidimyza, 21
Apiomerus pictipes, 26
Apis mellifera, 20
Argyrotaenia velutiana, 314
Aristolochia spp., 298
Artemesia spp., 140
Arthrobotrys:
  dactyloides, 133, 152
  musiformis, 132
  oligospora, 132
Artogeia rapae, 3, 106
Asclepias spp., 30, 247, 287–288
  currassivica, 289
  eriocarpa, 30
  humistrata, 290
  speciosa, 30
Asobara:
  gahani, 148

Asobara (*Continued*)
  rufescens, 67, 148–149
  tabida, 67, 148–149
Aspergillus:
  flavus, 114, 130
  repens, 130
Aster umbellatus, 280
Atta cephalotes, 103, 135, 140
Aureolaria spp., 288
Autographa californica, 334

Bacillus:
  cereus, 103, 106–107, 143
  entomocidus, 106
  thuringiensis, 104, 106–107, 109–
    110, 215–219
Bacterium, 101
Battus spp., 276, 283–284
  philenor, 283, 285, 298
  polydamas, 285
Beauveria bassiana, 106–107, 108
Bembidion obtusidens, 15
Bessa harveyi, 69
Besseya spp., 288
Biosteres longicaudatus, 146, 152
Blissus leucopterus, 106
Bombyx mori, 106
Botrytis allii, 131
Brachymeria intermedia, 186
Bufo:
  americanus, 259–260
  fowleri, 259–260
Bursaphelenchus fungivorus, 131

Callosamia promethea, 298
Calotropis gigantea, 289–290
Campoletis sonorensis, 69, 175–177,
    179, 182–185, 188, 211
Camponotus, 23
  abdominalis floridanus, 284
  ligniperda, 29
Cardiochiles nigriceps, 73
Cassia spp., 72
Castanea dentata, 141
Castilleja spp., 280, 288
  sulphurea, 288

Castnia spp., 283
Catalpa, 15, 101
  speciosa, 17, 18
Ceratocystis:
  encastanea, 135
  microspora, 135
  ulmi, 143
  wageneri, 141
Ceratomia catalpae, 17
Chaetomium:
  globosum, 131
  indicum, 131
Chelone spp., 245, 288
Chilo partellus, 74
Chlosyne:
  glabra, 299
  harrisii, 280, 299
  leanira, 280
Chrysochloa spp., 247
Chrysocus cobaltinus, 247
Chrysomela spp., 242–245, 255
  aenea, 255
  aenicollis, 20, 249–250, 262
  brunsvicensis, 237, 239, 265
  populi, 241, 255, 258, 261
  20-punctata, 259
  saliceli, 259
  scripta, 254, 258
  tremulae, 241–242, 258
  vigintipunctata costella, 256, 258
Chrysolina spp., 247
  brunviscensis, 246, 250–251
  dydimata, 246
  herbacea, 261
  hyperici, 245, 246, 261
  varians, 246
Chrysopa, see *Chrysoperla*
Chrysoperla spp., 69
  carnea, 3, 15, 21
Chrysophtharta spp., 247
Colinus virgianus, 259–260
Colletotrichum:
  graminicola, 114
  trifolii, 140
Collops vittatus, 15
Coptotermes formosanus, 101

Cossus cossus, 35
Costelytra zealandica, 113, 142, 150
Cotesia:
  congregata, 211–214, 186
  glomeratus, 4
  marginiventris, 74, 211
Crematogaster clara, 258
Crioceris spp., 240
Croton spp., 72
Cucumus sativus, 259
Cucurbita:
  andreana, 244, 259
  maxima, 244, 259
  pepo, 251
Culex pipiens fatigans, 142
Cycas spp., 282
Cyclas formicarum, 113
Cyptolestes ferugineus, 129, 130
Cyrilla racemiflora, 19

Dactylaria:
  confusus, 26
  pyriformis, 133
  thaumasia, 133
Danus spp., 31
  chrysippus, 291–293
  gilippus, 291, 293
  plexippus, 29, 247, 275, 277, 287–291, 298
Delia antiqua, 137, 152
Dendroctonus frontalis, 21, 103, 143, 152
Desmodium spp., 72
Diabrotica spp., 245, 256
  balteata, 244
  hirsutum, 251
  undecimpunctata, 13, 251
  undecimpunctata howardi, 244, 251, 259–260, 318
  virgifera, 244, 259
Diaeretiella rapae, 69, 75
Diaphania hyalinata, 3
Diatraea saccharalis, 74
Dibolia borealis, 237, 245
Diplacus spp., 288
Ditylenchus dipsaci, 132

Drino bohemica, 69
Drosophila spp., 67, 135–137, 139
  ampelophila, 135–136, 153
  fenestrarum, 137
  hydei, 148
  melanogaster, 135–136, 147, 149, 153
  phalerata, 137, 147–148
Dryas julia, 281
Dryocosmus dubiosus, 107

Encarsia formosa, 77
Endothia parasitica, 135, 141
Entamoeba histolytica, 134
Epilachna:
  borealis, 318
  varivestis, 24
Erigeron spp., 70, 72
Escherichia:
  aerogenes, 134
  coli, 134
Eucarcelia rutilla, 70
Eucelatoria, 69
Eumaeus spp., 276, 282–283
  atala, 283–284
  childrenae, 283
Euphydrayas spp., 275, 276, 279–280, 288, 290, 292
  anicia, 279, 289, 292–294
  chalcedona, 291, 292
  editha, 292
  gellettii, 291, 292
  phaeton, 280, 289, 292, 299
Euplectrus plathypenae, 211–212

Fannia canicularis, 139, 148, 152
Formica subintegra, 25
Fusarium:
  graminearum, 114
  monoliniforme, 130
  oxysporum pisi, 131
  roseum, 140
  tricinctum, 140

Galerucella lineola, 248–250
Gastrolina depressa, 244, 245, 248

Gastrophysa:
  cyanea, 239
  depressa, 256
  viridula, 247, 257, 263
Geocoris, 21, 67
Geranium carolinianum, 69, 178
Gliocladium roseum, 131
Gomphocarpus fruticosus, 289
Gossypium:
  anomalum, 185
  arboreum, 185
  barbadense, 184
  capitis-viridis, 185
  gossypoides, 185
  hirsutum, 175, 182, 184
  laxum, 185
  raimondii, 185
  thurberii, 185
  trilobum, 185
  turnerii, 185

Habrolepis rouxi, 186
Hades noctula, 283
Hansenula holstii, 143, 152
Heliconius spp., 281–282
  erato, 281
  melpomene, 281
  numata, 281
  sara, 281
Heliothis spp., 69
  virescens, 73, 175, 183, 188, 315, 316
  zea, 21, 24, 67–70, 72, 74–75, 108, 210–211
Helminthosporium savitum, 129
Hippodamia convergens, 252
Hylastinus obscurus, 135, 139
Hylemya antiqua, 115
Hymenaea courbaril, 103, 140
Hypericum spp., 245, 250
Hyphantria cunea, 106
Hyposoter:
  annulipes, 211–213
  exiguae, 211, 214

Ibalia leucospoides, 145
Icterus abeillei, 36, 295

Ichthyrua inclusa, 262–263
Ips paraconfusus, 143, 150
  grandicollis, 143
Iridomyrmex humilis, 25, 254, 258
Itoplectis conquisitor, 69

Juglans:
  atala, 283–284
  mandshurica var. sieboldiana, 258
  nigra, 101
Juniperus virginiana, 101
Junonia coenia, 290

Klebsiella spp., 138, 152
  pneumoniae, 138
Kleidotoma dolichocera, 148

Labidomera clivicollis, 247
Laetilia coccidivora, 26
Laphygma, see *Spodoptera*
Laser niger, 256, 258
Lema spp., 240
Leptinotarsa, *decemlineata*, 106, 237, 246, 261
  janeta, 239
Leptopilina:
  boulardi, 136, 146, 148, 153
  clavipes, 147–149
  heteroma, 136, 146–147, 149, 152
  janeta, 239
Leptothorax:
  acervorum, 25
  kutteri, 25
Lestrimellita limao, 25
Lilioceris spp., 240
Limenitis arthemis astyanax archippus, 275, 292
Liriodendron tulipifera, 101
Liriomyza, 5
Lixophaga diatraeae, 73–74
Lochmaea capreae, 248–250
Lomamyia latipennis, 26
Lonicera involucrata, 291
Lopidea instabile, 281
Lotus corniculatus, 285
Lupinus texensis, 178

Lycopersicon, 18
Lydella grisescens, 74
Lygus lineolaris, 72
Lymantria dispar, 262–263

Macrophya nigra, 299
Magnolia grandiflora, 101
Malacosoma:
    disstria, 106
    neustria, 35
Manduca sexta, 106, 206, 209–212,
    216–217
Mattesia grandis, 105
Medetera bistriata, 15
Megaponera fuetens, 21
Megaselia halterata, 137
Meris alticola, 300
Microplitis croceipes, 69, 74–75
    demolitor, 75
Monolinia fructicola, 146, 152
Monacrosporium rutgeriensis, 133
Morus sp., 101
Musca:
    domestica, 139, 152, 320
    stabulans, 139, 152
Myrmecaphodius excavaticollis, 25
Myrmica rubra, 243, 258, 259–261
Myzus persicae, 320

Nasonia vitripennis, 127
Neacoryphus bicrucis, 281
Neoterpes graefiaria, 300
Neotylenchus linfordi, 131
Nepeta cataria, 33
Nephila spp., 278, 280
Nicotiana spp., 18
Nigrospora sphaerica, 129, 152
Nilaparvata lugens, 24
Nosema:
    otiorrhychi, 105
    pyrausta, 105, 107–108
Nymphalis antiopa, 262–263

Odynerus nidulator, 255
Oncopeltus fasciatus, 320

Onchomys spp., 36
    torridus, 34
Ophiobolis graminis, 131
Opius fletcheri, 70
Oreina spp., 247
Orius tristicolor, 4
Orizaephilus:
    mercator, 129
    surinamensis, 129
Ornithoptera priamus, 285
Orthocarpus spp., 288
Ostrinia nubilalis, 74, 105, 107, 114
Otiorrhynchus ligustici, 105

Pachliopta aristolochiae, 285
Panagrellus redivivus, 131, 132, 133,
    152
Papilio:
    glaucus, 109, 298
    hector, 285
    polyxenes, 109, 298, 319
    troilus, 298
Paramecium spp., 134
Parides spp., 276, 283–284
    anchises, 283
    neophilus, 283
Paropsis spp., 247
Passiflora spp., 281–282
Pedicularis spp., 288
Penicillium:
    camemberti, 130
    digitatum, 146
Penstemon spp., 288, 300
Penstemon breviflorus, 291
Peridroma saucia, 334
Perimyscus maniculatus, 259–260
Perisoreus canadensis, 283
Peristenus pseudopallipes, 70, 72
Phaenocarpa canaliculata, 148
Pheidole bicornis, 17
    dentata, 23
Phenolepis spp., 257
Philanthus triangulum, 20
Phoruioru.
    tibialis, 255

Phoratora (*Continued*)
   vitellinae, 241, 244, 245, 248–250,
      255, 258–259, 263
Phormia regina, 148
Pichia:
   bovis, 152
   pinus, 143, 152
Pieris rapae see *Artogeia rapae*
Pimpla ruficollis, 70
Piper cenocladum, 17
Plagiodera versicolora, 244, 248–250,
      252, 255, 259, 262–263
Platanus occidentalis, 101
Poa annua, 336
Podisus spp., 24
   maculiventris, 3, 15
Popillia japonica, 336
Populus spp., 244–245
   sieboldi, 258
   trichocarpa, 258
Pratylenchus penetrans, 132
Pristiphora erichsonii, 72, 106
Pseudomonas spp., 140
   reptilivora, 142
Pseudotrichia ardua, 135, 140
Pseudotsuga menziesii, 101, 107
Ptilocerus ochraceus, 26
Pyrenochaeta terrestris, 131
Pyrrhalta luteola, 256

Quercus stellata, 101

Reticulitermes:
   hesperus, 26
   flavipes, 25, 101
Rhabditis oxycerca, 131, 152
Rhizoctonia solani, 131, 140
Rhyacionia buoliana, 70
Rhyssa persuasoria, 145
Romalea spp., 31–34
   guttata, 29, 31
Rumex spp., 247, 257

Saccharomyces cerevisiae, 129, 131,
      135, 146–147, 149, 152

Salix spp., 242, 244–245, 248, 258, 260
   babylonica, 258
   caprea, 241
   lasiolepsis, 249
   nigricans, 250, 258, 263
   orestera, 249
   planifolia, 249
   purpurea, 259
Salyavata variegata, 26
Sassafras albidum, 101
Scaptomyza pallida, 137, 148–149
Schizaphis graminum, 94
Scolytis:
   multristriatus, 143
   ventralis, 102
Scrophularia spp., 288
Seirarctia echo, 276, 286
Senecio spp., 247
   smallii, 281
Sirex juvencus, 145
   noctilio, 145
Solanum spp., 18, 107
   berthaultii, 24
Solenopsis invicta, 23, 24
Speyeria diana, 298
Spodoptera:
   eridania, 113, 315, 316, 321
   exigua, 106, 203, 210, 211, 214
   frugiperda, 74, 211–212, 321
   littoralis, 106, 321
Sporothrix spp., 143
Symmorphus cristatus, 255
Syrphis corollae, 21

Taeniopoda eques, 34, 36
Tenodera aridifollis senenesis, 244,
      259–260
Tenthredo:
   grandis, 299
   olivacea, 260
Tetraopes sp., 247
Thanasimus dubius, 21
Thuja occidentalis, 101
Timarcha spp., 240
Trialeuroides vaporariorum, 77

Triatoma infestans, 327, 333
Tribolium confusum, 129, 152
Trichoderma lignorum, 130
Trichogramma spp., 68, 69, 72
  pretiosum, 3, 69
Trichoplusia ni, 15, 21, 206, 210, 215–
  218
Trichopsenius frosti, 25
Trichosporium symbioticum, 102
Trichothecium roseum, 133, 153
Trigona fulviventris, 26
Troides spp., 283, 285
  aeacus, 285
Tuberolachnus salignus, 257
Tyrophagus putrescentiae, 133, 153

Utetheisa bella, 286

Venturia canescens, 127
Vespula maculifrons, 15
Viburnum cassinoides, 107

Womersia strandtmanni, 26

Xamia spp., 282
Xyleborus ferruginneus, 103
Xysticus spp., 255

Zea mays, 114
Zerynthia polyxena, 285
Zigadenus venenosus, 19

Subject Index

Acetaldehyde, 146, 152, 153
Acetylcholinesterase, 323
Aflatoxin G1, 114
Alcohols, 115, 137, 138, 146, 153
Aldehydes, 99, 115, 25
Alfalfa looper, 334
Alkaloid(s), 18, 99, 106, 111, 112,
    206–219, 246, 276, 277, 280, 282
Alkyldisulfide, 138
Allelochemical(s):
    definition, 12–14, 126, 150–151
    developmental effects, 113, 188–189,
        206–210, 212–214
    microbial, 127–128, 139
Allomone(s), 12, 13, 23–37, 38, 126,
    132, 133, 240
Allyl isothiocyanate, 39, 69
American chestnut blight, 135, 141
Amino acid(s), 100, 165
    non-protein, 99, 165
Anabasine, 18
Anthraquinone(s), 26, 239, 256
Anthrones, 239, 256
Antimone, 13, 18–19

Ants, 17, 21–26, 34, 103, 140, 173,
    254, 256, 283–284
Aphids, 17, 21, 24, 69, 75, 94, 100,
    113, 320
Apneumones, 127
Aposematism, 253, 265, 283, 285, 300
Aposymbiosis, 101
Argentine stem weevil, 113
Aristolochic acids, 276, 277, 284–285
Aromatic acids, 99
Arylphorin, 321
Assasin bug, 25, 26
Azoxyglycosides, 282

Bacteria, 98, 100, 104, 106, 109, 215–
    219
Bark beetle(s), 13, 15–16, 73, 102,
    143, 150
Bee wolf, 20
Benzaldehyde, 238, 241, 256, 258
Benzoquinone, 32
p-benzoquinone, 32
Biological control, 66, 75–77, 79, 190
    191

Biotechnology, 39
Bisabolene, 184–185
Bisabolol, 180, 184–185
Bluebonnet, 178
Bluegrass billbug, 113, 116
Boll weevil, 331–332, 339–340
Broccoli, 336
Browerian mimicry, *see* Mimicry, automimics

Cabbage looper, 206
Calotropin, 278
Camphene, 106
Canavanine, 107
Cardenolide(s), 29–31, 238, 239, 247, 261, 277, 285, 288–289, 291, 295, 321
Cardiac glycosides, 165, 203, 275
Carene, 102
Carminic acid, 26
Carrot, 333
Caryophyllene, 15, 69, 103, 106, 180, 185, 188, 203, 339
Catalpol, 278, 288–289, 292–293
Catechol, 32
Catnip, 33
Cellulases, 115
Cellulose, 100
Chemical plant defense, apparency/unapparency theory, 202, 203
Chitinases, 108
Cholesterol, 101
Collard, 69, 75
Colorado potato beetle, 114, 221, 246
Competition, 262–264
  mediation by microorganisms, 91
Conditioned stimulus, 35
Confused flour beetle, 129
Copaene, 179, 180–181, 182
Corn, 67–68, 69, 72, 74–75, 114
Cost and benefits, 150–151, 153, 240, 242–243, 257, 260, 293–294
Cotton, 69, 72, 75, 175, 176, 178, 180, 183–184, 187, 189, 331, 339
Coumarin(s), 99, 334
Cowpea, 69, 337

Cucumber beetle, 253, 318
Cucurbitacin(s), 237, 239, 244, 245, 250, 253–254, 259–260, 318
Cuticle, pathogen infection, 105, 108
Cyanide, 238, 252
Cyanogenic compounds, 99, 285
Cyanogenic glycosides, 275, 276, 282
Cycads, 282
Cycasin, 276, 277, 283–284
Cytochrome P-450, 105, 116, 324–325, 327, 328–336, 339–340
Cytoplasmic polyhedrosis virus, 105

Deterrency/toxicity dichotomy, 260
Detoxication:
  enzymes, 115, 116, 320
  genetic mutation, 325
  and induction, 333
  metabolic, 100, 326
2,4-Dichlorophenol, 32–33
Digestive enzymes, 114
n-Dipropyl disulfide, 138
Diseased plants, colonization of, 139–140
Dispersal, 16
Diversification, 3
DNA, 314, 317
Douglas fir tussock moth, 101
Drosophila, 67, 135

Ecdysone 20-monoxygenase, 325
Ectosymbiosis, 102
Egyptian armyworm, 3
Endosymbionts, 100
Endotoxin, 104, 109
Enemies hypothesis, 72–73
Epoxide hydrolases, 324
Ethanol, 137, 146, 152, 153
Ethyl acetate, 138, 146, 152
European corn borer, 114
European pine shoot moth, 70
Extrafloral nectaries, 14, 19, 22–23

Fall armyworm, 74, 113, 322, 336
Fat body, 105, 321
Fatty acids, 99, 152, 239

Fechone, 107, 129
Feeding:
    damage, attraction to, 15, 73, 175
    damage and plant defense, 113, 220
    deterrency, 112, 319
    preference, 113
    site selection, 135
Fermentation, 135, 146
Fireflies, 26–27
Flagellate, 101
Flavonoids, 99, 206, 282
Food location, 133
Fragilin, 243
Frass, chemical cues, 21, 73–145
Frontalin, 15, 21
Fungi, 98, 102, 104, 129–134
    endophytes, 112
Furanocoumarin(s), 98, 108, 319, 336

Gall(s), 27
Generalist:
    herbivores, 33–34, 167, 202, 205–
        210, 215, 221, 285, 339
    natural enemies, 27, 211–214
Glucosidase, 241, 244
Glucosinolates, 99, 203
Glutathione-S-transferases, 105, 324–
    325, 326, 328, 332, 334, 340
Gossonorol, 179, 180
Gossypol, 186–189, 339
Grand fir, 102
Grasshopper(s), 29, 38
Greenhouse whitefly, 77
Green leaf volatiles, 115
Gut:
    necrosis, 109
    permeability to xenobiotics, 203
    pH, 203
    wall, 108
Gypsy moth, 262

Heliocide, 187
Helmicellulases, 115
Hemigossypolone, 185, 186–187
Hemocytes, 110
Honeybee, 19, 20, 25

Honeydew, 17, 19, 21, 37
Host habitat location, 66–67, 149, 183
Host plant selection, 66, 69, 251–252
Host/prey:
    acceptance, 66
    location, 66, 73
    regulation, 66
    searching, 66
    selection, 66, 73, 252
    specificity, 67
    suitability, 66
House cricket, 113
Housefly, 320, 324, 326–328, 333, 339
Humulene, 180, 185
Hydroquinone, 32
Hypericin, 237, 245–246, 251–252, 265

Inducible defense(s), 22, 110, 219–
    222, 251
Induction, 249, 314, 316, 336
    and resistance, 333–340
Information, 126
Inquilines, 25
Insecticides, 105, 107–108, 314, 320–
    321
Intercropping, 38
Ipomeamarone, 113
Ipsdienol, 143
Ipsenol, 143
Iridoid(s), 280
Irodoid glycoside(s), 17, 19, 237, 245,
    265, 276, 279, 288–289, 291–293,
    299–300
Irritant(s), 253, 319
Isobornyl acetate, 107

Japanese beetle, 113, 336
Juglone, 238, 244–245, 256, 258

Kairomone(s), 12, 13, 19–23, 37, 73–
    75, 126, 131, 133
Kdr housefly, 339
Ketones, 115
Kidney bean, 333
Knock-down resistance, 323

Large milkweed bug, 320–321, 327
Learning, 137, 149, 260, 317
Leptinotarsin, 239
Leucoanthocyannins, 108
Lignification, 114
Limonene, 102, 107, 175, 180
Linamarin, 278, 282, 285
Lipophilic binding, 324
Lipophorin, 321
Lotaustralin, 282, 285
Lubber grasshopper, 31–36

Malpighian tubules, 101, 105
Methoxybenzaldehyde, 32
Methyl eugenol, 38
Methyl ketone, 18
Microbial allelochemicals, 127–128, 139
Microbivores, 128
Microhabitat(s), 137, 147–148
Microorganisms, 93, 126
  aquatic, 141–142
Microsporidia, 105
Milkweeds, 29, 277
Mimicry, 28, 274–275, 280, 283, 286–293, 295–300
  automimics, 31, 274, 277, 297–298
  Batesian, 31, 274, 281, 296–298
  Mullerian, 274, 281, 285, 296–298, 300
Mite(s), 130, 133
Monarch butterfly, 29–31, 247, 275, 287–288, 293, 298
Monoculture, 70–73
Mosquitoes, 142
Mustard oils, 69
Mycentangia, 102
Mycophagus nematodes, 131
Myrcene, 102, 106, 175, 180, 184

Nectar thieves, 18, 19
Nematode, 130
Nematophagous fungi, 132, 133
Nepeta lactone, 33
Nicotine, 18, 206–219, 320, 334, 337
Nitidulid beetles, 101

Nitriles, 99
Nitrocompounds, 99
Nitrogen-fixation, 100, 111
Nornicotine, 18
Nuclear polyhedrosis virus, 106
Nutrient/allelochemical interaction, 111

Ocimene, 180, 187
Okra, 69
Onion flies, 115, 137
Ovarian development, 70, 103
  of parasitoids, 70
Oviposition behavior, 137, 166, 289–291
  of parasitoids, 70
Oviposition site(s), 128, 166
  selection, 135, 142
Oxalic acid, 247

Palatability spectrum, 31
Parsley, 336
Parsnip, 336
Pectinases, 115
Pentamethylbenzene, 316
Peppermint, 334–335
Phellandrene, 107
Phenol(s), 32, 99, 142
Phenolglucosides, 20, 241, 242, 248–250, 265, 280
Phenol β-glucosyl transferases, 324
Phenolics, 32
Phenylethylesters, 252
Pheromone(s), 37, 103, 126, 279
  aggregation, 73, 143
  alarm, 17, 24
  epideictic, 263
  sex, 142
Phlox, 336
Phytoalexin(s), 112, 113, 114
Pinene, 102, 106–107, 175, 180, 334, 339
Piperonyl butoxide, 328
Plagiolactone, 244
Plant breeders, 38–39, 77
Pollination, 17, 18

Polyacetylenes, 99
Polyculture(s), 70–73
Poplar tentmaker, 262
Population cycles, 110
Propaganda substance, 25
Propane thiol, 115
Propenylcysteine sulfoxide, 115
N-propyl sulfide, 115
Proteases, 108
Protein, hemolytic, 321
Protozoa, 98, 100–101, 104, 105
Pulegone, 334
Pyrethroid(s), 330
Pyrrolizidine alkaloids, 31, 276–277, 280–281, 286, 295

Qualitative defenses, 202–203, 216, 220
Quantitative defenses, 202–203, 214, 220
Quinine, 99
Quinolines, 99
Quinones, 32, 99

Raimondal, 187
Redbanded leafroller, 314, 331
Red scale, 186
Reflex bleeding, 239, 240, 256
Resistance, 313, 316
  associational, 70, 78
  behavioral, 317, 318–320
  biochemical, 317, 321–323
  and induction, 333–340
  metabolic, 318, 324–333
  physiological, 317, 320–321
  plant, 24, 38, 77, 92, 116, 184, 189–190
  target site, 323–324
Resistance management, 3
Rhizobial symbionts, 116
Rickettsiae, 98, 100, 104
Romalenone, 32
Rutin, 205–210, 213–219

Salicin, 20, 239, 241, 243, 245–246, 249, 253, 258–259, 262

Salicortin, 243–244, 249
Salicylaldehyde, 238, 241, 243–245, 253–254, 256, 258, 260, 262–264
Salidrosid, 249
Saponins, 18, 280
Saprophytic microbes, 115
Sarmentogenin, 246
Sassafras, 336
Scopoletin, 339
Semiochemicals, 126
Senecionine, 278
Sequestration, 13, 29, 34, 39, 241, 246, 247, 253–254, 274, 276, 278–279, 282, 291, 293, 320, 323
Sod webworm, 113
Soil microorganisms, 130
Sorghum, 72, 74
Southern armyworm, 113, 315, 322, 333–334, 336, 339
Soybean, 69, 72, 75, 337
Spathulenol, 179, 182
Specialized herbivores, 167, 205–210, 215, 221, 248, 281, 339
  natural enemies, 211–212, 216
Spiders, 24, 25, 255, 278, 281, 295
Squash beetle, 318
Stalk-rotting fungus, 114
Steroids, 103
Stored products, 128
Swallowtail, 109, 285, 319, 320
Sweet potato, 113
Symbionts, 17, 116
Synomone(s), 12, 13–17, 68–73, 126, 183

Tannin(s), 99, 106–107, 109–110, 165, 203, 282
Termite(s), 21, 25–26, 100
Terpenes, 16, 32, 99, 101, 102, 176, 178, 183, 184, 187, 203, 238, 243–244, 252, 256, 260, 262, 334
Terpenoid(s), 13, 15, 103, 179
Tetramethyl pyrazine, 138
Thujone, 107
Tobacco, 178, 206

Tobacco budworm, 113, 315, 331–333, 339–340
Tobacco hornworm, 113, 186, 206, 320
Tobacco mosaic virus, 114
Tomatine, 18, 210
Tomato, 67–68, 69
Tremulacin, 243, 249
Trenching, 318
Trichomes, 18, 77, 165, 241
2-Tridecanone, 18
Trophic level, 1–6, 172–174
Turnip, 336

UDP glucosyl transferases *see* Phenol β-glucosyl transferases
Umbelliferone, 339
Unconditioned stimulus, 35
Unpalatability, 274–286, 288, 291, 295–297
Urates, 100

Variation:
   herbivore, 31–36, 250, 288–291
   plant chemistry, 31–34, 35, 166, 219–220, 248, 252, 264
Variegated cutworm, 334–335, 339
Velvetbean caterpillar, 339
Verbenol, 103, 143
Viruses, 104, 105
Vitamins, 100

Wild geranium, 178

Xanthotoxin, 320, 336

Yeast(s), 100, 129, 135–136, 143, 149
Ylangene, 179, 182